KEITH SELKIRK

LONGMAN MATHEMATICS HANDBOOK

the language and concepts
of mathematics explained

YORK PRESS
Immeuble Esseily, Place Riad Solh, Beirut.

LONGMAN GROUP UK LIMITED
Burnt Mill, Harlow, Essex.

First published 1991

ISBN 0 582 02161 8

Illustrations by Charlotte Kennedy and Philip Bushell
Phototypeset in Britain by MS Filmsetting Limited, Frome, Somerset, England

Produced by Longman Group (FE) Ltd
Printed in Hong Kong

Contents

How to use the handbook

This handbook contains nearly 3000 words used in mathematics. These are arranged in groups under the main headings listed on pp. 3–4. The entries are grouped according to the meaning of the words to help the reader to obtain a broad understanding of the subject.

At the top of each page the subject is shown in bold type and the part of the subject in lighter type. For example, on pp. 92 and 93:

92 · SEQUENCES / TYPES

SEQUENCES / MONOTONIC SEQUENCES · **93**

In the definitions, the words used have been limited so far as possible to about 1500 words that are in common use, taken from the *New Method English Dictionary* (fifth edition) by M. West and J. G. Endicott (Longman 1976).

In addition to the entries in the text, the handbook has two useful appendixes which are detailed in the Contents list and are to be found at the back of the handbook.

1. To find the meaning of a word

Look for the word in the alphabetical index at the end of the book, then turn to the page number listed.

In the index you may find words with a number or part of speech at the end. These only occur where the same word appears more than once in the handbook in different contexts. For example, **tangent**

tangent[1] is the ratio of a particular pair of sides of a triangle

tangent[2] is a particular line which meets a curve or arc

The description of the word may contain some words with arrows in brackets (parentheses) after them. This shows that the words with arrows are defined nearby.

(↑) means that the related word appears above or on the facing page;

(↓) means that the related word appears below or on the facing page.

A word with a page number in brackets after it is defined elsewhere in the handbook on the page indicated. Looking up the words referred to may help in understanding the meaning of the word being defined.

In some cases more than one meaning is given for the same word. Where this is so, the first definition given is the more (or most) common usage of the word. At the end of the definitions, you will find listed any symbols, any alternative terms (also known as . . ., similarly . . .), as well as further page references which may prove helpful in understanding the word and distinguishing it from similar words.

The explanation of each word usually depends on knowing the meaning of a word or words above it. For example, on p. 143 the meaning of *infinitesimal calculus*, *increment*, *decrement*, *derivative*, and the words that follow, depends on the meaning of *calculus* and *infinitesimal* which appear above them. Once the earlier words have been read, those that follow become easier to understand. The illustrations have been designed to help the reader understand the definitions, but the definitions are not dependent on the illustrations.

2. To find related words

Look at the index for the word you are starting from and turn to the page number shown. Because this handbook is arranged by ideas, related words will be found in a group on that page or one nearby. The illustrations will also help to show how the words relate to one another.

For example, words relating to matrices are on pp. 83–90. On p. 83 *array* and *matrix* are followed by words describing the different elements of matrices with illustrations showing these elements; p. 84 explains and illustrates various types of matrices and products with entries on, e.g. *symmetric matrix* and *sum of two matrices*; p. 85 gives further types of matrices and pp. 86–88 gather together words relating to determinants; and so on to the end of the section.

3. As an aid to studying or revising

The handbook can be used for studying or revising a topic. For example, to revise your knowledge of topology, you would look up *topology* in the alphabetical index. Turning to the page indicated, p. 238, you would find *topology*, *topological space*, *topological equivalence*, *topological invariant*, and so on; on p. 239 you would find *Jordan curve, network, graph theory,* and so on; on p. 240 you would find *circuit*, *collapse*, etc.

In this way, starting with one word in a topic you can revise all the words that are important to this topic.

4. To find a word to fit a required meaning

It is almost impossible to find a word to fit a meaning in most books, but easy with this handbook. For example, if you had forgotten the word for the various pairs of angles produced when a line intersects parallel lines, all you would have to do would be to look up parallel in the alphabetical index and turn to the page indicated, p. 176. On reading the definition and subsequent ones, you would find *vertically opposite angles*, *alternate angles* and *corresponding angles*, and a diagram showing them all.

5. Abbreviations used in the definitions

abbr	abbreviated as	i.e.	*id est* (that is to say)
adj	adjective	lim	limit
adv	adverb	n	noun
arg	argument	p.	page
conj	conjunction	pl	plural
det	determinant	pp.	pages
e.g.	*exempli gratia* (for example)	sing.	singular
etc	*et cetera* (and so on)	v	verb
exp	exponential	=	the same as

COMMON DERIVATIVES

$$\frac{d}{dx}(x^n) = nx^{n-1}$$

$$\frac{d}{dx}(\sin x) = \cos x$$

$$\frac{d}{dx}(\cos x) = -\sin x$$

$$\frac{d}{dx}(\tan x) = \sec^2 x$$

$$\frac{d}{dx}(\cot x) = -\operatorname{cosec}^2 x$$

$$\frac{d}{dx}(\sec x) = \sec x \tan x$$

$$\frac{d}{dx}(\operatorname{cosec} x) = -\operatorname{cosec} x \cot x$$

$$\frac{d}{dx}(\ln x) = \frac{1}{x}$$

$$\frac{d}{dx} e^x = e^x$$

$$\frac{d}{dx} a^x = a^x \ln a \ (a>0)$$

$$\frac{d}{dx}(\sinh x) = \cosh x$$

$$\frac{d}{dx}(\cosh x) = \sinh x$$

If $y = uv$

$$\frac{dy}{dx} = \frac{d}{dx}(uv) = u\frac{dv}{dx} + v\frac{du}{dx}$$

If $y = \frac{u}{v}$

$$\frac{dy}{dx} = \frac{d}{dx}\left(\frac{u}{v}\right) = \frac{v\frac{du}{dx} - u\frac{dv}{dx}}{v^2}$$

COMMON INTEGRALS

The arbitrary constant is omitted and $a>0$.

$$\int x^n \, dx = \frac{1}{n+1} x^{n+1} \ (n \neq -1)$$

$$\int \frac{1}{x} \, dx = \ln|x| = \begin{cases} \ln x & (x>0) \\ \ln(-x) & (x<0) \end{cases}$$

note: $x \neq 0$

$$\int \sin x \, dx = -\cos x$$

$$\int \cos x \, dx = \sin x$$

$$\int \tan x \, dx = \ln|\sec x|$$

$$\int \cot x \, dx = \ln|\sin x|$$

$$\int \sec x \, dx = \ln|\tan(\tfrac{1}{2}x + \tfrac{1}{4}\pi)|$$

$$= \ln|\sec x + \tan x|$$

$$\int \operatorname{cosec} x \, dx = \ln|\tan \tfrac{1}{2}x|$$

$$\int \ln x \, dx = x \ln x - x \qquad (x>0)$$

$$\int \sin^2 x \, dx = \tfrac{1}{2}(x - \tfrac{1}{2}\sin 2x)$$

$$\int \cos^2 x \, dx = \tfrac{1}{2}(x + \tfrac{1}{2}\sin 2x)$$

$$\int \frac{1}{x^2+a^2} \, dx = \frac{1}{a}\tan^{-1}\frac{x}{a}$$

$$\int \frac{1}{x^2-a^2} \, dx = \frac{1}{2a}\ln\left|\frac{x-a}{x+a}\right|$$

$$\int \frac{1}{\sqrt{(x^2+a^2)}} \, dx = \sinh^{-1}\frac{x}{a}$$

$$\int \frac{1}{\sqrt{(x^2-a^2)}} \, dx = \begin{cases} \cosh^{-1}(x/a) & (x>a) \\ -\cosh^{-1}(-x/a) & (x<-a) \end{cases}$$

$$\int \frac{1}{\sqrt{(a^2-x^2)}} \, dx = \sin^{-1}\frac{x}{a}$$

If u, v are functions of x

$$\int u\frac{dv}{dx} \, dx = uv - \int v\frac{du}{dx} \, dx$$

mathematics (*n*) the study of space, number and pattern; the proof of results about them by reasoning. **mathematical** (*adj*).
pure mathematics the part of mathematics which is not directly concerned with the actual world.
applied mathematics the part of mathematics which is concerned with results of immediate practical use.
application (*n*) something of practical use related to a part of mathematics.
abstract (*adj*) not concerned with any real or practical application. **abstraction** (*n*).
symbol (*n*) a letter or sign used for a mathematical concept. **symbolize** (*v*).
notation (*n*) a set of symbols used in some part of mathematics.
index[1] (*n*) a small symbol written above or below the level of other symbols as in x^3, a_i. *See also* pp. 35, 115. **indices** (*pl*).
superscript (*n*) an index (↑) written above the usual level of symbols as in x^3.
subscript (*n*) an index (↑) written below the usual level of symbols as in a_i.
suffix (*n*) = subscript (↑).
prefix (*n*) a symbol written before another symbol or symbols to which it applies, e.g. the *f* in $f(x)$.
postfix (*n*) a symbol written after another symbol or symbols to which it applies, e.g. the $^\times$ in $a^\times$, the *r* in ^nC_r.
infix (*n*) a symbol written between the symbols to which it applies, e.g. the + in $a + b$.
dummy suffix a letter used as a suffix (↑) which does not appear when a notation is finally worked out, so that any other letter could be used instead, e.g. in the sigma notation (p. 91)

$$\sum_{i=1}^{4} i^2,$$ the *i* could be replaced by any other letter with the same result.
vinculum (*n*) a line drawn over a set of symbols which has the same effect as a pair of brackets (↓), e.g. $a - \overline{b + c} = a - (b + c)$.

brackets (*n. pl.*) pairs of symbols (,); [,]; {,} which show that those mathematical quantities between them should be taken as a single quantity which in many cases will be calculated first. Brackets are really the symbols [,], but the word is now used more generally and these symbols are called square brackets. The others are **parentheses** (,) and **braces** {,}. *See also* vinculum (↑).

convention (*n*) a rule which is generally agreed to avoid misunderstanding, often a rule about symbols, but also rules such as those about axes in coordinate geometry (*see* analytical geometry, p. 214). **conventional** (*adj*).

diagram (*n*) a picture used in studying a mathematical problem. *See also* figure (p. 168).

graph (*n*) (1) a diagram representing numbers or relations (p. 16), e.g. histogram (p. 112), cartesian graph (p. 215), arrow diagram (p. 17); (2) a diagram representing a network (p. 239). The word *graph* alone is often taken to be a cartesian graph. *See also* network (p. 239).

graph paper special paper with lines printed on it for drawing graphs.

isometric graph paper graph paper (↑) with a pattern of equilateral (p. 177) triangles often used for drawing views of three dimensional objects. *See also* isometry (p. 233).

plot (*v*) to draw points or curves by measurement on a cartesian (p. 215) or other graph.

sketch (*v*) to draw a rough graph, usually only showing ideas of special interest and without exact measurement. Also known as **trace**. *See also* p. 84.

definition (*n*) an exact statement giving an explanation of a mathematical concept used as a starting point for further work. Definitions are often best understood from examples. **define** (*v*).

implicit definition a definition which uses other words which must also be defined instead of using basic words which may be assumed (p. 251).

inductive definition (1) a definition based on examples and not usually allowed in mathematics. *See also* induction (p. 255); (2) = recursive definition (p. 10).

diagram

diagram used in solving the problems of boats A, B, C passing boats D, E, F in a canal tunnel, where the wider space can only take one pair of boats

graph

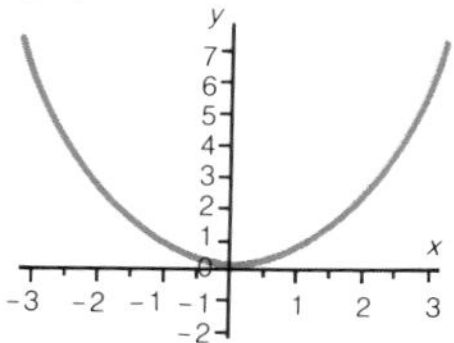

cartesian graph of $y = x^2$

graph paper

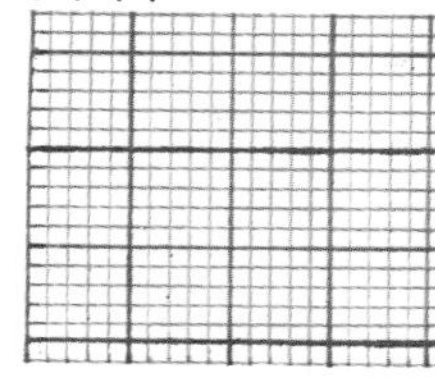

isometric graph paper

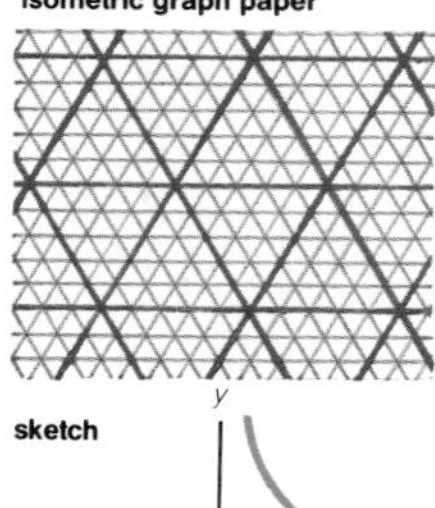

sketch

y
0
x

sketch graph of $y = \frac{1}{x}$ (not to scale)

recursive definition[1] a definition involving the non-negative integers (p. 24) and arranged in a similar way to proof by induction (p. 255). *See also* inductive definition (p. 9), p. 94.

ostensive definition a definition which is not formal (p. 289), often one based on ideas obtained from diagrams.

formal definition a definition which is based on a set of axioms (p. 250), and is used to build a theory of some part of mathematics.

undefined terms words which cannot be defined in a simpler way, e.g. *point, line*, and whose meaning can only be understood intuitively (p. 243).

set (*n*) a collection of different objects. There must be a rule to decide if any given object belongs to the set or not. *See also* extension (↓), intension (↓). Symbols: capital letters such as A, B, Z; the elements (↓) of the set listed between braces (p. 9).

element[1] (*n*) at one time an important part of a theory, now an object which belongs to a set. The symbol of set membership is $\in$, so that we can write, e.g., $a \in A$. *See also* p. 83. Symbols: small letters, e.g. a, b, z.

member (*n*) = element (↑), though this is better kept for an element of a sequence (p. 91). **membership** (*n*).

enumerate (*v*) to list the elements of a set.

extension[1] (*n*) the definition of a finite set by listing its members, e.g. $\{1, 2, 3, 4\}$ is the set of the first four positive integers. Also known as **enumeration**. *See also* pp. 252, 262.

intension (*n*) the definition of a set by stating properties of its elements, e.g. if n is an integer, then $\{n : n > 3\}$ is the set of all integers greater than 3.

equal (*adj*) two sets which have the same elements equal each other and are said to be equal sets. The word is also widely used for other pairs of things which can be regarded as the same, e.g. the elements of some equivalence classes (p. 19) such as the rational numbers (p. 26). **equality** (*n*), **equal** (*v*). Symbol: = .

element

if S is the set of
nations in South America,
b is Brazil and
t is Tanzania,
then $b \in S$
(b is a member of S),
$t \notin S$
(t is not a member of S)

equate (*v*) to make equal to.

unequal (*adj*) not equal. Also known as **inequal**.
inequality (*n*). Symbol: $\neq$.

set diagram a way of showing sets and how they relate by drawing closed curves or circles which may or may not cross one another. The elements of each set are placed inside the corresponding curve. Usually (but incorrectly) called Venn diagram (p. 249). *See also* union (p. 12), intersection (p. 13).

set diagram

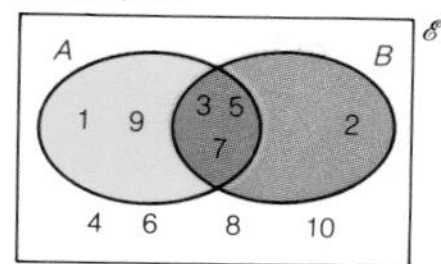

set diagram for first ten positive integers showing sets
$A = \{\text{odd numbers}\}$
$B = \{\text{prime numbers}\}$

Karnaugh map a type of table which can be used instead of a set diagram (↑).

Karnaugh map

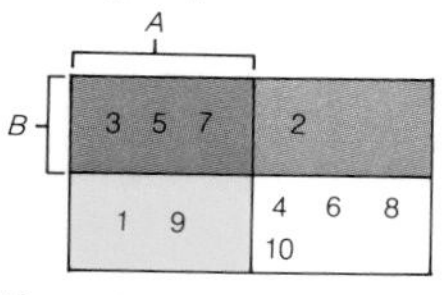

Karnaugh map for first ten positive numbers showing
$A = \{\text{odd numbers}\}$
$B = \{\text{prime numbers}\}$

empty set the set which has no elements. Also known as **null set**. Symbols: $\emptyset$ or $\{\}$.

empty (*adj*) having no elements.

non-empty (*adj*) containing at least one element.

universal set the set which contains all the elements which are being considered at any given time. In set diagrams (↑) it is shown by a rectangle around the whole diagram. Symbol: ℰ.

singleton (*n*) a set with only one element.

doubleton (*n*) a set with two elements. Also known as **doublet**. Similarly **triplet** (three), **quadruplet** (four), etc, using Latin prefixes (*see* Appendix 2, p. 294) and ***n*-tuplet** (*n* elements).

infinite (*adj*) never ending; greater than any number which may be defined.

infinity (*n*) the concept of an infinite number or set. Symbol: ∞.

finite (*adj*) not infinite, having a limit.

transfinite (*adj*) used to describe numbers which are infinite, e.g. *see* aleph null (p. 24), power of the continuum (p. 64).

count (*v*) to put the elements of a set into one : one correspondence (p. 52) with the positive integers one, two, three, ...

denumerable (*adj*) used of both finite and infinite sets whose elements can be put into one : one correspondence (p. 52) with the positive integers (p. 24), e.g. the rational numbers (p. 26). Also known as **countable**.

non-denumerable (*adj*) not denumerable (↑), e.g. the real numbers (p. 46). Also known as **uncountable**.

denumerable

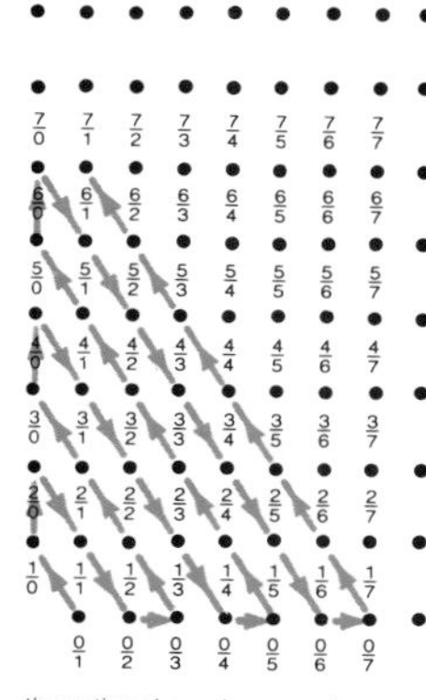

the rational numbers can be laid out as shown and can then be counted by the path whose start is shown. They are therefore denumerable. This cannot be done with the real numbers

subset (*n*) if every element of a set A is also an element of a set B, then A is said to be a subset of B, e.g. the integers are a subset of the rational numbers (p. 26). Symbols: $A \subseteq B$ or $B \supseteq A$.

proper subset if A is a subset (↑) of B and A is not equal to B, then A is called a proper subset of B. The word **proper** is used similarly in other places where equality is not allowed. Symbols: $A \supset B$ or $B \subset A$.

superset (*n*) if A is a subset (↑) of B, then B is a superset of A.

power set in a given set, the power set is the set of all subsets (↑) of the given set, including the null set (p. 11) and the given set itself.

inclusion (*n*) the concept of one set being a subset (↑) of another. $A \subset B$ is spoken as 'A is included in B' or as 'B includes A'. Inclusion and include are best used for subsets of a set, but some writers also use include instead of 'belong' for the elements of a set; this use is best avoided. Also known as **contain** which is used similarly.

exclude (*v*) used of two things which cannot both be true at the same time. *See also* mutually exclusive (p. 102). **exclusion** (*n*), **exclusive** (*adj*).

hereditary property a property of a subset (↑) which it has because the set which includes it also has that property, e.g. if A is a subset of B and B is a subset of C, then A is also a subset of C.

complement[1] (*n*) the complement of a set A is the set of elements not in the set A but in the universal set (p. 11) being considered. *See also* relative complement (↓), p. 25. Symbols: A' or $\bar{A}$.

complementation (*n*) the concept or process of obtaining a complement (↑).

union (*n, conj*) the union of two sets A and B is the set C whose elements belong either to A or to B or to both. Spoken as 'A union B' or 'A cup B'. Symbol: $C = A \cup B$.

cup (*conj*) = union (↑), so called because of the shape of the symbol (p. 8).

subset and proper subset

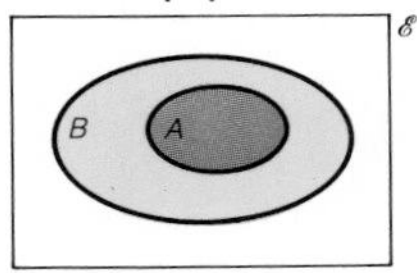

set diagram for $A \subset B$

if $A \subseteq B$, the space outside A but inside B may contain no elements

complement

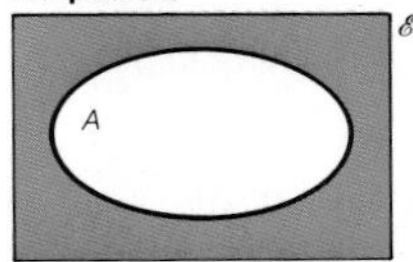

A' or $\bar{A}$, which is the complement of A, is in red

union

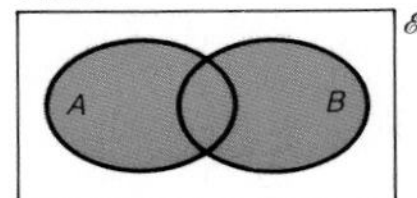

$A \cup B$ is in red

intersection

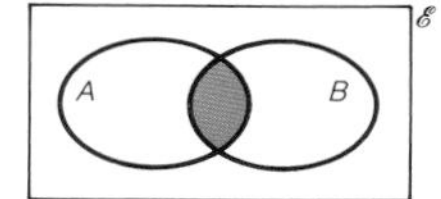

$A \cap B$ is in red

intersection (*n, conj*) the intersection of two sets *A* and *B* is the set *D* whose elements belong to both *A* and *B*. Spoken as '*A* intersection *B*' or '*A* cap *B*'. Also used widely in geometry of sets of points where lines, planes, regions etc meet. **intersect** (*v*). Symbol: $D = A \cap B$.

cap[1] (*conj*) = intersection (↑), so called because of the shape of the symbol (p. 8). *See also* p. 204.

symmetric difference

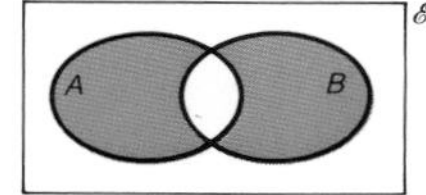

$A \triangle B$ is in red

symmetric difference of two sets *A* and *B*, the set *E* whose elements are either in *A* or in *B* but *not* in both. Symbol: $E = A \triangle B$.

relative complement

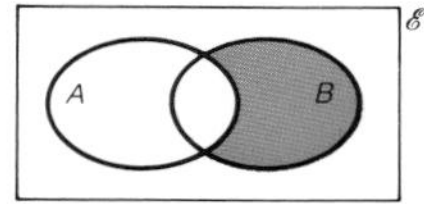

$B \setminus A$ is in red

relative complement of set *A* with respect to set *B*, the set *F* of all elements which are in *B* but not in *A*. Also known as **complement** of *A* with respect to *B*. *See also* complement (↑). Symbol: $F = B \setminus A$.

disjoint (*adj*) two sets *A* and *B* are disjoint if they have no common elements, that is $A \cap B = \emptyset$.

partition[1] (*n*) the division of a set into disjoint (↑) subsets (↑) so that each element of the set is in exactly one subset. *See also* p. 27.

dichotomy (*n*) a partition (↑) of a set into two disjoint (↑) subsets (↑). Similarly **trichotomy** into three disjoint subsets, **polytomy** into more than two disjoint subsets, e.g. the integers may be trichotomized into the positive integers, zero, and the negative integers. **dichotomize** (*v*), **dichotomous** (*adj*).

disjoint

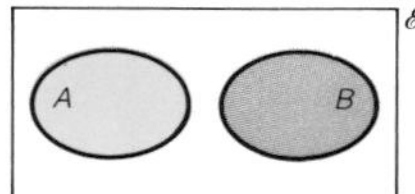

disjoint sets can be shown as non-intersecting (or shown with no elements in the intersection)

overlapping (*adj*) of sets which are not disjoint (↑).

cover (*v*) if the union (↑) of a set of subsets (↑) of a given set *S* contains a set *T*, then the set of subsets is said to cover *T*. **covering** (*n*).

class[1] (*n*) a set, usually a set whose members have some particular property in common. *See especially* equivalence class (p. 19). *See also* pp. 110, 187.

cover

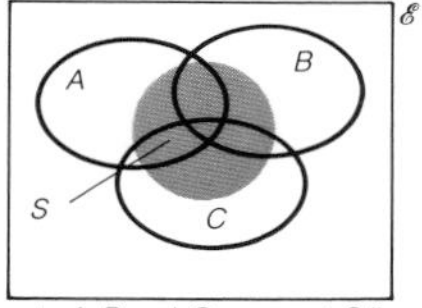

sets *A*, *B* and *C* cover set *S* (in red)

space (*n*) (1) the real world around us; that in which solid bodies or configurations are found; also often used for more than three dimensions; (2) a set having some or all of the properties of space as defined in the real world, e.g. vector space (p. 78), function space (p. 54), sample space (p. 99).

subspace (*n*) a subset (↑) of a space which is itself a space.

operation (*n*) a rule by which the elements of one or more sets can produce a new set of elements. In some cases the elements are sets themselves, e.g. as in union (p. 12). **operate** (*v*).

singulary operation an operation (↑) on the elements of only one set at a time, e.g. squaring (p. 34) a number, complementation (p. 12). Also known as **unary operation**.

binary operation an operation (↑) on the elements of two sets at a time, e.g. union (p. 12), intersection, addition, division.

ternary operation an operation on the elements of three sets at a time, e.g. scalar triple product (p. 82).

operator (*n*) (1) the symbol used for an operation (↑); (2) anything which operates on the elements of a set, especially in a singulary operation (↑).

operand (*n*) an element of a set on which an operation (↑) (especially a singulary operation, ↑) is made.

combine (*v*) to join; often used when two or more elements are combined by an operation (↑) to produce another element.

left-hand operator an operator (↑) placed on the left-hand side of the element on which it operates (↑) (usually in a singulary operation, ↑), e.g. the difference operator (p. 158). Similarly **right-hand operator**.

one-sided operator an operator (↑) which is either a left-hand or a right-hand operator (↑), but not both.

closed[1] (*adj*) used of a binary operation (↑) on any two elements of the same set which always gives another element of that set. The word *closed* is used differently for sets and intervals (p. 65). *See also* p. 66. **closure** (*n*).

unique (*adj*) the only one of its kind. Often used of a binary operation (↑) o for which, when elements a and a' are equal, and also elements b and b' are equal, then $a \circ b$ and $a' \circ b'$ are also equal, giving a unique result. Can be used similarly for other operations (↑). **uniqueness** (*n*).

closed

for the positive integers, addition is closed

$$6 + 8 = 14$$

subtraction is not closed

$$6 - 8 = -2$$

unique

for the rational numbers addition is unique since

$$\tfrac{1}{2} = \tfrac{2}{4} \text{ and } \tfrac{1}{3} = \tfrac{3}{9}$$

give

$$\tfrac{1}{2} + \tfrac{1}{3} = \tfrac{2}{4} + \tfrac{3}{9}$$

the operation of adding numerators is *not* unique since

$$1 + 1 \neq 2 + 3$$

(such operations are usually of little interest)

commutative (*adj*) used of a binary operation (↑) $*$ on two elements of the same set if for all elements a and b, $a * b = b * a$. Also known as **symmetric**, but this is best used for symmetric relations (p. 18). *See also* p. 232. **commutativity** (*n*), **commute** (*v*), **non-commutative** (*adj*).

anti-commutative (*adj*) used of a binary operation (↑) $\circ$ on two elements of the same set if for all elements a and b, $a \circ b = -(b \circ a)$, e.g. vector product (p. 82). **anti-commutativity** (*n*).

associative (*adj*) used of a binary operation (↑) $*$ on elements a, b and c when $a * (b * c) = (a * b) * c$. Since the operation can be done first on either a and b, or on b and c, such an operation may be written $a * b * c$. **associativity** (*n*).

idempotent (*adj*) used of a binary operation (↑) $\circ$ if for any element a, $a \circ a = a$, e.g. set union (p. 12), set intersection. **idempotence** (*n*).

distributive (*adj*) a binary operation (↑) $\circ$ is distributive over another binary operation, $*$, if for any three elements a, b and c, $a \circ (b * c) = (a \circ b) * (a \circ c)$. **distributivity** (*n*).

left distributive the law given by the equation under distributive (↑) is sometimes said to be left distributive. Similarly the **right distributive** law is given by $(b * c) \circ a = (b \circ a) * (c \circ a)$. In many cases the difference may be forgotten.

identity element an element I in a binary operation (↑) $\circ$ which for all elements a gives $I \circ a = a \circ I = a$, e.g. 0 for addition, 1 for multiplication, the universal set (p. 11) for set intersection, the empty set (p. 11) for set union (p.12). Also known as **neutral element**.

commutative

for the real numbers addition is commutative since

$$6 + 3 = 3 + 6$$

subtraction is not commutative since

$$6 - 3 \neq 3 - 6$$

associative

for the real numbers addition is associative since

$$6 + (2 + 5) = (6 + 2) + 5$$

subtraction is not associative since

$$6 - (2 - 5) \neq (6 - 2) - 5$$

distributive

for the real numbers multiplication is distributive over addition since

$$3 \times (4 + 7) = (3 \times 4) + (3 \times 7)$$

addition is not distributive over multiplication since

$$3 + (4 \times 7) \neq (3 + 4) \times (3 + 7)$$

set union and intersection are each distributive over the other

identity element

0 for addition

1 for multiplication

$\mathscr{E}$ (the universal set) for set intersection

$\emptyset$ (the empty set) for set union

subtraction has only a right inverse since

$a - 0 = a$ for all a but $0 - b \neq b$

identity operation a singulary operation (p. 14) which leaves every element on which it operates unchanged. Similarly **identity transformation**.

inverse elements elements a and b under a binary operation (p. 14) $*$ for which $a * b = I$ where I is the identity element (p. 15) for the operation $*$.

inverse operation an operation (p. 14) which 'undoes' a given operation, e.g. squaring (p. 34) and extracting the square root (p. 47) for positive real numbers (p. 46). *See also* p. 250. Similarly **inverse transformation**.

self-inverse (*n*) an element a in a binary operation (p. 14) $\circ$ with identity element I (p. 15) is self-inverse if $a \circ a = I$, e.g. -1 is a self-inverse element in multiplication. An operation is self-inverse if it is its own inverse operation (↑), e.g. 'take the reciprocal (p. 27) of'.

reciprocal operation a singulary operation (p. 14) which is its own inverse (↑), e.g. reflection (p. 234).

relation (*n*) a property of one or more sets of elements which is either true or not true. The idea is most easily seen in binary relations (↓). Also known as **relationship**. relate (*v*).

related[1] (*adj*) used of two or more elements with a true relation.

singulary relation this shows whether a property is true or false for each of a set of elements, e.g. whether an integer is even or not.

binary relation a relation between the elements of a set and the elements of another set (or the same set) which is either true or false, e.g. a positive integer A is or is not a multiple of another positive integer B. Other examples are 'is greater than' (p. 20), 'is equal to', 'is parallel to', 'is an element of'.

ternary relation a relation between the elements of three sets which is either true or false, e.g. the positive integer A is a common divisor (p. 26) of the two positive integers B and C.

quaternary relation a relation between the elements of four sets which is either true or false.

inverse elements

4 and $\frac{1}{4}$

for multiplication

$+6$ and -6

for addition

inverse operations

addition and subtraction are inverse operations because

$2 + 7 = 9 \quad 9 - 7 = 2$

domain and range

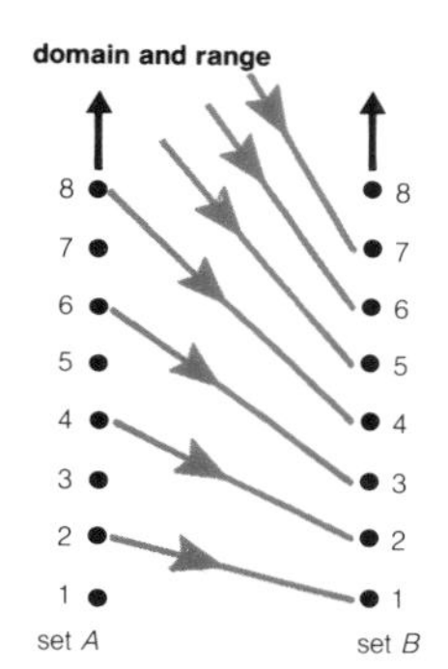

if sets A, B are the positive integers and the relation is 'is double', then the domain is the positive even integers, the range is all the positive integers

hold (*v*) a relation which is true is said to hold.

correspondence (*n*) two sets and a binary relation (↑) between some or all the elements of one set and some or all the elements of the other set.

corresponding (*adj*) used of pairs of elements between which a binary relation (↑) holds. *See also* p. 176. **correspond** (*v*).

domain (*n*) if two sets A and B are defined with a binary relation (↑) between them, then the subset (p. 12) of the first set A for which the relation is defined is called the domain of the relation. The word is often used with functions, and also for the set of values which a variable may take.

range[1] (*n*) if two sets A and B are defined with a binary relation (↑) between them, then the subset (p. 12) of the second set B for which the relation is defined is the range of the relation. The word is often used with functions. *See also* pp. 113, 167, 268.

inverse relation if a binary relation (↑) ρ gives the relation $a \rho b$, then the binary relation σ which gives the relation $b \sigma a$ is the inverse relation, e.g. 'is greater than' (p. 20) is the inverse relation of 'is less than' (p. 20).

arrow diagram a diagram or graph in which relations are shown by lines with arrow heads as in a flow chart (p. 69) and a Papygram (p. 18).

Papygram

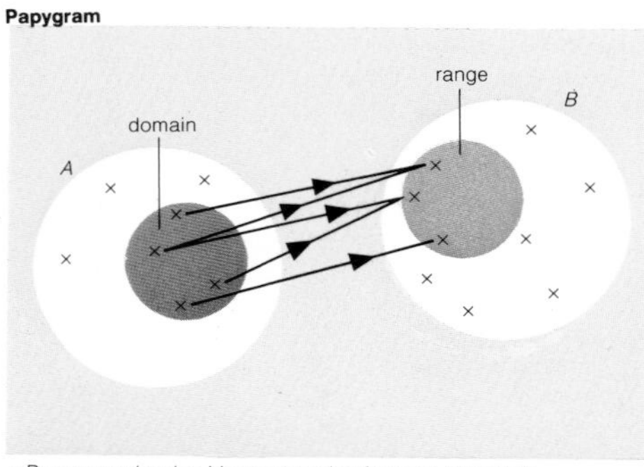

a Papygram showing binary operation from set A to set B

Papygram (*n*) an arrow diagram (p. 17) which is used to represent a binary relation (p. 16).

path diagram a diagram similar to an arrow diagram (p. 17), but not necessarily having arrows on the lines.

complementary relation for every relation there is a complementary relation which is true when the first relation is not true, and which is not true when the first relation is true, e.g. for the set of lines in the Euclidean plane (*see* Euclidean geometry, p. 171) 'is parallel to' and 'intersects'.

reflexive relation a binary relation (p. 16) ρ in which for every element a of a given set, $a \rho a$ is true, e.g. in the relation 'is a multiple of', every positive integer is a multiple of itself.

irreflexive relation a binary relation (p. 16) σ in which for every element a of a given set $a \sigma a$ is false, e.g. 'is less than' (p. 20), 'is perpendicular to'.

symmetric relation a binary relation (p. 16) ρ in which for all elements a and b the two expressions $a \rho b$ and $b \rho a$ are always either both true or both false. *Do not mix with* commutative (p. 15). The commonest example is the relation 'equals'.

antisymmetric relation a binary relation (p. 16) σ in which for all elements a and b, if the two expressions $a \sigma b$ and $b \sigma a$ are both true then $a = b$, e.g. the relation 'is greater than or equal to' (p.20). Also known as **asymmetric relation**.

transitive relation a binary relation (p. 16) ρ in which for all elements a, b and c, if both $a \rho b$ and $b \rho c$ are true, then $a \rho c$ is also true, e.g. in the relation 'is a divisor (p. 26) of', 3 is a divisor of 12 and 12 is a divisor of 84, means that 3 is a divisor of 84, and the relation is transitive.

intransitive relation a binary relation (p. 16) which is not transitive (↑), e.g. 'is the square of'.

equivalence relation a binary relation (p. 16) on a set which is reflexive (↑), symmetric (↑) and transitive (↑), e.g. equality (p. 10), similarity (p. 185), parallelism (p. 176).

reflexive relation

'is a factor of' on the set {2, 3, 4, 5, 6}

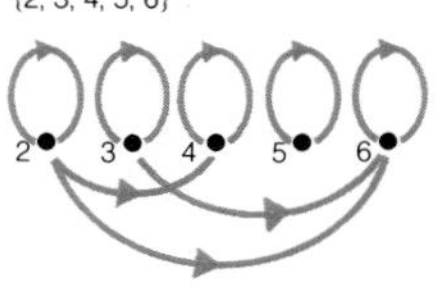

irreflexive relation

'is a proper factor of' on the set {2, 3, 4, 5, 6}

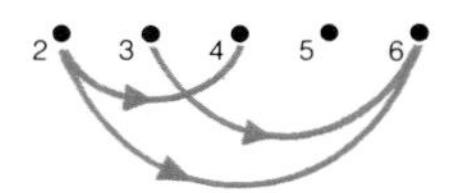

symmetric relation

(also reflexive relation) 'is equal to' on the set {2, 3, 4, 5, 6}

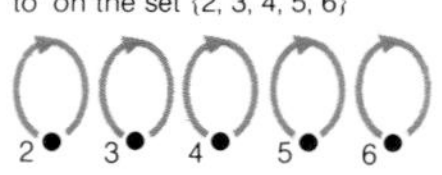

equivalence class a set whose elements form an equivalence relation (↑), usually a subset (p. 12) of a larger set, e.g. the set of fractions {2/3, 4/6, 6/9, ...}.

preordering (*n*) a binary relation (p. 16) on a set which is reflexive (↑) and transitive (↑), e.g. the relation between people which holds if and only if one is both not older and not taller than the other. This is not a partial ordering (↑) if there are two people of equal age and height.

dominance relation a binary relation (p. 16) on a set which is transitive (↑) and irreflexive (↑), e.g. 'less than' (p. 20).

partial ordering relation a binary relation (p. 16) on a set which is antisymmetric (↑) and transitive (↑) and which may or may not be reflexive (↑), e.g. 'is greater than or equal to' (p. 20), 'is less than or equal to' (p. 20) (whose symbols are sometimes used generally); 'is a multiple of', 'is a subset (p. 12) of'. Also known as **ordering relation**, but this can also be used for total ordering relation (↓).

total ordering relation a partial ordering relation (↑) ρ for which $x \rho y$ or $y \rho x$ whenever x and y are distinct (p. 60) elements of the domain (p. 17). It cannot be either symmetric (↑) or reflexive (↑). It can therefore be used to put the elements to which it applies in a unique (p. 14) order, as in the set of real numbers (p. 46), e.g. 'is greater than' (p. 20), 'is less than' (p. 20), 'is a proper subset (p. 12) of'. Also known as **linear ordering relation**, **simple ordering relation**, **serial relation**, **ordering relation**.

ordered set a set of elements with an ordering relation (↑), usually taken to be a total ordering relation (↑). Also known as **simply ordered set**, **chain**.

ordered pair an ordered set (↑) of two elements (usually related) often written (a, b). Also known as **number pair** when the elements are numbers.

ordered triple an ordered set (↑) of three elements. Also known as **number triple** when the elements are numbers. Similarly **ordered quadruple** for four elements.

order[1] (*n*) an arrangement, often used to give a number as in the order of a group (p. 72), which is the number of elements in the group; or the order of a derivative (p. 143), which is the number of times it has been derived (p. 143). *See also* pp. 72, 83, 190. **order** (*v*).

well-ordered set a totally ordered set (p. 19) for which each non-empty (p. 11) subset (p. 12) has a least element.

well-ordering theorem every set can be given an ordering relation (p. 19) for which it is well-ordered (↑), e.g. Farey sequence (p. 93).

cartesian product

if $A = \{1, 2, 3\}$ and $B = \{0, 1\}$
the cartesian product of sets A and B is
$\{(1, 0), (1, 1), (2, 0), (2, 1), (3, 0), (3, 1)\}$

cartesian product of two sets: the set of ordered pairs (p. 19) (a, b) where a and b are elements of the respective sets, e.g. the coordinates of a point in two dimensions. Also known as **cross product**. *Do not mistake for* cross product of two vectors (p. 82) or two matrices (p. 85). Symbol: $A \times B$, where A and B are the two sets.

greater than a total ordering relation (p. 19) which allows any two real numbers (p. 46) to be placed in order of size with the larger first. Also known as **more than**. Symbol: $>$.

less than a total ordering relation (p. 19) which allows any two real numbers (p. 46) to be placed in order of size with the smaller first. Symbol: $<$.

greater than or equal to the partial ordering relation (p. 19) derived from 'greater than' (↑) together with equality (p. 10). Symbols: $\geqslant$, $\geq$.

less than or equal to the inverse relation (p. 17) of 'greater than or equal to' (↑). Symbols: $\leqslant$, $\leq$.

max (*a*, *b*) the greater of the numbers a and b, also used for variables. *See also* maximum (p. 145).

min (*a*, *b*) the lesser of the numbers a and b, also used for variables. *See also* minimum (p. 145).

natural number one of the ordered set (p. 19) of numbers used in counting, called in English one, two, three, ... There are two binary operations (p. 14), addition and multiplication which are both closed (p. 14), unique (p. 14), commutative (p. 15) and associative (p. 15), and multiplication is distributive (p. 15) over addition. They can be defined using Peano's axioms (↓). Also known as **counting number**, **whole number**, but zero (↓) is sometimes included as one of the latter. *See also* integer (p. 24). Symbol for set of natural numbers: ℕ.

zero (*n*) (1) the numeral (↓) 0, also written Ø in computing (p. 42); (2) the identity element (p. 15) for addition (p. 22) which when added to any other element leaves it unchanged and when multiplied by any other element gives zero itself. An element of similar character in other sets. Also known as **nought** (especially the numeral 0), and **zero element** or **annihilator** (especially when the elements considered are not numbers).

successor (*n*) that which follows next, especially used of the next element to a given element in an ordered set (p. 19).

Peano's axioms a set of mathematical rules which is often used as a definition of the natural numbers (↑) and involves the idea of a successor (↑). They are:

- I 1 is a number;
- II the successor to any number is a number;
- III there are no two different numbers with the same successor;
- IV 1 is not a successor;
- V a property which belongs to the successor of every number which has the property, and which also belongs to 1, belongs to every number. (This allows the use of mathematical induction, p. 256).

numeral (*n*) a symbol for one of the natural numbers (↑) or zero. **numerical** (*adj*) (also refers to number).

digit (*n*) a symbol forming part of a numeral (↑), usually one of the ten digits 0, 1, 2, ..., 9. Also known as **figure**, **cipher**.

digital root a digit (p. 21) corresponding to each positive integer (p. 24) formed by repeatedly adding its digits until only one digit is left.

addition (*n*) the binary operation (p. 14) by which two natural numbers (p. 21) give a third number by counting on, which can be defined from Peano's axioms (p. 21) and is well known from simple arithmetic; also similar operations carried out on other sets of elements, especially numbers. It is commutative (p. 15) and associative (p. 15). Also known as **summation**. **add** (*v*), also known as **sum**, **additive** (*adj*), **plus** (*conj*), **add** (*conj*). Symbol: $+$.

digital root

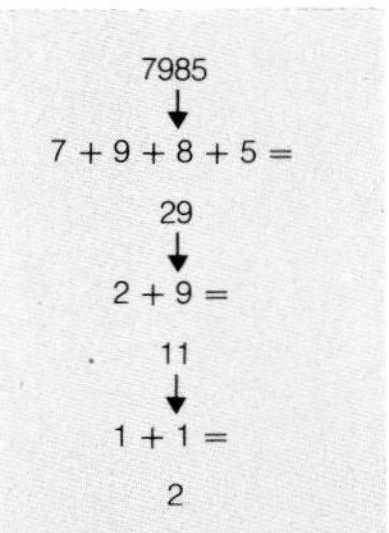

addition

in natural numbers	$2 + 5 = 7$
in integers	$(+7) + (-6) = (+1)$
in rational numbers	$\frac{1}{4} + \frac{2}{3} = \frac{11}{12}$
in vectors	$\begin{pmatrix} 5 \\ -1 \end{pmatrix} + \begin{pmatrix} -3 \\ -4 \end{pmatrix} = \begin{pmatrix} 2 \\ -5 \end{pmatrix}$

addend (*n*) one of the numbers (sometimes not the first number) in an addition.

augend (*n*) the first of the numbers in an addition.

sum (*n*) the result of an addition. Also known as **aggregate**, **total**.

total (*n*) the whole, the sum, everything being considered. Also known as **aggregation**.

addition

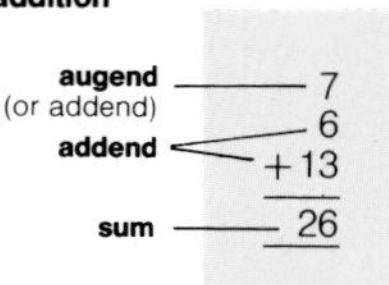

multiplication (*n*) the binary operation (p. 14) by which two natural numbers (p. 21) give a third number by repeated addition, which can be defined from Peano's axioms (p. 21) and is well known from simple arithmetic; also similar operations carried out on other sets of elements; especially numbers. It is associative (p. 15), distributive (p. 15) over addition, and usually (but not always) commutative (p. 15). **multiply** (*v*), **multiplicative** (*adj*), **times** (*conj*). Symbols: $\times$, . (a dot), and others in special cases, including juxtaposition (↓).

juxtaposition (*n*) placing next to; with two or more variables it is often used to show multiplication. **juxtapose** (*v*).

multiplication

in natural numbers

$$2 \times 5 = 10$$

in integers

$$(+7) \times (-6) = (-42)$$

in rational numbers

$$\tfrac{1}{4} \times \tfrac{2}{3} = \tfrac{1}{6}$$

in matrices

$$\begin{pmatrix} 2 & 1 \\ 1 & 3 \end{pmatrix}\begin{pmatrix} 1 & 2 \\ 3 & 4 \end{pmatrix} = \begin{pmatrix} 5 & 8 \\ 10 & 14 \end{pmatrix}$$

but

$$\begin{pmatrix} 1 & 2 \\ 3 & 4 \end{pmatrix}\begin{pmatrix} 2 & 1 \\ 1 & 3 \end{pmatrix} = \begin{pmatrix} 4 & 7 \\ 10 & 15 \end{pmatrix}$$

multiplication

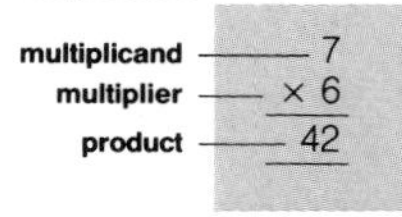

multiplicand (*n*) the first (or sometimes the second) element in a multiplication.

multiplier (*n*) the second (or sometimes the first) element in a multiplication.

product (*n*) the element which is the result of a multiplication.

short multiplication

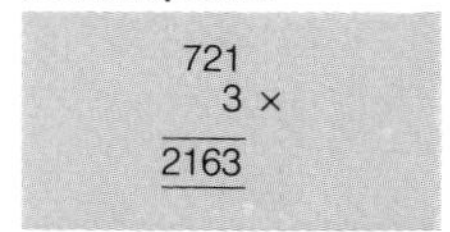

short multiplication a method of multiplying simple numbers. *See* diagram opposite.

long multiplication

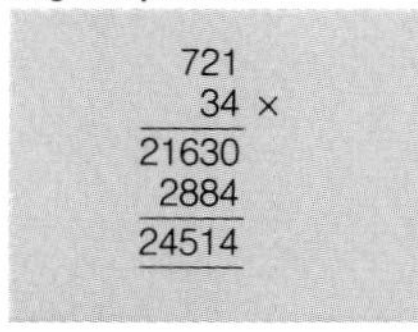

long multiplication a method of multiplying large numbers. *See* diagram opposite.

carrying figure

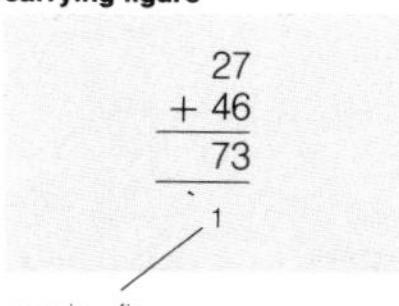

carrying figure

carrying figure a digit (p. 21) taken from one column (p. 83) to another in a calculation.

borrow[1] (*v*) sometimes used of the operation (p. 14) of taking a carrying figure (↑). *See also* p. 39.

premultiplication (*n*) multiplication in which the multiplier (↑) is placed on the left of the multiplicand (↑), (the difference is important when multiplication is not commutative, p. 15). Also known as **left multiplication**. Similarly in **postmultiplication** the multiplier is placed on the right of the multiplicand. Also known as **right multiplication**.

cancellation law a rule for simplifying (p. 292) additions or multiplications by taking out repeated numbers or variables, e.g.

if $a + c = b + c$, then $a = b$,

if $a \times c = b \times c$, then $a = b$,

except when $c = 0$ in multiplication. *See also* cancellation (p. 31).

one (*n*) (1) the numeral (p. 21) 1; (2) the identity element (p. 15) for multiplication, that element which when multiplied by any other element remains unchanged. Also known as **unity**, often when special force is to be given to the idea, and as **unit element** when the elements considered are not numbers. *See also* unit (p. 284).

cardinal numbers one of the natural numbers (p. 21) 1, 2, 3, . . . , especially when compared with the ordinal numbers (p. 24).

cardinality (*n*) the number of elements in a set, obtained by putting the elements into one : one correspondence (p. 52) with the positive integers.

aleph null the cardinality (p. 23) of the set of natural numbers (p. 21), and hence of any denumerably (p. 11) infinite set. Symbol: $\aleph_0$.

ordinal number one of the set of numbers showing order; in English first, second, third, fourth, ... **1st**, **2nd**, **3rd**, **4th**, ... (*abbr*).

positive integer one of the set of numbers $+1$, $+2$, $+3$, ..., (or $^+1$, $^+2$, $^+3$, ...), each formed from the natural numbers (p. 21) by placing $+$ or $^+$ before it. Sometimes defined as for the natural numbers. Symbol for set of positive integers: $\mathbb{Z}^+$.

non-negative integer one of the set of positive integers with the added element zero.

negative integer one of the set of numbers -1, -2, -3, ..., (or $^-1$, $^-2$, $^-3$, ...), each of which is the inverse element (p. 16) under addition of the corresponding positive integer $+1$, $+2$, $+3$, ...

integer (*n*) a positive integer (↑), a negative integer (↑) or zero. Every element of the set of integers has an inverse element (p. 16) under addition. **integral** (*adj*). Symbol for set of integers: $\mathbb{Z}$.

sign (*n*) (1) any symbol or character other than a letter or numeral; (2) often one of the two symbols $+$, $-$, (better written $^+$, $^-$) when used with a natural number (p. 21) to make a positive or negative integer (↑), e.g. $+6$, -3.71. *Do not mix with* the same symbols used for the operations (p. 14) of addition and subtraction when they are always written $+$, $-$. **signed** (*adj*).

signless (*adj*) having no sign.

plus (*adj*) the sign $+$ or $^+$. Also known as **positive**.

minus (*adj*) the sign $-$ or $^-$. Also known as **negative**.

positive (*adj*) greater than zero; having the symbol $+$ or $^+$ when showing a positive integer (↑).

negative (*adj*) less than zero; having the symbol $-$ or $^-$ when showing a negative integer (↑).

directed number a number with one of the signs $+$ or $-$ (or $^+$ or $^-$) used to show direction. The $+$ conventionally (p. 9) shows a direction to the right or upwards, or an anticlockwise (p. 174) turn. The sign $-$ shows the opposite.

subtraction

minuend	76
subtrahend	− 32
difference	44

subtraction (*n*) the inverse operation (p. 16) to addition, always possible in the set of integers and also many other sets of numbers, but not the natural numbers (p. 21). It is neither commutative (p. 15) nor associative (p. 15). **subtract** (*v*), also known as **take**, **take away**. **minus** (*conj*). Symbol: −.

minuend (*n*) the first element in a subtraction.

subtrahend (*n*) the second element in a subtraction.

difference (*n*) the result of a subtraction. **differ** (*v*).

absolute difference the result of a subtraction, forgetting any negative sign which may arise. Symbols: $a \sim b$, $|a - b|$.

complement[2] (*n*) the complement of a number *a* with respect to a number *b* is the number $b - a$, e.g. the complement of 32 with respect to 76 is 44. *See also* p. 12.

complementary addition a method of subtraction by writing down the complement (↑) of the subtrahend (↑) with respect to the minuend (↑), often used for subtraction 'in the head'. Also known as **counting on**.

subtraction three methods for 643 − 276

complementary addition

276	add	7	makes	283
283	add	60	makes	343
343	add	300	makes	643
	total	367		

decomposition

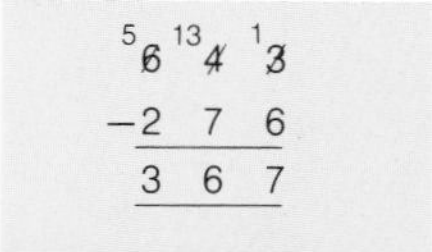

```
 5  13  1
 6   4   3
−2   7   6
 3   6   7
```

equal additions

```
     1   1
 6   4   3
 3   8
−2   7   6
 3   6   7
```

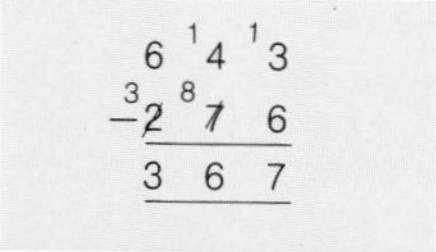

decomposition (*n*) a method of subtraction which uses carrying figures (p. 23) subtracted from the minuend (↑).

equal additions a method of subtraction which uses carrying figures (p. 23) added to the subtrahend (↑).

rational number one of the numbers used in arithmetic which can be written as one integer over another (a/b) where b is not equal to 0. The inverse operation (p. 16) to multiplication is always possible except when the multiplication is by zero. A full definition can be given by using ordered pairs (p. 19) of integers $\{a, b\}$ where b is not zero so that:

I $\{a, b\}$ is greater than, equal to or less than $\{c, d\}$ if $ad - bc$ is greater than, equal to or less than zero;

II $\{a, b\} + \{c, d\} = \{ad + bc, bd\}$;

III $\{a, b\} \times \{c, d\} = \{ac, bd\}$.

Seen in this way the rational numbers are a set of equivalence classes (p. 19) and are denumerable (p. 11). The rational numbers $\{a, 1\}$ behave exactly like the integers and are thus called by this word. The three rules given above are in fact the three well-known rules for fractions in arithmetic. *See also* irrational number (p. 46). Symbol for set of rational numbers: $\mathbb{Q}$.

rational number rules

I $\{3, 4\} > \{2, 3\}$, $\frac{3}{4} > \frac{2}{3}$ since $(3 \times 3) - (4 \times 2) > 0$

II $\{3, 4\} + \{2, 3\} = \{(3 \times 3) + (4 \times 2), 4 \times 3\} = \{17, 12\}$

$$\frac{3}{4} + \frac{2}{3} = \frac{(3 \times 3) + (4 \times 2)}{4 \times 3} = \frac{17}{12}$$

III $\{3, 4\} \times \{2, 3\} = \{3 \times 2, 4 \times 3\} = \{6, 12\} = \{1, 2\}$

$$\frac{3}{4} \times \frac{2}{3} = \frac{3 \times 2}{4 \times 3} = \frac{6}{12} = \frac{1}{2}$$

division (n) the inverse operation (p. 16) to multiplication. It is always possible in the set of rational numbers (↑) except for division by zero, and similarly in many other sets of numbers, though not in the integers. It is neither commutative (p. 15) nor associative (p. 15). **divide** (v). Symbols: ÷ or /.

dividend[1] (n) the first number in a division. *See also* p. 40.

divisor (n) the second number in a division. *See also* factor (↓).

division

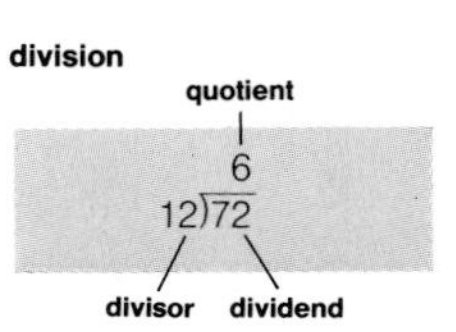

quotient (*n*) the result of a division, sometimes the integral part of that number.

divisible (*adj*) an integer a is divisible by another integer b if $a \div b$ is also an integer. Used similarly of expressions in algebra, e.g. $a^2 - b^2$ is divisible by $a - b$.

divisibility test a simple means of deciding if one positive integer (p. 24) is divisible by another.

remainder (*n*) when one positive integer (p. 24) a, is divided by another positive integer b, then $a = bq + r$ where q is the integral part of the quotient (↑) and the integer r which is less than b and may be zero is called the remainder, e.g. $46 \div 8$ gives 5 with remainder 6 since $46 \div 8 = (8 \times 5) + 6$. Used similarly in algebra, e.g. with polynomials.

multiple (*n*) when two positive integers (p. 24) can be divided without remainder, the dividend (↑) is said to be a multiple of the divisor (↑). Any positive integer is a multiple of unity and of itself. Used similarly in algebra for expressions.

factor (*n*) (1) any quantity which has an effect on another quantity, especially in statistics; (2) when two positive integers (p. 24) can be divided without remainder, the divisor (↑) is said to be a factor of the dividend (↑). Unity is a factor of all positive integers, and all positive integers are factors of themselves. Used similarly in algebra, especially with polynomials. Also known as **divisor**. **factorize** (*v*), **factorization** (*n*).

proper factor any factor of a number or expression other than the number or expression itself is a proper factor. Also known as **proper divisor**.

reciprocal (*n*) two non-zero numbers or expressions are reciprocal to each other if their product is unity, e.g. 8 and $\frac{1}{8}$ are reciprocals, as are $x^3 + 3$ and $1/(x^3 + 3)$. **reciprocal** (*adj*).

quotition (*n*) division regarded as repeated subtraction.

partition² (*n*) division regarded as giving out one at a time to a known number of places, as in dealing out a pack of cards. *See also* p. 13.

divisibility test

some common divisibility tests

by 2	last digit is even
by 4	last two digits divisible by 4
by 5	last digit is 0 or 5
by 3	sum of digits (or digital root) divisible by 3
by 9	sum of digits (or digital root) divisible by 9

factor

48 is a multiple of 12

12 is a factor of 48

or

12 is a divisor of 48

division

$48 \div 6 = 8$

quotition

from 48 matches take 6 at a time away so that there are enough for 8 people to have 6 each

partition

6 people each take one match at a time from 48 until each person has 8 matches

short division a method of dividing in simple cases.

long division a method of dividing in harder cases.

undetermined form an arrangement of symbols which is not defined. The commonest is $0 \div 0$, others are 0^0 and $\log_e 0$. Also known as **indeterminate form**.

fraction (*n*) loosely a rational number (p. 26), especially one written as the quotient (p. 27) of two integers, one which is not an integer or one which is between zero and unity.

solidus (*n*) the symbol / used to show division or a fraction as in $7/11 = 7 \div 11$ and $3/8 = \frac{3}{8}$.

ratio[1] (*n*) a quantity obtained by dividing one number by another, often written as the two smallest possible integers with a colon (:) between them, or as a rational number and unity with a colon between, e.g. 48 and 84 are in the ratio 4 : 7 or 1 : (7/4) or (4/7) : 1. *See also* p. 111.

inverse ratio the reciprocal (p. 27) of a given ratio, e.g. 4 : 7 and 7 : 4 are inverse ratios.

rate (*n*) (1) a ratio between two quantities with different units, as in kilometres per second, pounds per annum; (2) the reciprocal (p. 27) quantity to time.

proportion (*n*) two or more numbers or quantities which are always in the same ratio are said to be in the same proportion. **proportional** (*adj*). Symbol: $\propto$ for 'is proportional to'.

proportional division division of a quantity into parts which are in a given proportion.

proportional part a part in a proportional division.

inverse proportion used of two numbers or quantities, one of which is always in a constant ratio to the reciprocal (p. 27) of the other. Also known as **reciprocal proportion**.

mean proportional when *a* and *c* are given numbers, the number *b* such that $a/b = b/c$. It is a special case of the geometric mean (p. 114).

third proportional when *a* and *b* are given numbers, the number *c* such that $a/b = b/c$ is the third proportional to *a* and *b*.

short division

```
  544
7)3814(6   remainder
```

long division

```
     165
  23)3814
     23
     ---
     151
     138
     ---
      134
      115
      ---
       19   remainder
```

proportional division

84

divided in the proportions

4:2:1

gives

48, 24 and 12

mean proportional

12 is the mean proportional of 9 and 16 since

$\frac{9}{12} = \frac{12}{16}$ or $12^2 = 9 \times 16$

continued proportion

9, 12 and 16 are in continued proportion

third proportional

16 is the third proportional of 9 and 12

fourth proportional

15 is the fourth proportional of 4, 6 and 10 since $\frac{4}{6} = \frac{10}{15}$

unitary method

if a car travels 650 km on 50 litres of petrol, how far will it travel on 15 litres?

650 km ⟷ 50 litres
13 km ⟷ 1 litre
195 km ⟷ 15 litres

amicable numbers

divisors of 220	divisors of 284
1	1
2	2
4	4
5	71
10	+ 142
11	220
20	
22	
44	
55	
+110	
284	

continued proportion used of the numbers *a*, *b* and *c* in the definition of third proportional (↑) and mean proportional (↑).

fourth proportional when *a*, *b* and *c* are given numbers, the number *d* such that $a/b = c/d$ is the fourth proportional to *a*, *b* and *c*.

rule of three the rule for calculating a fourth proportional (↑) by multiplying the second and third numbers and dividing by the first.

unitary method a method of calculating proportions and inverse proportions (↑) by making one or more of the numbers or quantities in the proportion a unit amount.

aliquot part an exact fraction of a quantity or number which divides it without remainder, e.g. 45 minutes is an aliquot part of 3 hours.

cross multiplication a method of simplifying (p. 292) an equation by using the fact that if $a/b = c/d$, then $ad = bc$. *The use of the words is not advised* because of the dangers of misunderstanding in algebra.

even (*adj*) divisible by 2.

odd (*adj*) (of integers) not divisible by 2. *See also* odds (p. 101).

perfect number a positive integer which is the sum of all its divisors (p. 26) except itself, e.g. $28 = 1 + 2 + 4 + 7 + 14$, also 6 and 496.

abundant number a positive integer for which the sum of its divisors (p. 26) except for itself is greater than the number, e.g. 24, since $1 + 2 + 3 + 4 + 6 + 8 + 12 > 24$.

defective number a positive integer for which the sum of its divisors (p. 26) except for itself is less than that number, e.g. 15, since $1 + 3 + 5 < 15$.

amicable numbers pairs of positive integers each having the sum of its divisors (p. 26) apart from itself equal to the other integer.

arithmetic (*n*) the part of mathematics which deals with calculations using numbers and quantities, especially those calculations called the four rules of arithmetic (↓). **arithmetic** (*adj*).

number (*n*) loosely an element of one of the sets of integers, rational numbers (p. 26), real numbers (p. 46), complex numbers (p. 47), etc. **no.** (*abbr*), **numerical** (*adj*), **number** (*adj*).

number system (1) a system which is formed by a set of numbers, most often the positive integers (p. 24), the integers, the rational numbers (p. 26), the real numbers (p. 46) or the complex numbers (p. 47); (2) a system of numerals (p. 21) such as the Hindu-Arabic number system (p. 37).

number theory the part of mathematics which studies number systems (↑(1)) and their properties, sometimes including also some algebraic ideas.

extended number system a number system (↑) which has been made larger to allow another operation (p. 14) to be made, e.g. the positive integers (p. 24) are extended to the integers so that subtraction is always possible.

calculation (*n*) a problem in which numbers must be used to obtain the answer. **calculate** (*v*).

algorithm (*n*) a rule or method for solving a problem or calculation, e.g. the method of long multiplication (p. 23).

four rules of arithmetic the four operations (p. 14) of addition, subtraction, multiplication and division. Also known as **arithmetic operations**.

elementary operations the four rules of arithmetic; similar operations (p. 14) applied outside arithmetic.

number bonds the sums and products of single digit (p. 21) positive integers (p. 24) and their application to subtraction and division which must be known in order to use the algorithms (↑) for harder calculations, e.g. $7 + 4 = 11$, $3 \times 8 = 24$.

vulgar fraction a rational number (p. 26) written (apart from sign) as one positive integer (p. 24) divided by (or over) another, e.g. $\frac{7}{8}$, (5/3).

number bonds

$0+0=0,$	$0+1=1,$	$0+2=2,$	$\ldots,$	$0+9=9$
$1+0=1,$	$1+1=2,$	$1+2=3,$	$\ldots,$	$1+9=10$
•	•	•	•	•
$9+0=9,$	$9+1=10,$	$9+2=11,$	$\ldots,$	$9+9=18$

$0\times 0=0,$	$0\times 1=0,$	$0\times 2=0,$	$\ldots,$	$0\times 9=0$
$1\times 0=0,$	$1\times 1=1,$	$1\times 2=2,$	$\ldots,$	$1\times 9=9$
•	•	•	•	•
$9\times 0=0,$	$9\times 1=9,$	$9\times 2=18,$	$\ldots,$	$9\times 9=81$

numerator
denominator

$\frac{7}{8}$ — numerator — $\frac{5}{2}$
— denominator —

numerator (*n*) the upper or first number in a vulgar fraction (↑) corresponding to the dividend (p. 26) in a division.

denominator (*n*) the lower or second number in a vulgar fraction (↑) corresponding to the divisor (p. 26) in a division.

proper and improper fractions

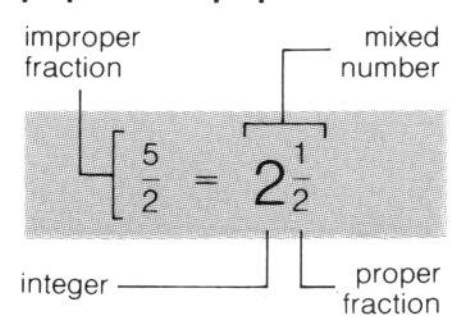

proper fraction a fraction in which (apart from sign) the numerator (↑) is less than the denominator (↑) and whose value thus lies between -1 and $+1$, e.g. $\frac{3}{4}$, $(-1/10)$.

improper fraction a fraction in which (apart from sign) the numerator (↑) is greater than the denominator (↑), e.g. $(17/8)$, $(-5/3)$.

mixed number an improper fraction (↑) written as an integer followed by a proper fraction (↑), e.g. $2\frac{1}{8}$, $-5\frac{3}{4}$.

unit fraction a fraction in which the numerator (↑) (apart from sign) is 1, e.g. $\frac{1}{4}$, $(-1/463)$.

complex fraction a fraction in which the numerator (↑) and the denominator (↑) are not both integers, e.g. $\frac{2\frac{3}{8}}{(7/12)}$.

equivalent fractions vulgar fractions (↑) which have the same value, that is which belong to the same equivalence class (p. 19) of rational numbers (p. 26), e.g. 2/5, 4/10, 6/15, ...

cancellation

$$\frac{\overset{3}{\cancel{6}}}{\underset{5}{\cancel{10}}}=\frac{3}{5},\quad \frac{\overset{\overset{3}{\cancel{6}}}{\cancel{48}}}{\underset{\underset{7}{\cancel{14}}}{\cancel{112}}}=\frac{3}{7}$$

cancellation (*n*) the expression of a vulgar fraction (↑) as the simplest equivalent fraction (↑) by dividing the numerator (↑) and the denominator (↑) by the same number; the similar process in algebra. *See also* cancellation law (p. 23). **cancel** (*v*).

decimal (*n*) a number where (apart from sign) the part which is a proper fraction (p. 31) is given as a set of digits (p. 21) placed after a decimal point (↓), the first digit after the decimal point showing tenths, the next showing hundredths, the next thousandths and so on, e.g. 3·67 = 3 + (6/10) + (7/100) and −79.482 = −79 − (4/10) − (8/100) − (2/1000). Also known as **decimal fraction**. **decimal** (*adj*).

decimal point the symbol used in a decimal; either a dot (.) or (·), or a comma (,).

finite decimal a decimal with a finite number of digits (p. 21). Also known as **terminating decimal**.

recurring decimal a decimal in which the digits (p. 21) after the decimal point (↑) do not end but repeat the same sequence (p. 91) for ever. All rational numbers (p. 26) form either finite decimals (↑) or recurring decimals. Also known as **periodic decimal**, **repeating decimal**. Symbol: a dot placed over the first and last digits to repeat, e.g. 1/3 = 0.3333333... written as $0.\dot{3}$, 1/60 = 0.0161616... written as $0.01\dot{6}$.

infinite decimal a decimal which is neither finite (↑) nor recurring (↑) and is therefore an irrational number (p. 46), e.g. the decimal forms of $\sqrt{2}$ = 1.414... (*see* square root, p. 47) and π = 3.142... (*see* pi, p. 192).

non-terminating decimal a decimal which is either infinite (↑) or recurring (↑).

prime (*n*) a positive integer (p. 24) which has exactly two factors, namely 1 and itself. One is not a prime since it has only one factor. There is no largest prime number. **prime** (*adj*).

prime number = prime (↑).

coprime (*adj*) used of two or more positive integers (p. 24) whose only common factor is unity, e.g. 25 and 28. Also used of polynomials without a common factor. Also known as **relatively prime**.

prime factor a factor of a positive integer (p. 24) which is a prime (↑) number. *See also* fundamental theorem of arithmetic (↓).

decimal point

37.638

USA, Britain

37·638

another British form, still common

37,638

many other countries

all mean

$\frac{37638}{1000}$ or $37\frac{638}{1000}$

composite number any positive integer other than unity and the primes (↑) and which thus has more than two factors.

fundamental theorem of arithmetic every composite number (↑) can be written as a product of prime factors (↑) in only one way apart from rearrangements. Also known as **unique factor theorem.**

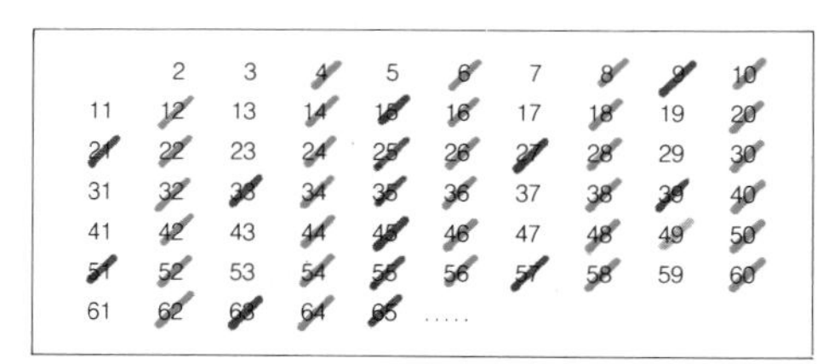

Eratosthenes' sieve
the multiples of 2 are first crossed out (red), then the remaining multiples of 3 (blue), of 5 (green), of 7 (yellow), of 11 (the first such multiple is 121) and so on. The primes are left

Eratosthenes' sieve a method of finding primes (↑) by taking a list of positive integers greater than 1 and crossing out in turn the multiples of each prime, leaving the primes themselves behind.

prime pairs pairs of primes (↑) which differ by 2, e.g. 17 and 19, 71 and 73.

highest common factor the largest positive integer which is a factor of every element of a given set of positive integers. Also known as **greatest common divisor** GCD, **greatest common measure** GCM. **HCF** (*abbr*).

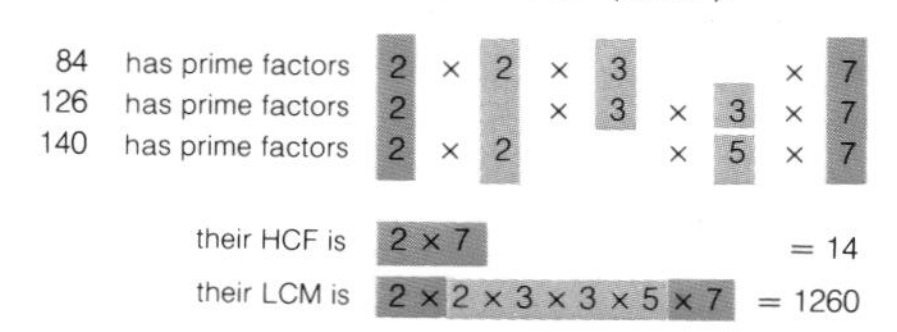

highest common factor
lowest common multiple

lowest common multiple the smallest positive integer which is a multiple of every element of a given set of positive integers. When used to find the denominator (p. 31) in the addition or subtraction of fractions it is known as the **lowest common denominator** LCD. **LCM** (*abbr*).

Euclid's algorithm an algorithm (p. 30) for the highest common factor (p. 33) of two numbers.

Euler function the Euler function of a positive integer (p. 24) is the number of positive integers less than the given integer which do not have a factor in common with it, e.g. the Euler function $\phi(12)$ of 12 is 4, the numbers 1, 5, 7, 11 being the only numbers less than 12 which do not have a factor in common with it. Also known as **totient function**.

Goldbach's conjecture a guess that every even positive integer (p. 24) greater than 2 is the sum of two primes (p. 32). It has not yet been either proved or disproved. *See also* conjecture (p. 254).

Mersenne prime a prime (p. 32) of the form $2^p - 1$ where p is prime, important in the history of number theory (p. 30).

Fermat number a positive integer (p. 24) of the form $p = 2^{2^n} + 1$ where n is a positive integer. (That is 2 raised to the power 2^n, plus 1.) Important in the construction (p. 186) of regular polygons (p. 177) having p sides, which is possible using ruler (p. 185) and compasses (p. 185) only when p is prime (p. 32).

polygonal number a positive integer (p. 24) which gives the number of spots in a pattern shaped like a regular polygon (p. 177), often called by the name of the particular polygon. Also known as **figurate number**.

triangular number a positive integer (p. 24) which gives the number of spots in a triangular pattern with an integral number of spots along each side. It gives the sequence (p. 91) 1, 3, 6, 10, 15, 21, ... Similarly **pentagonal number**, **hexagonal number** etc. *See also* square number (↓).

square number any number which is the product of an integer with itself, e.g. 0, 1, 4, 9, 16, ... Also known as **square** (*n*).

square[1] (*v*) to multiply any number or expression by itself. *See also* p. 181.

cube[1] (*n*) any number or expression which is the product of another integer with itself two more times, e.g. 0, ±1, ±8, ±27, ±64, ... *See also* p. 201. **cube** (*v*), **cubic** (*adj*).

Euclid's algorithm

for 36 and 96

$$
\begin{array}{l}
36)96(2 \\
\quad \underline{72} \\
\quad 24)36(1 \\
\qquad \underline{24} \\
\qquad 12)24(2 \\
\qquad\quad \underline{24} \\
\qquad\quad 0
\end{array}
$$

96 ÷ 36 gives 2 remainder 24
36 ÷ 24 gives 1 remainder 12
24 ÷ 12 gives 2 remainder 0
the HCF is the last non-zero remainder: 12

Goldbach's conjecture

4 = 2 + 2	10 = 5 + 5
6 = 3 + 3	= 3 + 7
8 = 3 + 5	12 = 5 + 7

and so on

Mersenne prime

$2^2 - 1 = 3$	$2^7 - 1 = 127$
$2^3 - 1 = 7$	$2^{11} - 1 = 2047$
$2^5 - 1 = 31$	$= 23 \times 89$

and so on

the first four are Mersenne primes

Fermat number

n = 0 gives $2^1 + 1 = 2$
n = 1 gives $2^2 + 1 = 5$
n = 2 gives $2^4 + 1 = 17$
n = 3 gives $2^8 + 1 = 257$
n = 4 gives $2^{16} + 1 = 65537$
and so on

polygonal numbers
the first four pentagonal numbers

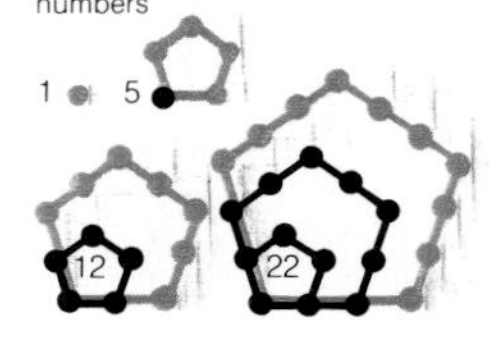

trapezoidal number

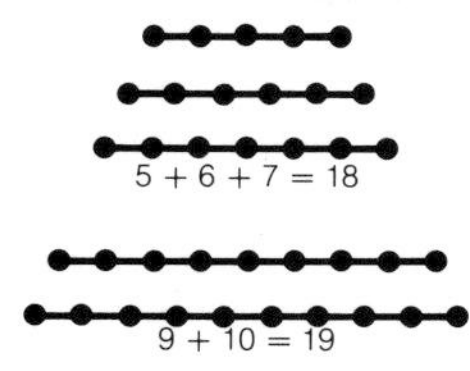

perfect square any number (especially an integer) or expression which is the exact product of another number or expression with itself, e.g. $9 = 3^2$, $2.56 = 1.6^2$, $a^2 + 2ab + b^2 = (a + b)^2$. Similarly **perfect cube**.

trapezoidal number a positive integer (p. 24) which is similar to a triangular number (↑) but which uses the shape of a trapezium (p. 181); it is any number which is the sum of consecutive integers.

pyramidal number similar to a triangular number (↑), but using a pyramid (p. 203) in three dimensions. For a triangular pyramid the first four are 1, $1 + 3 = 4$, $1 + 3 + 6 = 10$, $1 + 3 + 6 + 10 = 20$, each being formed from a sum of triangular numbers. For a square pyramid the first four are similarly 1, $1 + 4 = 5$, $1 + 4 + 9 = 14$, $1 + 4 + 9 + 16 = 30$, each being formed from a sum of square numbers.

index[2] (*n*) when an index is used as a superscript (p. 8) placed after a number or expression it most often shows a power. If it is a positive integer (p. 24) n it shows that unity is to be multiplied by the number or expression n times, e.g. $5^3 = 1 \times 5 \times 5 \times 5$; $a^2 = 1 \times a \times a$. If it is not a positive integer then it still obeys the index laws (↓). In particular $x^0 = 1$ for all non-zero x. *See also* pp. 8, 115. Also known as **exponent**. **indicial** (*adj*).

index laws if a is a number, and m, n are rational numbers (p. 26), then:

I $a^m \times a^n = a^{m+n}$;
II $a^m \div a^n = a^{m-n}$;
III $(a^m)^n = a^{mn}$.

Similar laws are also true for some other systems, e.g. in group (p. 72) theory. If m, n are real numbers (p. 46), then Cauchy sequences (p. 94) allow indices (↑) to be defined. Similarly if they are complex numbers (p. 47), Demoivre's theorem (p. 50) allows them to be defined.

index laws

$3^3 \times 3^4 = 3^{(3+4)} = 3^7$

$2^{1/2} \times 2^2 = 2^{(1/2+2)} = 2^{5/2}$

$1.5^{-3} \times 1.5^2 = 1.5^{-1}$

$3^3 \div 3^4 = 3^{(3-4)} = 3^{-1}$

$2^{1/2} \div 2^2 = 2^{(1/2-2)} = 2^{-3/2}$

$(6^{-1/2})^2 = 6^{(-1/2 \times 2)} = 6^{-1}$

$(6^{-3})^2 = 6^{(-3 \times 2)} = 6^{-6}$

power[1] (*n*) the index (↑) of a number or expression; the number or expression is said to be 'raised' to the given power. *See also* pp. 126, 270.

raise[1] (*v*) (to a power) *see* power (p. 35). *See also* p. 175.

involution[1] (*n*) the process of raising (↑) to a power. *See also* p. 55. **involve** (*v*).

involve (*v*) to concern, mix up with.

billion (*n*) in the USA, the number 10^9 = 1 000 000 000. In Britain now usually the same, but in older books, 10^{12}.

trillion (*n*) in the USA, the number 10^{12} = 1 000 000 000 000. In Britain now usually the same, but in older books, 10^{18}.

modular arithmetic an arithmetic obtained from the positive integers (p. 24) using only the remainders on division of each number by a fixed positive integer. Also known as **finite arithmetic**.

modulus[1] (*n*) the divisor (p. 26) *n* used in defining any given modular arithmetic (↑), which is then said to be *modulo n*. *See also* pp. 49, 140. **modulo** (*adj*).

congruent modulo *n* used of positive integers (p. 24) which have the same remainder on division by a positive integer *n*. **mod *n***. (*abbr*).

clock arithmetic similar to modular arithmetic (↑), except that instead of 0, the value of the divisor (p. 26) *n* is used. (Not just used for $n = 12$ as on a clock.)

residue (*n*) the remainder, usually in modular arithmetic (↑).

residue class an equivalence class (p. 19) of positive integers (p. 24) having the same residues (↑) under a given modular arithmetic (↑), e.g. 3, 11, 19, 27, ... is a residue class in arithmetic modulo (↑) 8.

quadratic residue if *p* is a prime (p. 32) number other than 2, then the residues (↑) of $1^2, 2^2, 3^2, \ldots, (p-1)^2$ when divided by *p* form the set of quadratic residues modulo (↑) *p* which has $\frac{1}{2}(p-1)$ elements.

Wilson's theorem if *p* is a prime (p. 32) number, then $(p-1)! + 1$ (*see* factorial, p. 97) is divisible by *p*.

Fermat's (little) theorem if *p* is a prime (p. 32) number and *p* is not a divisor (p. 26) of *a*, then $a^{p-1} - 1$ is divisible by *p*.

modular arithmetic
arithmetic modulo 8

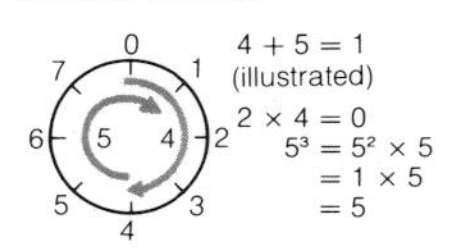

complete structure tables for addition and multiplication under arithmetic modulo 5

+	0	1	2	3	4
0	0	1	2	3	4
1	1	2	3	4	0
2	2	3	4	0	1
3	3	4	0	1	2
4	4	0	1	2	3

×	0	1	2	3	4
0	0	0	0	0	0
1	0	1	2	3	4
2	0	2	4	1	3
3	0	3	1	4	2
4	0	4	3	2	1

quadratic residue
in arithmetic modulo 7 residues can be found thus

$1^2 = 1$
$2^2 = 4$
$3^2 = (9 - 7) = 2$
$4^2 = (16 - 14) = 2$
$5^2 = (25 - 21) = 4$
$6^2 = (36 - 35) = 1$

the set has the three elements {1, 2, 4}

Wilson's theorem

if p = 7
$(7 - 1)! + 1 = (6.5.4.3.2.1) + 1$
$= 721$
which is divisible by 7

Fermat's (little) theorem

if p = 5, a = 11
$11^{5-1} - 1 = 14640$
which is divisible by 5

Euler's theorem

if a = 5, b = 12, so that
$\varphi(b) = 4$, then
$5^{\varphi(b)} = 5^4 = 624$
which is divisible by 12

Arabic digits

٠	0
١	1
٢	2
٣	3
٤	4
٥	5
٦	6
٧	7
٨	8
٩	9

decimal system

in 304.7

the 3 has value
$3 \times 10^2 = 300$

the 0 has value
$0 \times 10^1 = 0$

the 4 has value
$4 \times 10^0 = 4$

the 7 has value
$7 \times 10^{-1} = \frac{7}{10}$

the 0 is a place-holder showing zero value in the tens position

Fermat's last theorem there are no solutions in non-zero integers of the equation $x^n + y^n = z^n$ where n is an integer greater than 2. This is not a theorem, but a conjecture (p. 254) since it has not yet been proved or disproved.

Euler's theorem if a and b are coprime (p. 32) positive integers (p. 24), and $\phi(b)$ is the Euler function (p. 34) of b, then $a^{\phi(b)} - 1$ is divisible by b.

Hindu-Arabic number system the number system (p. 30 (2)) in common use using the ten digits (p. 21) 0, 1, 2, . . . , 9, with 0 or zero as a place-holder for numbers like 20, 100, 0.708. In some parts of the world the same system is used but with different symbols, e.g. the Arabic digits.

Arabic digits the digits (p. 21) corresponding to the digits 0 to 9 in the Hindu-Arabic number system (↑) and used in Arabic speaking countries.

decimal system a number system such as the Hindu-Arabic number system (↑) which counts in 10s and powers of 10. Also known as **denary system** which is less common, but more correct.

base[1] (*n*) the base of a number system (p. 30) is that number on which the system is built, as 10 is in the decimal system (↑). *See also* pp. 167, 179. Also known as **radix**.

exchange value the value of one symbol or position in a number system (p. 30) in relation to the next, e.g. the I and V in the Roman number system (p. 38) have exchange value 5; in the decimal system (↑) the exchange value is 10.

shift (*n*) the operation (p. 14) of moving a digit (p. 21) from one place to another in a numeral (p. 21); in the decimal system (↑) this changes its value by a power of 10. **shift** (*v*).

binary (*adj*) having two parts; having base (↑) 2 so that the numerals (p. 21) are 1, 10, 11, 100, 101, . . . , instead of 1, 2, 3, 4, 5, ... The binary system is important in computing (p. 42) and it can be extended to deal with fractions in a similar way to decimals. Sometimes incorrectly known as **bicimal**.

hexadecimal

the hexadecimal numbers
0, 1, 2, 3, 4, 5, 6, 7, 8, 9, A, B, C, D, E, F, 10, 11, ... 19, 1A ...
are used in computing

bit (*n*) abbreviation of binary (p. 37) digit (p. 21). It is either 0 or 1 and is used in computing.

ternary (*adj*) having three parts; having base 3. Similarly **quaternary** (4), **quinary** (5), **octal** or **octonary** (8), **duodecimal** or **duodenary** (12), **hexadecimal** or **hexadenary** (16).

sexagesimal (*adj*) having base 60 as in some measurements of time and angle.

binary code a way of describing things using binary (p. 37) digits (p. 21), which define a set of dichotomies (p. 13).

nim (*n*) a game for two people played with heaps of matches from which players take some or all in turn. Binary (p. 37) numbers can be used to help to find ways of winning the game.

place value the value a digit (p. 21) has because of its place in a numeral (p. 21), e.g. in 546 the digit 4 has value 40.

place-holder a digit (p. 21) which has zero place value (↑), written as 0 in the Hindu-Arabic number system (p. 37).

Babylonian number system a number system (p. 30 (2)) used by the Babylonians with mixed base of 10 and 60 from which the methods of measuring time and angle used today are probably obtained.

Egyptian number system ancient Egyptian number system (p. 30 (2)) using base 10 and repeated digits. There were also symbols for unit fractions (p. 31) and for 2/3; other fractions had to be expressed as sums of different unit fractions.

Greek number system a number system (p. 30 (2)) used in ancient Greece, the letters of the alphabet meaning the numbers 1 to 9, 10 to 90 in tens and 100 to 900 in hundreds. There is no zero.

Roman number system a number system (p. 30 (2)) used by the ancient Romans and up to the present day with repeated letters and a mixed base of 5 and 10.

Babylonian number system

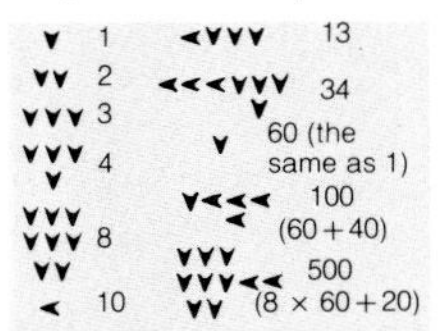

Egyptian number system

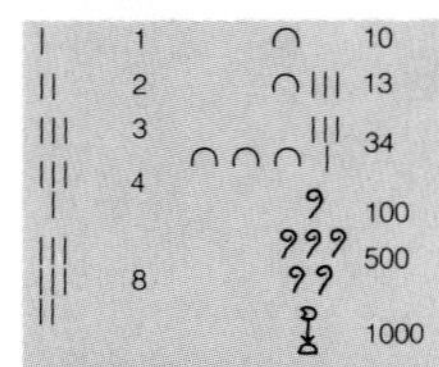

Greek number system

α	1	κ	20	τ	300
β	2	λ	30	υ	400
γ	3	μ	40	φ	500
δ	4	ν	50	χ	600
ε	5	ξ	60	ψ	700
ϝ	6*	ο	70	ω	800
ζ	7	π	80	ϡ	900*
η	8	ϙ	90*	ια	100
θ	9	ρ	100	ιγ	13
ι	10	σ	200	λδ	34

*older letters not in the usual Greek alphabet

Roman number system

I	1	XVI	16
II	2	XIX	19
III	3	XXXIV	34
IV	4	L	50
V	5	LXVII	67
VII	7	LXXIX	79
IX	9	XCIII	93
X	10	C	100
XIII	13	D	500
XIV	14	M	1000

a single symbol of less value placed before one of greater value shows that it must be subtracted

IV = 5 − 1 = 4
VI = 5 + 1 = 6

percentage (*n*) the number of parts for each hundred. **percent** (*abbr*). Symbol: %, e.g. 38% = 38/100.

profit (*n*) the gain obtained from lending, buying, selling or other use of money. **profit** (*v*).

investment (*n*) the lending of money for future profit. **invest** (*v*).

interest (*n*) the profit (↑) obtained from the lending of money; a share in a business.

borrow[2] (*v*) to be lent, obtain the use of something which must be returned. *See also* p. 23.

loan (*n*) that which is lent, often money.

capital (*n*) the total sum of money which is invested (↑) in a company, or owned by a person; an amount of money lent or borrowed (↑).

principal (*n*) an amount of money lent or borrowed (↑) for interest (↑).

rate of interest the amount of interest (↑) charged when money is lent, usually given as a percentage (↑) of the principal (↑).

period[1] (*n*) a length of time generally, often the time for which money is lent. *See also* pp. 133, 276.

simple interest interest (↑) which stays at a fixed percentage (↑) of the principal (↑) no matter how long the money is lent for.

compound interest interest (↑) which is added to the principal (↑) after a given time (perhaps six months or a year) and on which interest is then also paid.

share (*n*) a part, often a part of the capital (↑) of a company.

stock (*n*) money lent to a government or company at a fixed rate of interest (↑), and which may then be sold to other people at its market value (p. 40).

shareholder (*n*) a person who owns shares in a company. Similarly **stockholder**.

nominal value (1) the value given to a share or stock (↑) on which the interest (↑) is calculated, but which usually differs from the market value (p. 40) and sometimes from the issue price (p. 40) as well; (2) the given price of any article on which a discount (p. 40) or premium (p. 40) is calculated. Also known as **nominal price**.

simple interest

$$I = \frac{PRT}{100}$$

where *P* is the principal
R is the rate percent
T is the time of loan

compound interest

$$I = P\left(1 + \frac{R}{100}\right)^T$$

where *P* is the principal
R is the rate percent
T is the time of loan
and must be measured in the units of time at which the interest is compounded

issue price the price at which a share or stock (p. 39) is first sold.
market value the price at which any article, especially a share or stock (p. 39) can be sold.
dividend[2] (*n*) the money paid out by a company to a shareholder (p. 39) as part of a company's profit (p. 39). *See also* p. 26.
yield (*n*) the dividend (↑) on a share or the interest (p. 39) on a stock (p. 39) given as a percentage of its market value (↑).
net[1] (*adj*) (1) calculated without the costs of buying included; (2) (of weight) not counting the weight of the packing or container (e.g. a box). *See also* p. 202. Also written **nett**.
gross (1) (*adj*) in total, with the cost of buying included; (2) (of weight) counting the weight of the packing or container; (3) (*n*) one hundred and forty-four of any article.
tare (*adj*) of the weight of the packing or container, the difference between the gross (↑) and net (↑) weights.
discount (*n*) an amount of money taken off the nominal value (p. 39) of an article, most often given as a percentage and sometimes given for immediate payment.
premium (*n*) an amount of money added on to the nominal value (p. 39) of an article, most often given as a percentage.
account (*n*) a report, often about money, e.g. the money left by a person at a bank.
deposit (*n*) an amount of money lent to a bank or similar company; an amount of money left with a seller of goods to show that it is intended to buy them.
balance (*n*) the amount of money in an account (↑) at a bank or similar company.
overdraft (*n*) a balance (↑) which is negative, so that money is owed.
hire purchase the buying of goods by borrowing (p. 39) money at interest (p. 39), the money then being paid back in regular amounts.
instalment (*n*) an amount of money paid regularly, as in hire purchase (↑).
tax (*n*) money paid to the government of a country by its people.

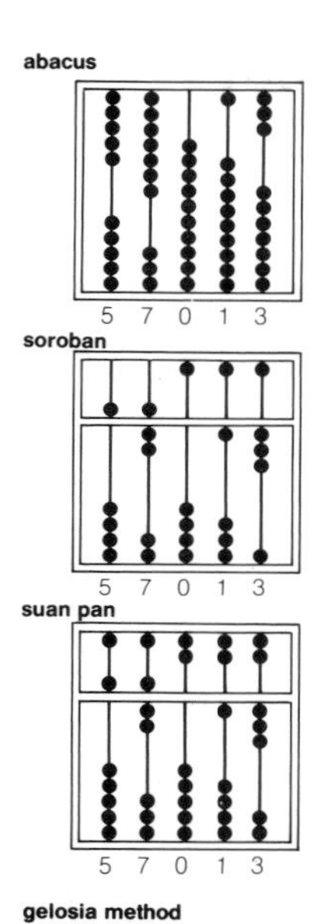

duty (*n*) a sort of tax (↑), especially one on articles such as oil or wine.

poundage (*n*) a payment calculated on each pound of money in places where the unit of money is the pound.

counting board a marked board on which counters are placed to do calculations.

abacus (*n*) a frame with balls on wires used for doing calculations, especially in Asian countries. In the simplest form used in the USSR there are ten balls on each wire. **abacuses** (*pl*), **abaci** (*pl*).

soroban (*n*) a Japanese form of abacus (↑) in which each wire has two parts, the first having four balls, and the second having one ball counting as five balls in the first part.

suan pan a Chinese form of abacus (↑) in which each wire has two parts, the first having five balls, and the second having two balls, each counting as five balls in the first part.

tally

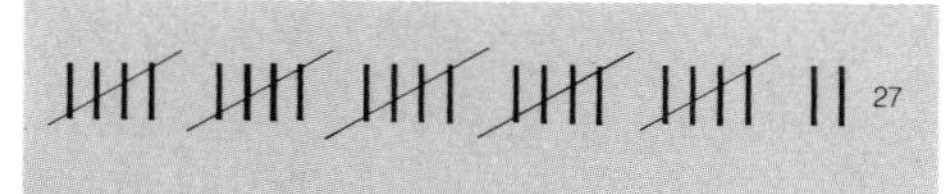

tally (*n*) a mark on a piece of paper or wood used for counting. **tally** (*v*).

gelosia method

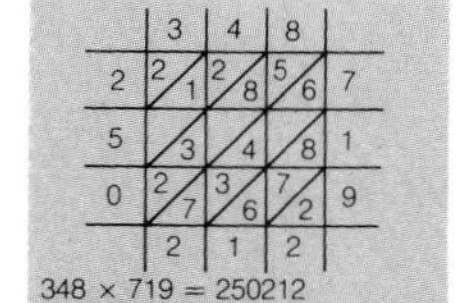

gelosia method a way of multiplying numbers by using an array (p. 83) of squares cut by diagonal (p. 177) lines. Also known as **grating method**, **lattice method**.

Napier's bones a way of multiplying numbers obtained from the gelosia method (↑) in which the numbers are written on vertical pieces of bone or other material. Also known as **Napier's rods**.

nomogram

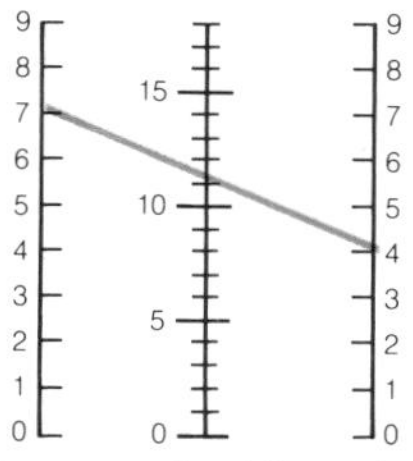

a nomogram for addition and subtraction showing 7 + 4 = 11 and 11 − 7 = 4

nomogram (*n*) a set of marked lines (which may not be parallel or even straight) across which a straight edge may be placed to make a calculation or solve a formula. Also known as **alignment chart**.

nomography (*n*) the drawing and use of nomograms (↑).

slide rule an instrument used for calculating, in which two sets of marked edges slide against each other. Most often used for multiplying and dividing using a logarithmic scale (p. 140).

Gunter scale an early form of slide rule (↑) using a logarithmic scale (p. 140) and dividers (p. 185).

calculator (*n*) a person or machine which calculates, now usually a small machine worked by electricity which can be carried in the pocket.

programmable calculator a calculator (↑) which has special keys so that it can also be used as a simple computer (↓).

graphical calculator a calculator (↑) which can also be used to plot simple graphs.

reverse Polish notation a notation where an operation (p. 14) is written after the two elements to which it applies, used in a few calculators (↑) where the two numbers are entered followed by the operation to be performed on them, e.g. 3, 6, x, gives 18. It avoids the use of brackets (p. 9).

Polish notation a notation where the operation (p. 14) is written before the two elements to which it applies.

computer (*n*) any machine used for calculating, now usually a machine worked by electricity which can be used for long and difficult calculations and for processing data, as well as having other uses. **computing** (*n*), **compute** (*v*).

digital computer a computer (↑) where calculations are made in discrete (p. 64) amounts, usually given in binary (p. 37) form. Most present day computers work in this way.

analogue computer a computer (↑) where all calculations are made in a continuous (p. 63) way, e.g. by the turning of toothed wheels or by changing electrical forces, so that the idea of the real number line (p. 46) is copied. A slide rule (↑) is an example of an analogue computer. Also written **analog computer**.

slide rule

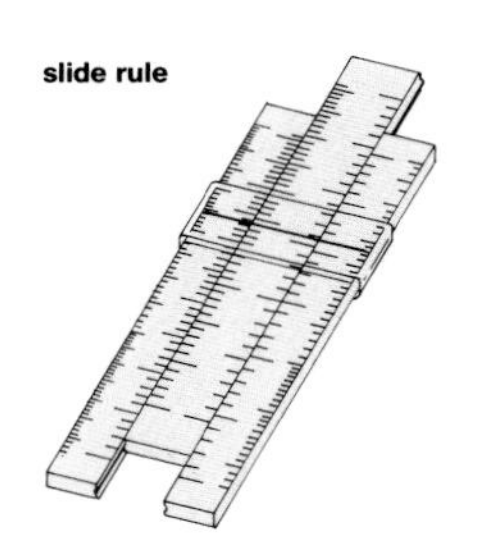

calculator

reverse Polish notation

$(6 + 2) \times 14 \div (7 - 3) = 28$

can be written

$6, 2, +, 14, \times, 7, 3, -, \div = 28$

error (*n*) the difference between the true value and the calculated value in a result caused because exact measurement or calculation is not possible. *It does not mean the same as* mistake.

approximation (*n*) a value given which contains some small amount of error (↑); a value given by the first few terms in a convergent series (p. 94). **approximate** (*v*), **approximate** (*adj*). Symbols: $\approx$, $\doteqdot$ for 'is approximately equal to'.

linear approximation a straight line which approximates to a curve or set of points; the corresponding idea in equations.

accuracy (*n*) the amount of error (↑), the nearness of a particular value to the true value, sometimes given as a ratio of the error to the result, e.g. an answer which is given as 900 may have an error of 1 : 100 and so the correct result would lie between 891 and 909. **accurate** (*adj*).

absolute error the error (↑) given as an exact quantity and not as a ratio or percentage.

relative error the error (↑) given as a ratio to the total amount. *See also* accuracy (↑).

percentage error the error (↑) given as a percentage of the total amount.

random error an error (↑) which happens in a random way.

systematic error an error (↑) in measurement which happens in a regular pattern such as an inaccurate instrument of measurement, and can thus sometimes be allowed for.

rounding error an error (↑) caused by using too few decimal places (p. 44) in a calculation. Also known as **rounding-off error**.

truncation error an error (↑) similar to rounding error (↑) caused by taking too few terms in a sequence (p. 91).

precision (*n*) the nearness of a set of values to the true value or to each other (*compare* accuracy, ↑). It is concerned with dispersion (p. 117) rather than location (p. 113).

tolerance (*n*) the limit of error (↑) of measurement allowed in practical work.

calculus of errors the study of errors (↑) and tolerances (↑).

iteration (*n*) a repeated procedure (p. 290) which allows better and better approximations to an answer to be made. **iterative** (*adj*).

cobweb curve an iterative (↑) method of obtaining an approximate solution to an equation $f(x) = 0$ by plotting $y = f(x) + x$ and $y = x$ and obtaining better and better approximations to the correct result in a spiral (p. 230) pattern on the graph.

decimal places a number is given to *n* decimal places when there are *n* digits (p. 21) after the decimal point (p. 32). Thus 3.142 has three decimal places. A number such as 0.7301 is 0.730 to three decimal places, but 0.7305 is given as 0.731, the final '5' being rounded (↓) up by convention (p. 9).

significant figures a number is given to *n* significant figures when there are *n* meaningful digits (p. 21) in it, e.g. 3740, 0.730, 0.000438 and 3.02 are all given to three significant figures (though the first could also be given to four). Zeros at the start of numbers are ignored, and zeros at the end of integers are ambiguous (p. 254); they may or may not be significant.

truncate (*v*) to cut short; used of series (p. 91), solid figures, and numbers. In the last case it is used when numbers are approximated to a given number of decimal places (↑) or significant figures (↑), by replacing non-zero final digits (p. 21) by zeros, e.g. 3748 by 3740. *See also* p. 203. **truncated** (*adj*), **truncation** (*n*).

round[1] (*v*) used when numbers are approximated to a given number of decimal places (↑) or significant figures (↑), by replacing non-zero final digits (p. 21) by zeros and increasing the final digit by 1 when this gives a closer number, e.g. 3748 becomes 3750. When the result is smaller this is *rounding down*, when larger *rounding up*. If these are equally close, it is conventional (p. 9) to round up, except sometimes in statistics, when rounding is done to the nearest even digit, e.g. 3745 to 3740, but 3755 to 3760. *See also* p. 192. Also known as **round off**. **rounding** (*adj*), **rounding off** (*adj*).

cobweb curve

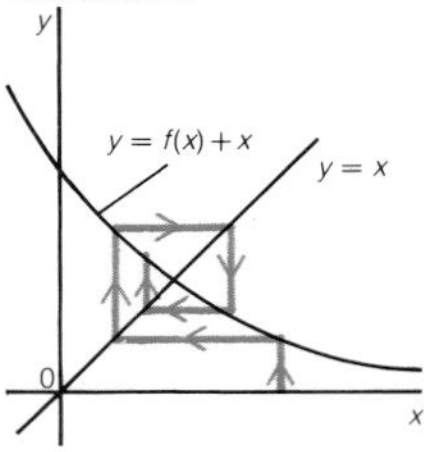

the 'cobweb' produced by the approximations shown in red to the intersection of $y = f(x) + x$ and $y = x$

order of magnitude a rough approximation to the size of a number (often to one significant figure, ↑, or to the nearest power of 10); also used of two functions whose ratio remains finite as the variable tends to infinity and which are said to be of the same order of magnitude. If $f(x)/g(x)$ tends to 0 as x tends to infinity, then f has a smaller order of magnitude than g. If $f(x)/g(x)$ tends to any other finite value, the orders of magnitude are equal. Symbol: in *both* cases $f = 0(g)$.

scientific notation a notation in which a number is written as another number multiplied by some positive or negative power of 10, e.g. 3740 can be given as 37.4×10^2 and 0.000374 can be given as 0.374×10^{-3}. Also known as **floating point notation**.

standard form a notation in which a number is given in scientific notation (↑) as $a \times 10^n$ where $1 \leqslant a < 10$ and n is an integer, e.g. 3.74×10^3, 3.74×10^{-4} are the standard forms of the examples which are given under scientific notation. Also known as **standard index form**.

interpolation

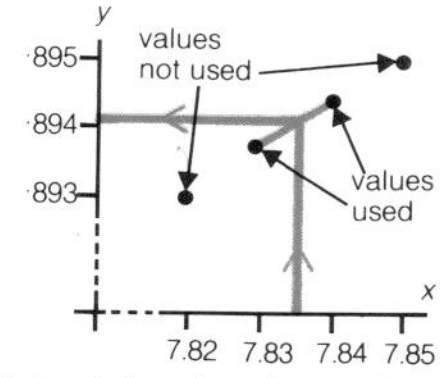

interpolation of $y = \log_{10} x$ using values from tables. The estimate is $\log_{10} 7.836 = 0.8941$

interpolation (n) the estimation (p. 109) of the value of a function at a given point from the values for one or more points on each side, e.g. given that $\log_{10} 7.83 = 0.8938$ and $\log_{10} 7.84 = 0.8943$, by interpolation $\log_{10} 7.836 = 0.8941$. This is linear interpolation by proportional parts (p. 28) using one point on each side; more difficult methods can be used to give better results. **interpolate** (v).

extrapolation

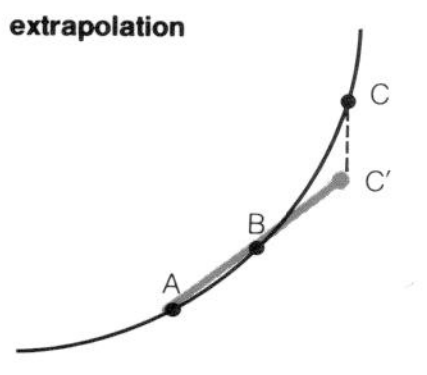

the dangers of using two points A and B on a curve to extrapolate a third C′ which is nowhere near the point C on the curve which it is used to estimate

extrapolation (n) the same idea as interpolation (↑) but used to find values outside the range of the given values, e.g. in the example under interpolation the given values can be used to estimate (p. 109) $\log_{10} 7.846$. *This procedure is usually considered to be unwise.* **extrapolate** (v).

collocation polynomial a polynomial of degree n which meets a curve at $n + 1$ given points and can be used to find approximate values of the function given by the curve in the range (p. 113) of the chosen values.

real number line position of π

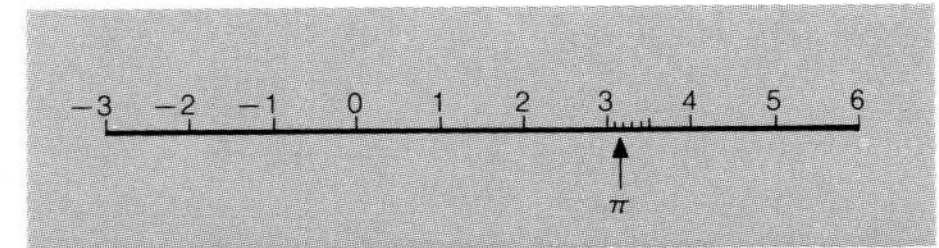

π lies on the real number line between 3.1 and 3.2. It can be defined by a sequence of upper bounds:

4, 3.2, 3.15, 3.142, 3.1416, 3.14160, ...

or by a sequence of lower bounds:

3, 3.1, 3.14, 3.141, 3.1415, 3.14159, ...

real number every number which can be written as a decimal (including a non-terminating decimal, p. 32), whether positive or negative. This includes the roots (↓) of all positive numbers, e.g. $\sqrt{2}$. The full definition is difficult but can be seen through the idea of completeness (p. 95), using either the Cauchy sequence (p. 94) or the Dedekind cut (↓). The real numbers form a complete ordered field (p. 76). Symbol for set of real numbers: $\mathbb{R}$.

real number line a continuous line on which every point stands for a number, usually imagined to be straight across the page. No other points are possible on the line, so it shows the completeness (p. 95) of the real numbers (↑).

irrational number any real number (↑) which is not a rational number (p. 26).

irrational exponents indices (p. 35) or exponents which are irrational numbers (↑). They must be defined by a sequence (p. 91) of approximations.

Dedekind cut a partition (p. 13) of the rational numbers (p. 26) into two sets which can be used in defining the real numbers (↑). Also known as **Dedekind section**.

algebraic number any number which is a solution of a polynomial equation with rational (p. 26) coefficients (p. 61). They may be real (↑) or complex numbers (↓), e.g. 3, -5, 6.71, $7 + \sqrt{3}$, $3 + 4i$.

algebraic integer an algebraic number (↑) which is a root (↓) of a polynomial equation in which the coefficient (p. 61) of the highest power is 1 and the other coefficients are all integers.

transcendental number a number which is not an algebraic number (↑), e.g. π (pi, p. 192), 3π, e (the exponential constant, p. 138), e^2.

surd (*n*) an irrational (↑) algebraic number (↑), usually one given in the form of a root (↓), e.g. $\sqrt{5}$, $3 + \sqrt[3]{7}$.

commensurable (*adj*) (1) having ratios which are rational numbers (p. 26); (2) being quantities which can be measured in the same units. Similarly **incommensurable**.

continued fraction a fraction of the form

$$a_1 + \cfrac{1}{a_2 + \cfrac{1}{a_3 + \dots}}, \text{ often written as}$$

$$a_1 + \frac{1}{a_2 +}\frac{1}{a_3 +}\dots, \text{ where the sequence}$$

(p. 91) $a_1, a_2, a_3, \dots$, may or may not end.

root[1] (*n*) the opposite of a power (p. 35), e.g. if $5^4 = 625$, then 5 is a fourth root of 625. Symbol for the positive *n*th root of *a* is $\sqrt[n]{a}$. *See also* p. 58.

square root the second root (↑), e.g. the square roots of 49 are +7 and −7, since $(+7)^2 = (-7)^2 = 49$. Symbol: $\sqrt{}$.

cube root the third root (↑), e.g. the cube root of $x^3 + 3x^2 + 3x + 1$ is $x + 1$.

extraction (*n*) the process of finding a root (↑). Also known as **evolution**.

radical (*n*) a root (↑), especially one which is not an exact number; an expression containing a root.

rationalize (*v*) to make rational, especially used of the process of expressing the denominator (p. 31) of a fraction so that it does not contain a root (↑). **rationalization** (*n*).

complex number a number of the form $a + ib$ where *a* and *b* are real numbers (↑), and *i* (sometimes written as *j*) is the square root (↑) of −1. A more mathematical definition is given by one of the set of ordered pairs (p. 19) of real numbers {a, b} for which:

I $\{a, b\} = \{c, d\}$ if and only if (p. 246) $a = c$ and $b = d$;

II $\{a, b\} + \{c, d\} = \{a + c, b + d\}$;

III $\{a, b\} \times \{c, d\} = \{ac - bd, ad + bc\}$.

Symbol for set of complex numbers: $\mathbb{C}$.

commensurable

similarly incommensurable

3.142 and 2.718 are commensurable

π and e, 3 and π, 3 and $\sqrt{7}$ are incommensurable

3 metres and 7 feet are commensurable

3 metres and 7 grams are incommensurable

continued fraction

$$1 + \frac{1}{2 + \frac{1}{3}} = 1 + \frac{6}{7} = \frac{13}{7}$$

rationalize

the denominator of $\frac{1}{(2+\sqrt{3})}$

$$\frac{1}{2+\sqrt{3}} = \frac{2-\sqrt{3}}{(2+\sqrt{3})(2-\sqrt{3})} = \frac{2-\sqrt{3}}{4-3} = 2-\sqrt{3}$$

complex number

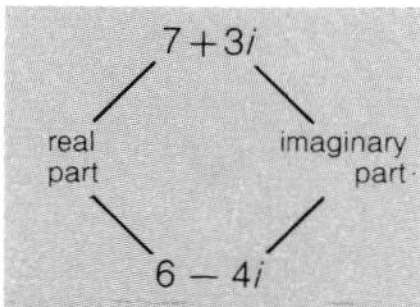

real part of a complex number the real number (p. 46) a which is part of the complex number (p. 47) $a + ib$ where b is also real.

imaginary part of a complex number the number, b, which is part of the complex number (p. 47) $a + ib$ where a and b are both real.

conjugate complex numbers pairs of complex numbers (p. 47) $a + ib$ and $a - ib$ whose product is $a^2 + b^2$, which is a real number (p. 46), e.g. $7 + 3i$ and $7 - 3i$. Also known as **conjugate numbers**.

realize (*v*) to make real, especially to multiply a complex number (p. 47) by its conjugate complex number (↑) to make it real, often used with the denominator (p. 31) of a complex fraction (p. 31), the numerator (p. 31) also being multiplied by the same number. *Compare* rationalize (p. 47). **realization** (*n*).

Argand diagram the representation of a complex number (p. 47) $z = a + ib$ or $\{a, b\}$, (where a and b are real numbers, p. 46) by the cartesian coordinates (p. 214) (a, b).

conjugate complex numbers

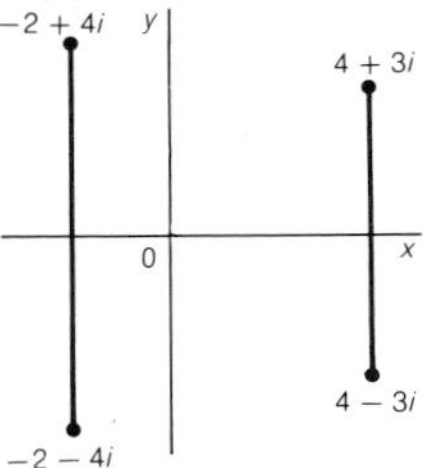

two pairs of conjugate complex numbers $4 \pm 3i$ and $-2 \pm 4i$

realize the denominator of $\dfrac{1}{2 + 3i}$

$$\frac{1}{2+3i} = \frac{2-3i}{(2+3i)(2-3i)} = \frac{2-3i}{2^2+3^2} = \frac{1}{13}(2-3i)$$

Argand diagram
real and imaginary axes

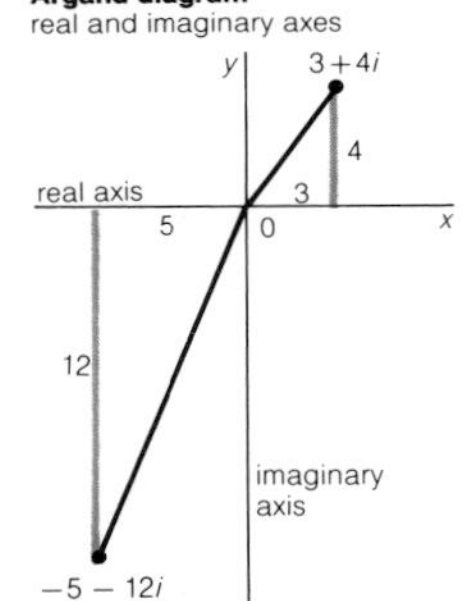

$3 + 4i$ and $-5 - 12i$ on the Argand diagram

affix (*n*) in the Argand diagram (↑), the affix of the complex number (p. 47) $z = a + ib$ is the point $P\,(a, b)$ representing z.

real axis the x-axis (p. 215) of the Argand diagram (↑) whose points form the real number line (p. 46).

imaginary axis the y-axis (p. 215) of the Argand diagram (↑).

complex plane the plane drawn in the Argand diagram (↑) which represents the set of all complex numbers.

extended complex plane the complex plane (↑) together with a point which is the image (p. 51) of the origin under the transformation (p. 54) $w = 1/z$. This transformation is then a bijection (p. 52) of the extended complex plane onto itself.

ideal complex point the additional point in the extended complex plane (↑). Also known as **point at infinity**. *See also* p. 218.

real (*adj*) (1) as in real number (p. 46), related to the real part of a complex number (p. 47); (2) in geometry used of points, lines, etc which exist in the everyday world.

imaginary (*adj*) (1) related to the imaginary part of a complex number (p. 47); (2) in geometry, imaginary is used to describe points, lines, etc which cannot exist in the everyday world, such as a tangent to a circle from a point inside it.

Argand diagram
moduli and arguments

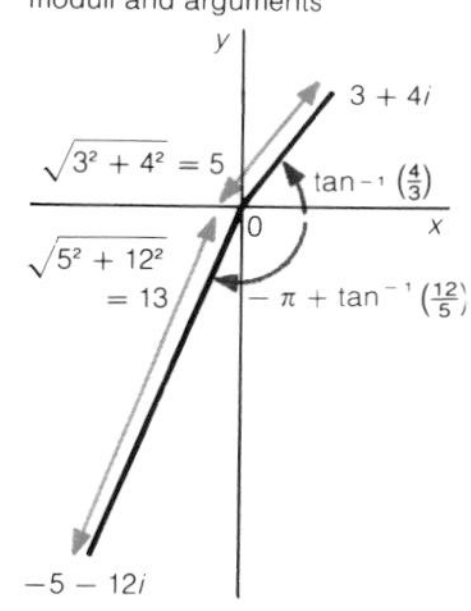

Argand diagram with the complex numbers 3 + 4*i* and −5 − 12*i* showing their moduli (red) and the principal values of their arguments (blue)

modulus[2] (*n*) the modulus of the complex number (p. 47) $a + ib$ is $\sqrt{(a^2 + b^2)}$, which is the distance of the point (a, b) on the Argand diagram (↑) from the origin. For real numbers (p. 46) this gives the number without a sign and is written, e.g., $|a|$ so that $|+7| = |-7| = 7$. *See also* pp. 36, 140. Also known as **absolute value**.

argument[1] (*n*) (1) the value of a function $f(x)$ for any given value of x; (2) the angle in the Argand diagram (↑) between the positive real axis (↑) and the line OP joining the origin O to the point P (a, b) which represents the complex number $a + ib$; the same angle used as the angular coordinate in polar coordinates (p. 216). *See also* principal value (p. 57), p. 243. Also known as **amplitude**.

modulus-argument form the complex number (p. 47) $a + ib$ written as $r\cos\theta + ir\sin\theta$, where r is its modulus (↑) and θ is its argument (↑). In the Argand diagram (↑) this is represented by the point (r, θ) in polar coordinates (p. 216).

cis (*n*) cis θ is short for $\cos\theta + i\sin\theta$. *See* modulus-argument form (↑).

Euler's formula[1] $\cos\theta + i\sin\theta = \text{cis}\,\theta = e^{i\theta}$. If $\theta = \pi$, the well known formula $e^{i\pi} = -1$ is obtained. *Do not mistake for* Euler's formula for polyhedra (p. 202), or π (p. 149).

exponential form of a complex number the form $re^{i\theta}$. *See* Euler's formula (↑), modulus-argument form (↑).

Demoivre's theorem

if $n = 2$

$(\cos\theta + i\sin\theta)^2 = \cos 2\theta + i\sin 2\theta$ gives $\begin{cases}\cos^2\theta - \sin^2\theta = \cos 2\theta \\ 2\sin\theta\cos\theta = \sin 2\theta\end{cases}$

Demoivre's theorem $(\cos\theta + i\sin\theta)^n = \cos n\theta + i\sin n\theta$, where n is an integer. This can be written $(\text{cis}\,\theta)^n = \text{cis}\,n\theta$. If n is a rational number (p. 26) p the theorem becomes cis $p\theta$ is one value of $(\text{cis}\,\theta)^p$. Also known as **de Moivre's theorem**.

roots of unity the nth roots of unity are the n solutions in complex numbers (p. 47) of the equation $z^n = 1$ which lie at the vertices of a regular n-sided polygon in the Argand diagram (p. 48).

roots of unity

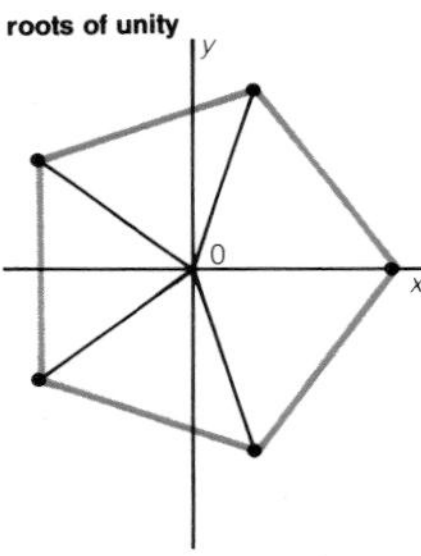

the complex fifth roots of unity form a regular pentagon on the Argand diagram

cyclotomic equation the equation $z^n = 1$ which gives the n roots of unity (↑).

primitive root of unity a root of unity (↑) whose lowest positive power which is equal to unity is the nth power, e.g. if $1, \omega, \omega^2, \ldots \omega^5$ are the sixth roots of unity, then the lowest powers of each which are equal to unity are given by 1^1, $(\omega)^6$, $(\omega^2)^3$, $(\omega^3)^2$, $(\omega^4)^3$ and $(\omega^5)^6$. The primitive roots are ω and ω^5, and there are two of them, which is the Euler function (p. 34) of 6.

principal root of an equation in complex numbers (p. 47): that solution for which the argument (p. 49) has the smallest non-negative value.

complex exponents indices (p. 35) or exponents which are themselves complex numbers (p. 47). If w, z are complex, then z^w may be defined as e to the power $w \operatorname{Ln} z$ where $\operatorname{Ln} z$ is the natural logarithm (p. 139) of the complex number, z. This expression is not a function (↓) as it has many values and the index laws (p. 35) then become difficult to use.

Gaussian integer a complex number (p. 47) in which both the real part (p. 48) and the imaginary part (p. 48) are integers, e.g. $7 - 4i$.

Gaussian prime a prime (p. 32) which cannot be factorized into two Gaussian integers (↑), e.g. 3 and 7 have no factors in Gaussian integers, but 5 is not a Gaussian prime because $5 = (1 + 2i)(1 - 2i)$.

function (*n*) a binary relation (p. 16) in which for every element of the first set, one and only one element of the second set is defined. *Warning:* older books use different definitions, but this is now the usual one (*see* so-called many-valued function, p. 56). Symbol: $y = f(x)$ where x, y are elements of the first set (the domain, p. 17), and the second set (the range, p. 17) respectively and f is the mapping (↓). Other letters than f may be used.

mapping (*n*) the rule, process or operation (p. 14) which defines a function. The word *mapping* is often loosely used instead of the word *function*, especially in geometry. Also known as **map**. **map** (*v*). Symbol: f or sometimes $\longrightarrow$ or $\longmapsto$.

codomain (*n*) a set which includes every element of the second set in a function. The range (p. 17) of a function must be a subset (p. 12) of every possible codomain. Also known as **target**. *See also* domain (p. 17).

object (*n*) an element of the domain (p. 17) of a function which is to be mapped (↑); a subset (p. 12) of such elements such as a set of numbers or a geometrical curve.

image (*n*) the result of mapping (↑) an object (↑) using a given function.

function machine an imaginary machine into which the elements of the domain (p. 17) of a function can be put and which then gives out their images (↑).

function box the part of the function machine (↑) which holds the function.

composition (*n*) joining together, e.g. of functions to give a single function, and of arcs of networks (p. 239) joined end to end. **compose** (*v*).

mapping

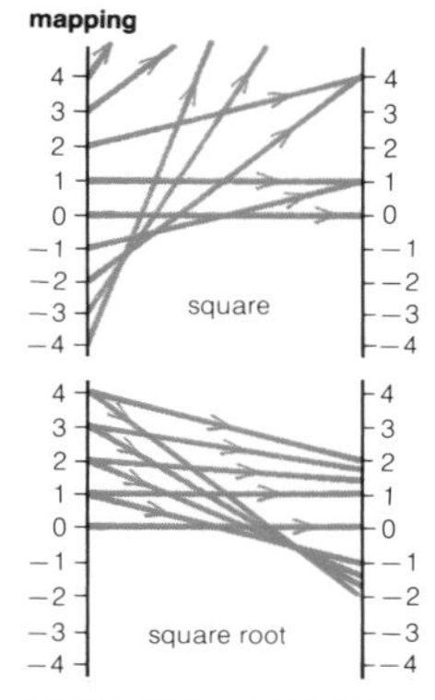

with domain the real numbers, 'square' gives a mapping, 'square root' does not since each positive number has two square roots and each negative number has none

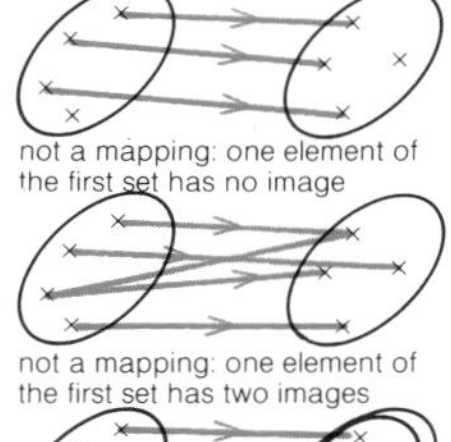

not a mapping: one element of the first set has no image

not a mapping: one element of the first set has two images

mapping: every element of domain has exactly one image

function machine

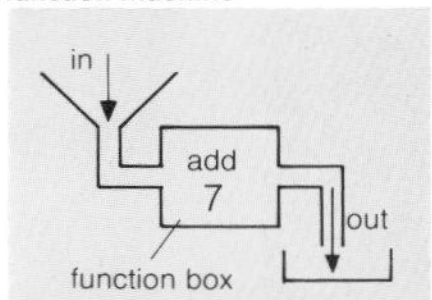

the function machine 'add 7', if −3 is fed in, 4 is given out

composition

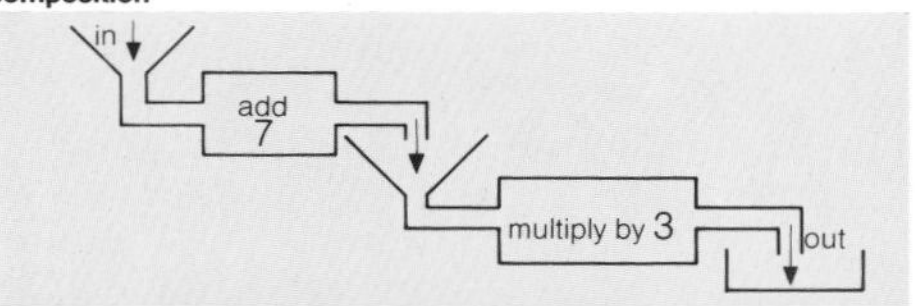

the composite function 'add 7 and multiply by 3' shown by two function machines. If the order of the machines is changed, a different composite function is obtained

composite function the single function produced by the composition (p. 51) of two or more functions. If *f* and *g* are functions, the composite function *f* followed by *g* is written $g\{f(x)\}$ or $gf(x)$ where *x* is an element of the domain (p. 17) of *f*, or sometimes as $g \circ f$.

surjection (*n*) a mapping (p. 51) where the codomain (p. 51) which is being considered and the range (p. 17) are the same. Also known as **mapping onto**. **surjective** (*adj*).

injection (*n*) a mapping (p. 51) where each element of the domain (p. 17) has a different image (p. 51). Also known as **mapping one: one into**. **injective** (*adj*).

bijection (*n*) a mapping (p. 51) which is both a surjection (↑) and an injection (↑). Also known as **one: one mapping**, **one to one mapping**. **bijective** (*adj*).

inverse function the function obtained when the set of images (p. 51) of a given function becomes the domain (p. 17) of the new function, and the original elements become the new codomain (p. 51). This is only possible when the first mapping (p. 51) is a bijection (↑), e.g. in the real numbers (p. 46) the inverse function of 'doubling' is 'halving'. Symbol: $x = f^{-1}(y)$ where $y = f(x)$ is the original function.

one: one correspondence a correspondence (p. 17) having two sets together with a bijection (↑) or one: one mapping. It is a function. Also known as **one to one correspondence**, **biuniform correspondence**, **biunique correspondence**. Symbol: ↔ between the symbols for the two sets.

many: one correspondence a correspondence (p. 17) in which more than one element of the first set may be related to a given element of the second set. It is a function. E.g. the squares of the real numbers (p. 46) where, e.g., $6^2 = 36$ and $(-6)^2 = 36$. Also known as **many to one correspondence**. *See* diagram on opposite page.

surjection

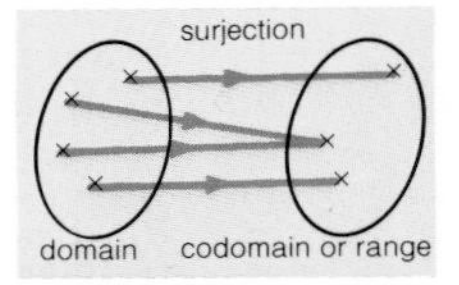

injection

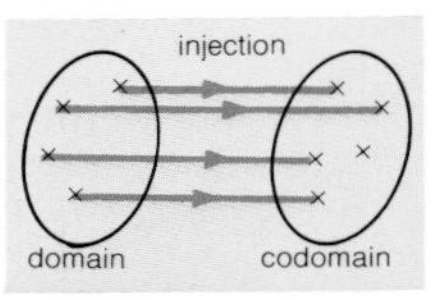

bijection

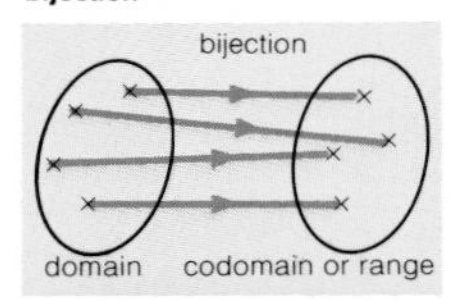

if the arrows are reversed the inverse function is obtained. The whole diagram shows a one: one correspondence

many:one correspondence

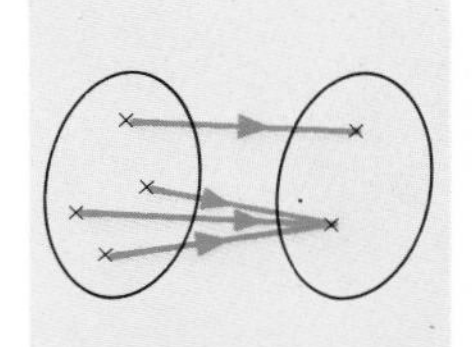

one:many correspondence

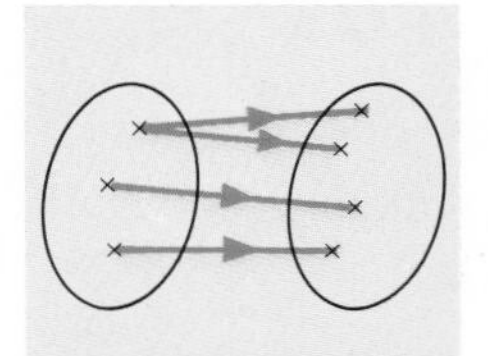

many:many correspondence

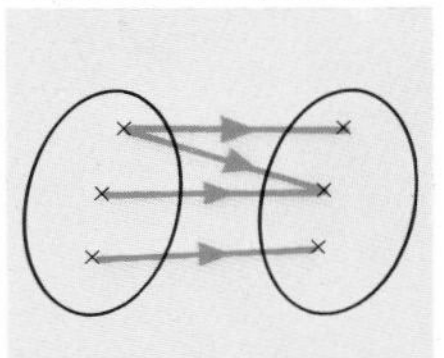

one: many correspondence a correspondence (p. 17) in which more than one element of the second set may be related to a given element of the first set. This is not now regarded as a function, e.g. the square root (p. 47) of a positive real number (p. 46), since each real number has two square roots. Also known as **one to many correspondence**.

many: many correspondence a correspondence (p. 17) in which more than one element of the first set may be related to any given element of the second set, and more than one element of the second set may be related to any element of the first set. This is not now regarded as a function. Also known as **many to many correspondence**.

homography (*n*) a one: one correspondence (↑) which can be written in the form

$y = (ax + b)/(cx + d)$ where $ad \neq bc$.

Often used of ranges (p. 167) of points or pencils (p. 167) of lines which are transformed (p. 54) onto themselves. *See also* bilinear transformation (p. 55). **homographic** (*adj*).

double point[1] in a homographic (↑) range (p. 167), one of the two points which transforms (p. 54) into itself. *See also* p. 189. Also known as **united point**.

elation (*n*) a homography (↑) in which the two double points (↑) are coincident (p. 61).

homology (*n*) the relation in a homography (↑). **homologous** (*adj*).

equinumerous (*adj*) having the same number of elements; used of sets whose elements can be put into complete one: one correspondence (↑). Also known as **equipotent**.

homomorphism (*n*) a mapping (p. 51) which keeps the algebraic structure (p. 57) of the domain (p. 17). *Do not mix with* homeomorphism (↓). Also known as **morphism**. **homomorphic** (*adj*).

epimorphism (*n*) a homomorphism (↑) which is also a surjection (p. 52). **epimorphic** (*adj*).

monomorphism (*n*) a homomorphism (↑) which is also an injection (p. 52). **monomorphic** (*adj*).

isomorphism (*n*) a homomorphism (↑) which is also a bijection (p. 52). **isomorphic** (*adj*).

endomorphism (*n*) a homomorphism (↑) whose codomain (p. 51) is a subset (p. 12) of its domain (p. 17). **endomorphic** (*adj*).

automorphism (*n*) a homomorphism (↑) whose domain (p. 17) and codomain (p. 51) are the same. **automorphic** (*adj*).

kernel (*n*) the subset (p. 12) of elements in a homomorphism (↑) which map (p. 51) onto the neutral element (p. 10).

homeomorphism (*n*) an isomorphism (↑) whose mapping (p. 51) is continuous or without 'breaks', e.g. the mapping from the position of the hands of a clock onto the time is not a homeomorphism because time is not continuous at 12 o'clock. *Do not mix with* homomorphism (↑). **homeomorphic** (*adj*).

bivariate (*adj*) having two variables.

multivariate (*adj*) having more than one variable.

bivariate function a function in which the elements of the domain (p. 17) have two variables, e.g. $f(x, y) = x^2 + y^2$ is a bivariate function of x and y. Similarly **multivariate function**.

real-valued function a function whose range (p. 17) is a subset (p. 12) of the real numbers (p. 46).

function space a space (p. 13) which is a set of functions defined for two other sets of elements.

transformation (*n*) an operation (p. 14) which changes a set of elements into another set of elements, used generally of mappings (p. 51) in both algebra and geometry, e.g. taking the logarithm (p. 139) of the positive real numbers (p. 46); translation (p. 233) and inversion (p. 195) in geometry. *See also* projection (p. 237). **transform** (*v*).

monomorphism

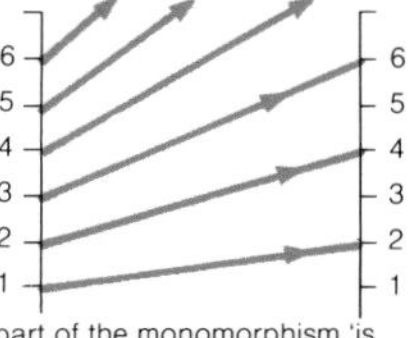

part of the monomorphism 'is half of' which maps the set of positive integers into itself. Note that 4 is half of 8, 5 is half of 10 and so on. It is also an automorphism

endomorphism

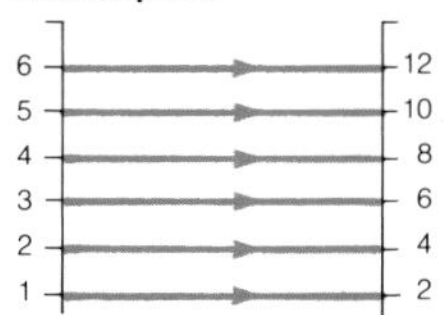

part of the endomorphism 'is half of' which maps the set of positive integers onto the set of even positive integers. It is also an isomorphism

transform (*n*) that which has been transformed (↑), the image (p. 51) or set of images in a mapping (p. 51).

bilinear form an equation of the type

$$cwz + dw + az + b = 0$$

where a, b, c, d are real numbers (p. 46), c is non-zero and w and z are elements of the set of complex numbers (p. 47), which gives a bilinear transformation (↓), and is linear (↓) in both w and z.

bilinear transformation if a, b, c, d are real numbers (p. 46), w, z are elements of the set of complex numbers (p. 47) and $ad \neq bc$; a transformation (↑) such as

$$w = (az + b)/(cz + d)$$

which is a one: one correspondence (p. 52) and is linear (↓) in both w and z.

linear transformation if a, b, d are real numbers (p. 46) and w and z are elements of the set of complex numbers (p. 47), a transformation (↑) of the form $w = (az + b)/d$ where d is non-zero. Such a transformation maps (p. 51) lines into lines and planes into planes.

Joukowski's transformation if a is a positive real number (p. 46) and w and z are elements of the set of complex numbers (p. 47), a transformation (↑) of the form $w = z + (a^2/z)$ (of importance in fluid mechanics, p. 280). It is a conformal (p. 235) transformation.

involution[2] a mapping (p. 51) which when repeated gives the identity (p. 57) transformation (↑), leaving every element unchanged, e.g. taking the reciprocal (p. 27) for non-zero real numbers (p. 46); reflection (p. 234) in geometry. *See also* p. 36.

mates (*n.pl.*) any two elements which map (p. 51) onto each other in an involution (↑).

invariant (*adj*) used of an element or set of elements which remain unchanged in a transformation (↑), e.g. unity in the mapping (p. 51) 'take the reciprocal (p. 27) of'.

self-corresponding (*adj*) used of elements in a transformation (↑) which transform into themselves, i.e. remain invariant (↑).

algebra (*n*) the part of mathematics which studies sets of symbols and the rules for operations (p. 14) on them. **algebraic** (*adj*). *See also* other definitions with algebraic.

expression (*n*) a set of symbols or numbers expressing a mathematical idea, usually one which does not contain the symbol for equality (p. 10), e.g. $(A \cup B) \cap (A \cup C)$.

variable (*n*) a symbol (or sometimes an expression) which can take various values. These values are the domain (p. 17) of the variable. Most often the letters x, y, z and t are used for variables, but many others are used as well. *See also* p. 104. **vary** (*v*), **variation** (*n*).

constant (*n*) that which does not vary; a symbol used for a number whose value is fixed but which may not be known.

real variable a variable whose domain (p. 17) is a subset (p. 12) of the real numbers (p. 46).

complex variable a variable whose domain (p. 17) is a subset (p. 12) of the complex numbers (p. 47).

conjugate functions if $u + iv$ is a function of the complex variable (↑) $x + iy$, then u and v are conjugate functions.

joint variation variation which depends on two or more separate variables which vary independently of each other, e.g. velocity (p. 258) varies jointly with distance and time.

single-valued expression an expression which has only one value for each value of the domain (p. 17); it defines a function.

many-valued expression an expression which has more than one value for some or all values of the domain (p. 17).

single-valued function now called just a function and no longer used.

many-valued function not now used *and incorrect*, since it is not a function as function is now defined. Use many-valued expression (↑). It is still found in older books.

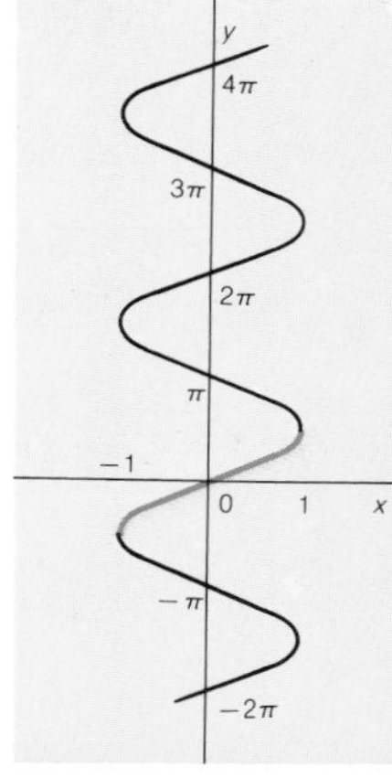

$y = \mathrm{Sin}^{-1}x$ is many-valued for $-1 \leqslant x \leqslant 1$
$y = \sin^{-1}x$ (in red) is the principal branch of this giving the principal values of y and is single-valued
$y = \sin^{-1}x$ is a function but $y = \mathrm{Sin}^{-1}x$ is not

equation

$x^2 + 3x + 2 = 0$

is an equation with roots or solutions $x = -1$ and $x = -2$

identity

$x^2 + 3x + 2 \equiv (x+1)(x+2)$

is an identity, true for all possible x

principal value of a many-valued expression (↑): a single value of the expression chosen for each element of the domain (p. 17) which thus allows a function to be defined, e.g. for the argument (p. 49) θ of a complex number (p. 47) it is that value of θ for which $-\pi < \theta \leqslant \pi$, where θ is measured in radians (p. 133). A useful convention (p. 9) is then to define Arg z as the argument of z, and to define arg z as its principal value, arg z then being a function of z. *See also* inverse trigonometric functions (p. 137), relations (p. 137).

principal branch of a many-valued expression (↑): the graph of the set of its principal values (↑).

secondary value a value of a many-valued expression (↑) which is not its principal value (↑).

constraint (*n*) an outside property which limits the values of an expression, or which limits the movement of a system in mechanics. Also known as **restraint**.

free variable a variable which does not have a constraint (↑) on its domain (p. 17).

abstract algebra the study of the laws of sets of elements with particular properties such as groups (p. 72) or fields (p. 76).

algebraic structure the set of laws for any given algebra and the results which can be obtained from them.

equation (*n*) two expressions which are equal and are joined by the symbol = . The equality (p. 10) is most often only true for some elements in the set of possible values or meanings of the symbols, but *see also* identity (2) (↓).

identity (*n*) (1) that transformation (p. 54) which leaves an element or set of elements unchanged; (2) an equation which is true for all elements of the sets of possible meanings of the symbols within a given domain (p. 17). **identical** (*adj*). Symbol for (2): ≡.

conditional equation an equation which is not an identity (↑).

formula (*n*) an important equation, often an identity (↑), which is used in further work in mathematics or in science.

root[2] (*n*) an element of the set of possible values of the symbols for which an equation is true, often used when that element is a number, e.g. 3 and -1 are roots of $x^2 - 2x - 3 = 0$. *See also* p. 47.
satisfy (*v*) the root of an equation (↑) is said to satisfy that equation.
solution (*n*) generally the answer to a problem, often the root or roots of an equation (↑). **solve** (*v*).
soluble (*adj*) able to be solved. Also known as **solvable**. **solubility** (*n*).
trivial solution a solution to an equation which is immediately clear but which is of little use or interest, e.g. $\sin x = x/2$ has a trivial solution $x = 0$.
zero of a function a value of the domain (p. 17) of a function for which the function is zero. If the function is defined by an equation, it is a root of the equation (↑), e.g. -1 is a zero of the function $y = x^2 + 3x + 2$.
unknown (*n*) not known, often a symbol in an equation whose value has to be found.
explicit (*adj*) used of a function defined as an equation $y = f(x)$ where x is an element of the domain (p. 17) and y is the corresponding element of the range (p. 17). Used similarly with relations, derivatives (p. 143) and other definitions.
implicit (*adj*) not explicit (↑), but resulting from what is given. Used of functions given in the form $f(x, y) = 0$ where values y of the range (p. 17) are not given at once in terms of values x of the domain (p. 17). Used similarly with relations, derivatives (p. 143) and other definitions.

explicit

$$y = 3x + 7$$

y is given explicitly in terms of x

implicit

$$3x - y + 7 = 0$$

y is given implicitly in terms of x

inequation (*n*) two expressions which are joined by an inequality (unequal, p. 11). Multiplication or division by a negative number changes the direction of the inequality. *See also* greater than (p. 20), less than (p. 20).
conditional inequality an inequality (unequal, p. 10) which is true only for some possible values of its variables, e.g. $x^2 + y^2 > 2xy$ if and only if x and y are not equal.

inconsistent equations

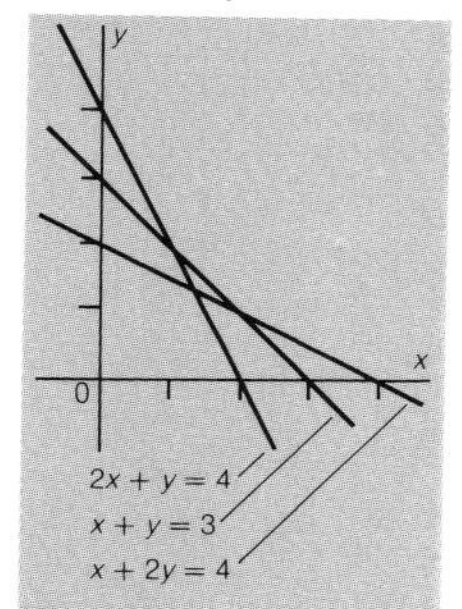

$x + y = 3$
$2x + y = 4$
$x + 2y = 4$
are inconsistent since no value of x and y satisfies all three at the same time

equivalent expression

$$\frac{x+3}{6} + \frac{x-2}{7} = \frac{x-3}{2}$$

$$14(x+3) + 12(x-2) = 42(x-3)$$

$$16x = 144$$

are all equivalent equations

term

in $x^2 + 3x + 2$

$3x$ is the 'second term' or the 'term in x'

consistent equations a set of equations which has a solution or solutions satisfying all of them at the same time.

inconsistent equations a set of equations which has no solution satisfying all of them at the same time. Also known as **incompatible equations**, **self-contradictory equations**.

indeterminate (*adj*) used of an equation or system of equations which has an infinite set of solutions, e.g. $x + y = 1$.

Diophantine equation an indeterminate (↑) equation in which only solutions in integers are allowed.

independent variable in an explicit (↑) equation $y = f(x)$, the variable x.

dependent variable in an explicit (↑) equation $y = f(x)$, the variable y. Also known as **subject**.

equivalent expression an expression which is another way of giving the same facts as a given expression. Similarly **equivalent equation**.

transformation of an equation changing an equation into an equivalent equation (↑).

changing the subject used of an equation which is explicit (↑) in one variable when it is to be made explicit in another variable. *See also* dependent variable (↑).

transposition (*n*) changing the subject (↑) of an equation; moving a term from one side of an equation to another. *See also* p. 84. **transpose** (*v*).

term (*n*) (1) a word with a particular meaning or definition; (2) one of a set of algebraic expressions (p. 60) or numbers; part of an algebraic expression which is completely separated from the rest of the expression by the symbols for plus (p. 24) (+) or minus (p. 24) (−); a member of a sequence (p. 91); one of the elements whose sum makes up a series (p. 91). *See also* sequence (p. 91), in terms of (p. 289).

like terms terms in an expression containing the same powers or same functions of the same variables which can be added by adding their coefficients (p. 61), e.g. in $5x + 4xy - 7y - xy$, the $4xy$ and the $-xy$ are like terms and can be replaced by $3xy$. Similarly **unlike terms** which cannot be so added.

polynomial (*n*) an algebraic expression (↓) in which each term is a constant, a positive integral power of a variable or a product of powers of such variables, possibly multiplied by a constant, e.g. $3x^4 + 4x^3 - 2x + 1$.

monomial (*n*) a polynomial having only one term.

binomial (*n*) a polynomial having two terms, e.g. $x^2 + 3$, but not $(x + 3)^2$. It is also used of a function, usually a power, of a binomial, as in binomial theorem (p. 98), e.g. the binomial expansion (↓) of $(x + 3)^2$ is $x^2 + 6x + 9$. Similarly **trinomial** having three terms, **multinomial** having more than two terms.

algebraic expression most often a polynomial expression or one which when put equal to zero is an equivalent equation (p. 59) to a polynomial equation (↓), e.g. $3x^2 + (1/x)$. Sometimes used to mean any expression using symbols which can take the place of numbers.

polynomial equation an equation which is a polynomial equated to zero.

algebraic equation an equation which is an equivalent equation (p. 59) to a polynomial equation (↑), or sometimes an algebraic expression (↑) equated to zero.

algebraic function an expression in two or more variables which is an algebraic equation (↑), and which can be written in a form where one variable is a function of the others.

transcendental function a function which is not an algebraic function (↑), e.g. the trigonometric functions (p. 132).

degree[1] (*n*) the highest power present in a polynomial. If there is more than one variable, then the powers in each term (p. 59) are added. *See also* pp. 73, 173.

repeated roots roots (p. 47) of a polynomial equation (↑) which correspond to repeated factors of the polynomial. Also known as **equal roots**, **multiple roots**. *See also* distinct (↓).

distinct (*adj*) different, separate, not the same. Used of roots (p. 47) which are not repeated (↑), and of points, lines, etc in geometry.

polynomial

$$3x^5 + 4x^2 - 7x + 2$$

is a polynomial of degree five, the coefficient of x^2 is 4 and of x is -7, the leading coefficient is 3

$$x^3y^2 + 3x^2y + y^4$$

is a polynomial in x and y of degree five

coefficient (*n*) (1) a quantity in an expression, used particularly for a number by which a variable or unknown (p. 58) is multiplied; (2) a measure which is used in science, statistics or in mechanics, e.g. coefficient of friction (p. 265).

leading coefficient the coefficient (↑) of the highest power in a polynomial.

monic polynomial a polynomial whose leading coefficient (↑) is 1.

literal equation an equation in which the coefficients (↑) are unknown constants shown by letters, e.g. $ax + by + c = 0$ has unknown constants a, b and c.

linear combination an expression containing only the first powers of variables in it, e.g. $a(x^2 + 1) + b(x^2 - 1)$ is a linear combination of $x^2 + 1$ and $x^2 - 1$.

homogeneous

$x^3y^2 + 3x^3yz + y^3z^2$ is homogeneous of degree five since each term has fifth power

$x^3 + \frac{y^4}{x+y} = y^3$ is homogeneous since it can be written as an equivalent homogeneous equation

homogeneous[1] (*adj*) having each term of the same degree, so that the sum of the powers of the variables is the same for each term. Used also of algebraic expressions (↑) or functions which can be arranged in this way. E.g. $x^3 + 3x^2y + 2y^3$.

expansion (*n*) the rewriting of an expression as an equivalent expression (p. 59) in a longer but simpler form, often with the removal of brackets (p. 9) or as a sum of separate terms, e.g. $(a + b)^3$ has the expansion $a^3 + 3a^2b + 3ab^2 + b^3$; $\sin(A + B)$ has the expansion $\sin A \cos B + \cos A \sin B$. **expand** (*v*).

algebraic dependence a variable y has algebraic dependence on another variable x if y is a root (p. 58) of a given polynomial in x. Also used of algebraic systems (p. 292), e.g. the rational numbers (p. 26) are algebraically dependent on the integers since every rational number is a solution of $ax + b = 0$, where a and b are integers. Similarly **algebraic independence**.

linear polynomial a polynomial whose highest degree is 1. Similarly **linear equation**, e.g. $3x + 4y = 7$.

quadratic polynomial a polynomial whose highest degree is 2. Similarly **quadratic equation**, e.g. $x^2 + 3x + 2 = 0$ and $x^2 + 3xy + 4y^2 + 2x + 1 = 0$. Similarly **cubic** (highest degree 3), **quartic** or **biquadratic** (4) and **quintic** (5) **polynomials** and **equations**.

elementary function usually taken to be a function which is a polynomial, a logarithmic (p. 139), exponential (p. 138), circular or hyperbolic (p. 141) function, or a composition (p. 51) of any of these.

Jordan factorial a polynomial of degree n with the form $x(x - 1) \dots (x - 2)(x - n + 1)$.

reciprocal equation a polynomial equation (p. 60) which gives an equivalent equation (p. 59) when the variable x is replaced by $(1/x)$, e.g. $x^4 + 4x^3 + 6x^2 + 4x + 1 = 0$.

remainder theorem if a polynomial $f(x)$ is divided by $x - \alpha$, then the remainder is $f(\alpha)$, e.g. if $2x^3 + 4x^2 - 9x + 5$ is divided by $x - 2$, the remainder is $2.2^3 + 4.2^2 - 9.2 + 5 = 16 + 16 - 18 + 5 = 19$.

factor theorem if $x - \alpha$ is a factor of a polynomial $f(x)$, then $f(\alpha) = 0$. A special case of the remainder theorem (↑). Also known as **quotient test**.

detached coefficients coefficients (p. 61), usually in polynomials, written without the variables so that calculations may be done more easily.

synthetic division a method of dividing a polynomial by a factor $x - \alpha$ using detached coefficients (↑).

nested multiplication if a polynomial such as $a_0x^4 + a_1x^3 + a_2x^2 + a_3x + a_4$ is written as $x\{x[x(a_0x + a_1) + a_2] + a_3\} + a_4$ its value may then be easily calculated by nested multiplication for a given value of x.

fundamental theorem of algebra every polynomial which is not a constant has at least one zero in the complex number (p. 47) system. Also known as **d'Alembert's theorem**, **Gauss's theorem**.

synthetic division

quotient — dividend

$$\begin{array}{r|l} & 2x^2 + 8x + 7 \\ x - 2 & 2x^3 + 4x^2 - 9x + 5 \\ & 2x^3 - 4x^2 \\ \hline & 8x^2 - 9x \\ & 8x^2 - 16x \\ \hline & 7x + 5 \\ & 7x - 14 \\ \hline & 19 \end{array}$$

divisor

remainder → 19

$$\frac{2x^3 + 4x^2 - 9x + 5}{x - 2} = 2x^2 + 8x + 7 + \frac{19}{x - 2}$$

using synthetic division this is written

$$\begin{array}{rrrr} 2 & 4 & -9 & 5 \\ & -4 & -16 & -14 \\ \hline & 8 & 7 & 19 \end{array}$$

these numbers (except for the last column) are detached coefficients

nested multiplication

$2x^3 + 4x^2 - 9x + 5$

can be written as

$x[x(2x + 4) - 9] + 5$

and if $x = 2$ we have

$2[2(4 + 4) - 9] + 5 = 19$

Descartes' rule of signs

$f(x) = x^5 - x^4 - 13x^3 + 13x^2 + 36x - 36$

has three changes of sign and three positive roots $x = 3$, $x = 2$ and $x = 1$

$f(-x) = -x^5 - x^4 + 13x^3 + 13x^2 - 36x - 36$

has two changes of sign and two positive roots $-x = +2$ and $-x = +3$

Descartes' rule of signs in any polynomial $f(x)$ with real (p. 49) coefficients (p. 61) arranged in powers of the variable starting at the highest, the number of positive roots (p. 58) equals the number of changes of sign of the coefficients or is less than this by an even number. Similarly for $f(-x)$ with the negative roots.

reducible polynomial a polynomial which factorizes into polynomials of a lower degree, e.g. $2x^3 + 4x^2 - 9x - 14 = (x - 2)(2x^2 + 8x + 7)$.

irreducible polynomial one which is not reducible (↑), e.g. $x^2 + 1$.

rational function a function which is obtained by dividing one polynomial by another.

proper rational function a rational function (↑) in which the degree of the numerator (p. 31) is less than the degree of the denominator (p. 31).

partial fractions a proper rational function (↑) may be written uniquely as a sum of fractions in which the denominators (p. 31) are polynomials of degree 1 or 2, or their powers, and the numerators (p. 31) are of lower degree than the denominators (without their powers, if any). Each of these fractions is a partial fraction. Used in integration and power series (p. 149) expansions (p. 61), e.g.

$$\frac{2x^3 + 11x^2 + 16x + 13}{(x^2 + 1)(x + 3)^2} = \frac{x + 1}{x^2 + 1} + \frac{1}{x + 3} + \frac{1}{(x + 3)^2}$$

cover-up rule

$$\frac{4x - 10}{(x - 1)(x - 2)(x - 3)} \equiv \frac{A}{x - 1} + \frac{B}{x - 2} + \frac{C}{x - 3}$$

A can be found by covering up $x - 1$ and putting $x = 1$ so that

$$A = \frac{4 - 10}{(-1)(-2)} = -3$$

B can be found by covering up $x - 2$ and putting $x = 2$ so that

$$B = \frac{(4 \times 2) - 10}{(1)(-1)} = 2$$

similarly $C = 1$ and the partial fractions are

$$\frac{-3}{x - 1} + \frac{2}{x - 2} + \frac{1}{x - 3}$$

cover-up rule a simple way of finding the coefficients (p. 61) of certain partial fractions (↑) by covering up a given factor in the denominator (p. 31) and substituting (p. 69) the value given by that factor equated to zero in the rest of the expression.

continuous (*adj*) without a break, usually used of functions of a real variable (p. 56). A function f is continuous at a value a if the limit of $f(x)$ as x tends to a is $f(a)$, so $\lim_{x \to a} f(x) = f(a)$, e.g. all polynomials, $y = \sin x$. Functions may be continuous for only some values of x. **continuity** (*n*).

discontinuous (*adj*) not continuous at at least one point, e.g. $y = (1/x)$ is discontinuous only at $x = 0$; $y = \tan x$ is discontinuous at $x = \pm(\pi/2), \pm(3\pi/2), \pm(5\pi/2), \ldots$ **discontinuity** (*n*).

removable discontinuity a point at which a function may be made continuous by redefinition, e.g. $y = (\sin x)/x$ has such a point at (0, 1) which can be removed by redefining the function as $y = \sin x/x$ when $x \neq 0$ and $y = 1$ when $x = 0$.

jump discontinuity a discontinuity (↑) where the left limit (↓) and the right limit (↓) differ, as in a step function (p. 166).

discrete (*adj*) not continuous at any point; having separate values, e.g. the integers.

continuum (*n*) a set which is infinite and continuous everywhere, e.g. the points of a line interval (↓). Such a set is not denumerably (p. 11) infinite and hence has cardinality (p. 23) greater than aleph null (p. 24).

continuum hypothesis the hypothesis (p. 251) which asks if the cardinal (p. 23) of the continuum (↑) is the next cardinal after aleph null (p. 24). This is in fact a matter of choice, and is thus a separate axiom (p. 250).

power of the continuum used of a set with the cardinality (p. 23) of the continuum (↑).

tend (*v*) to become very close to. If for any positive number, ε, no matter how small, there is a real number (p. 46) δ such that $|f(x) - L| < \varepsilon$ when $0 < |x - \alpha| < \delta$ we say that $f(x)$ tends to L as x tends to α, written as $f(x) \rightarrow L$ as $x \rightarrow \alpha$. Used of two graphs or functions which tend to each other as well as to a given value, e.g. the hyperbola (p. 221) $y = (1/x)$ tends to $y = 0$ as x tends to plus (p. 24) or to minus (p. 24) infinity. *See also* converge (p. 94) which can have the same meaning.

limit[1] (*n*) of a function: the number L defined under tend (↑). Also used similarly in geometry, e.g. the tangent to a circle may be regarded as the limit of a chord (p. 187) as the two points of intersection tend towards each other. *See also* p. 94. **limiting** (*adj*). Symbols: $\lim_{x \rightarrow a} f(x) \rightarrow L$.

dense

real numbers are
everywhere dense
continuous

rational numbers are
everywhere dense
not continuous

integers are
nowhere dense
not continuous

Zeeman catastrophe machine

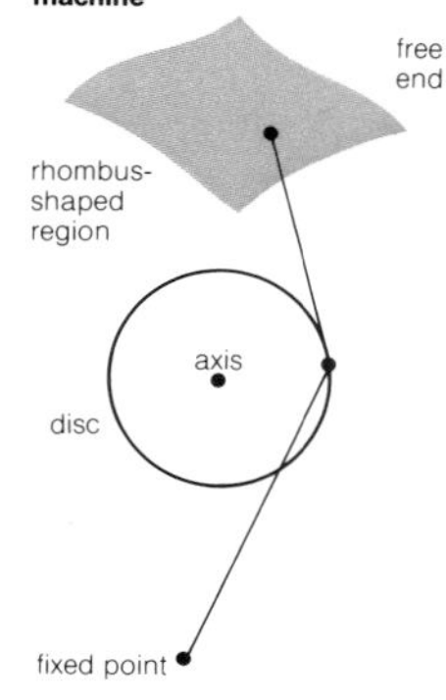

interval

open and closed intervals

diagram for open interval on a straight line

diagram for closed interval on a straight line

left limit a limit when x tends to α from below. Similarly **right limit**. In some cases these are not the same, e.g. in step functions (p. 166).

cluster point suppose there is a set of points (or real, p. 46, or complex numbers, p. 47) such that for another point P there is always a member of the set within a given distance from P no matter how small that distance is, then P is a cluster point of that set (P need not actually be a member of the set), e.g. the set $\{0.3, 0.33, 0.333, 0.3333, \ldots\}$ has a cluster point whose value is one third. Also known as **limit point**, **accumulation point**.

dense (*adj*) used of an ordered set (p. 19) having a third element between any two given elements.

nowhere dense used of a set of points (or numbers) which has no cluster points (↑).

catastrophe (*n*) a discontinuity (↑) in the values of a function, especially one which models (p. 290) real events.

Zeeman catastrophe machine a machine which shows the idea of a catastrophe (↑), having a disc (p. 192) rotating about a fixed axis with two equal pieces of rubber fixed to the edge; the other end of one has a fixed point, the free end of the second can be placed in different places; if these are inside an approximately rhombus-shaped (p. 181) region, there are two positions of equilibrium (p. 260), otherwise there is one. The disc makes sudden jumps as the free end is moved about.

interval[1] (*n*) a set of continuous values which is part of a line or curve, or for which a function is defined. *See also* p. 111.

open (*adj*) used of sets, intervals or neighbourhoods (p. 95). Not including the end or boundary (p. 201) points, and so defined by relations such as 'greater than' (p. 20), 'less than' (p. 20). In topology (p. 238) which has no metric (p. 169) it is used for sets and subsets (p. 12) which are an arbitrary (p. 287) collection of objects. Symbols: (a, b), or $)a, b($ for the open interval $a < x < b$, but some writers use (a, b) for closed (p. 66) interval.

closed[2] (*adj*) used of sets, intervals, neighbourhoods (p. 95). Including the end points, and so defined by relations such as 'greater than or equal to' (p. 20), 'less than or equal to' (p. 20). *Do not mistake for* closed as used with operations (p. 14). Symbols: $[a, b]$, or (a, b) for the closed interval such that $a \leqslant x \leqslant b$, but some writers use (a, b) for open (p. 65) interval.

half-closed (*adj*) closed at one end and open (p. 65) at the other.

nested intervals a sequence (p. 91) of intervals where the elements of each successive (p. 292) interval form a subset (p. 12) of the elements of the interval before it.

Bolzano-Weierstrass theorem any infinite set of points in a closed interval contains at least one cluster point (p. 65).

nested intervals

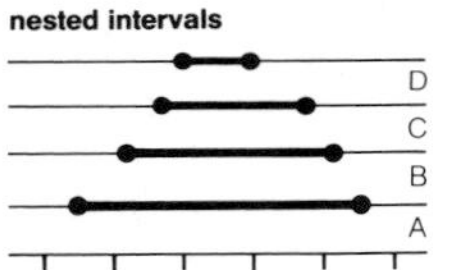

a sequence of nested intervals A, B, C, D related to the real number line

bound

in the sequence $\frac{1}{2}, \frac{3}{4}, \frac{7}{8}, \frac{15}{16}, \frac{31}{32}, \ldots$
1, 2, 3, 4, 5 are all upper bounds, 1 is the least upper bound (LUB)
$0, \frac{1}{4}, \frac{1}{3}, \frac{1}{2}$ are all lower bounds, $\frac{1}{2}$ is the greatest lower bound (GLB)

upper bound any number which is greater than or equal to all the elements of a set of real numbers (p. 46). *Note:* some books use upper bound for least upper bound (↓). Similarly **lower bound**.

bound (*n*) either an upper or lower bound (↑). Also known as **boundary value**. *See also* boundary (p. 201). **bound** (*v*), **bounded** (*adj*).

least upper bound the least element of the set of upper bounds (↑) of a given set. Also known as **supremum**. **LUB** (*abbr*), **sup** (*abbr*). Similarly **greatest lower bound**, or **infimum**. **GLB** (*abbr*), **inf** (*abbr*).

saltus (*n*) of any function in a given interval, the difference between its least upper (↑) and greatest lower bounds (↑) in that interval. Also known as **oscillation**. *See also* p. 275.

span[1] (*n*) of an area or volume, the least upper bound (↑) of the lengths of line segments (p. 167) whose end points lie in the area or volume. *See also* p. 80.

decreasing function **increasing function**

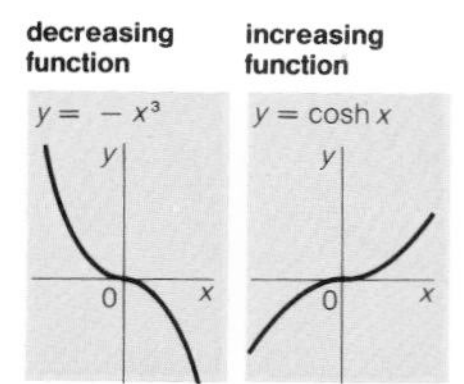

functions which decrease or increase or have zero gradient everywhere

monotonic decreasing function **monotonic increasing function**

$y = -x^3 - x$ $y = e^x$

even function $y = x^2(x^2 - 1)$

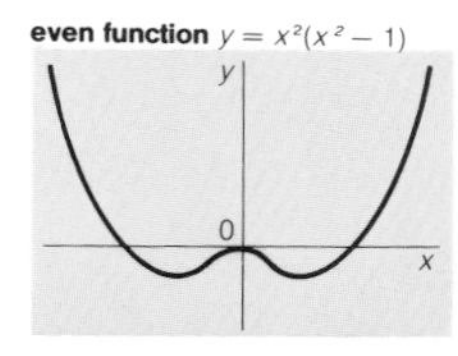

odd function $y = x^3 - x$

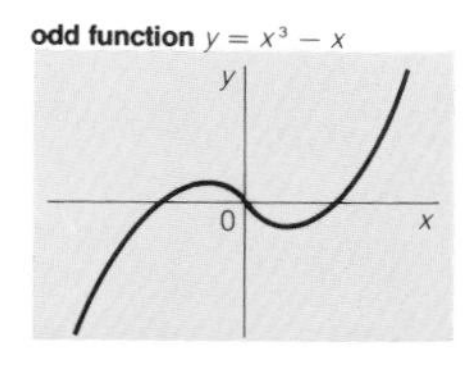

parameter
the parabola $y^2 = 4x$ can be given by the parameter t where $x = t^2$, $y = 2t$

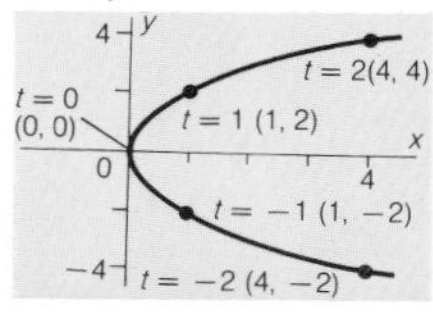

decreasing function a real-valued function (p. 54) is decreasing in an interval of its domain (p. 17) if, when $x_2 > x_1$, $f(x_2) \leqslant f(x_1)$ for all values x_1 and x_2 of x in the interval. If the derivative (p. 143) $f'(x)$ exists, it is negative or zero at all points in the interval. Similarly **increasing function**.

monotonic decreasing function defined as for a decreasing function (↑) except that $f(x_2) < f(x_1)$. If the derivative exists, it is negative at all points in the interval. Also known as **strictly decreasing function**. Similarly **monotonic increasing function**, **strictly increasing function**.

symmetric function (1) a function of two or more variables which remains the same function if any two of the variables are interchanged, e.g. $f(x, y, z) = x^3 + y^3 + z^3 + (x + y)(y + z)(z + x)$. The symmetric functions of the roots (p. 58) α and β of a quadratic equation (p. 68) $ax^2 + bx + c = 0$ are given by $\alpha + \beta = (-b/a)$ and $\alpha\beta = c/a$. Similar expressions may be found for polynomials of higher degree; (2) used to describe functions which are odd or even. *This use is better avoided.*

symmetric equation the equation of a symmetric function (↑), which remains the same if any two variables are interchanged, e.g. $x^2yz + xy^2z + xyz^2 = 0$.

even function a function with a graph which has line symmetry (p. 232) about the y-axis (p. 215) and for which $f(-x) = f(x)$.

odd function a function with a graph which has rotational symmetry (p. 232) of even order about the origin and for which $f(-x) = -f(x)$.

parity (*n*) whether a number is even or odd; whether a function is an even function (↑) or an odd function (↑), e.g. $y = x^2 + 3$ has even parity.

parameter[1] (*n*) a variable which is used only to give a definition of the variables of interest, e.g. the parameter θ may be used to define the circle $x^2 + y^2 = a^2$ by writing $x = a \cos \theta$, $y = a \sin \theta$. *See also* phase space (p. 78), p. 110. **parametric** (*adj*).

parametric equations those where the variables of interest are given as explicit (p. 58) functions of a parameter or parameters (p. 67).

quadratic form an expression of the type $ax^2 + bx + c$ where a is non-zero and a, b and c are real numbers (p. 46); also a binary quadratic form (↓) and similar general expressions of degree 2 in more than two variables. Similarly **quadratic equation**.

binary quadratic form a homogeneous (p. 61) expression of the type $ax^2 + 2hxy + by^2$, where a, h and b are real numbers (p. 46).

quadratic equation formula the general form of solution of the quadratic equation (↑) $ax^2 + bx + c = 0$, which is $x = \{-b \pm \sqrt{(b^2 - 4ac)}\}/2a$, e.g. the solutions of $3x^2 + 5x + 2 = 0$ are given by $x = \{-5 \pm \sqrt{(25 - 24)}\}/6 = -(2/3)$ or -1.

discriminant (n) any expression helping to show the difference between two or more things, e.g. the part of the quadratic equation formula (↑) $b^2 - 4ac$ which, when a, b and c are real (p. 49), shows the nature of the roots (p. 58) of the equation (if greater than zero, the roots are real and distinct, p. 60, equal to zero they are real and coincident, p. 61, less than zero they are complex and distinct); also the determinant (p. 86)
$$\begin{vmatrix} a & h & g \\ h & b & f \\ g & f & c \end{vmatrix}$$
which, when zero, shows that the conic (p. 220) $ax^2 + 2hxy + by^2 + 2gx + 2fy + c = 0$ must be a degenerate conic (p. 221) or line pair.

positive definite always positive, e.g. quadratic forms (↑) $ax^2 + bx + c$ in which $a > 0$ and $b^2 - 4ac < 0$ which are positive for all values of the variable, and thus have no real roots (p. 47). Similarly **negative definite** (here $a < 0$ and $b^2 - 4ac < 0$).

sum of roots the sum of roots of the polynomial $a_0x^n + a_1x^{n-1} + a_2x^{n-2} + \dots + a_n = 0$ is $-(a_1/a_0)$.

product of roots the product of roots of the polynomial $a_0x^n + a_1x^{n-1} + a_2x^{n-2} + \dots + a_n = 0$ is (a_n/a_0) multiplied by -1 if n is odd.

discriminant

$3x^2 + 5x + 2 = 0$
$\Rightarrow a = 3 \quad b = 5 \quad c = 2$
$\Rightarrow b^2 - 4ac = 25 - 24 = 1$
$\Rightarrow$ two distinct real roots
$3x^2 + 5x + 9 = 0$
$\Rightarrow a = 3 \quad b = 5 \quad c = 9$
$\Rightarrow b^2 - 4ac = 25 - 108 = -83$
$\Rightarrow$ two distinct complex roots
further $3x^2 + 5x + 3$
is positive definite

sum of roots and product of roots

$3x^2 + 5x + 2 = 0$
has sum of roots $= -\frac{5}{3}$
product of roots $= \frac{2}{3}$

$4x^4 - 6x^3 + 3x - 2 = 0$
has sum of roots $= -(\frac{-6}{4}) = \frac{3}{2}$
product of roots $= \frac{-2}{4} = -\frac{1}{2}$

simultaneous equations
simultaneous linear equations in two unknowns

$$\begin{cases} 3x + 4y + 3 = 0 \\ x - y + 1 = 0 \end{cases}$$

substitution replace y in the first equation by $(x + 1)$ giving

$$3x + 4(x + 1) + 3 = 0$$
$$\Rightarrow x = -1$$

elimination multiply the second equation by 4 and add the two equations to give

$$3x + 4y + 3 + 4x - 4y + 4 = 0$$
$$\Rightarrow 7x + 7 = 0 \quad \Rightarrow x = -1$$

in both cases y is found by substitution

undetermined coefficients

if $ax + by + 1 = 0$
passes through $(-1, 1)$ and $(3, 4)$, then
$$\begin{cases} -a + b + 1 = 0 \\ 3a + 4b + 1 = 0 \end{cases}$$
allowing a and b to be found

flow chart

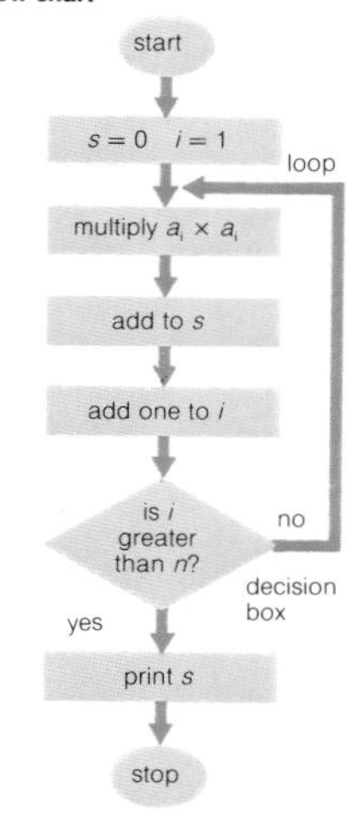

flow diagram to find $s = a_1^2 + a_2^2 + \cdots + a_n^2$

completing the square a method of solving a quadratic equation (↑) by putting the variables into a form which is a perfect square (p. 35), e.g. $3x^2 + 5x + 2 = 0$ can be written as $[x + (5/6)]^2 = 1/36$.

Ferrari's method a method of solving quartic (highest degree 4) equations similar in idea to completing the square (↑).

Cardan's solution a method of solving cubic (highest degree 3) equations which have only one real (p. 49) root (p. 47).

simultaneous equations a set of equations having variables in common. Also known as **system of equations**.

substitution (*n*) the replacement of part of an expression by another expression which is equal to the part removed. **substitute** (*v*).

elimination (*n*) removing a variable from a set of simultaneous equations (↑), often by adding or subtracting multiples of the separate equations, or sometimes by substitution (↑). **eliminate** (*v*).

degrees of freedom the number of independent elements in a set of simultaneous equations (↑) or in a mechanical (p. 259) system; the number of independent pieces of data in statistics.

independent equations equations which are not equivalent (p. 247) and cannot be transformed (p. 55) into each other.

undetermined coefficients a method of finding the coefficients (p. 61) of an equation by setting up simultaneous equations (↑) with these coefficients as the unknowns and using known values of the variables which satisfy the equations.

flow chart an arrow diagram (p. 17) which uses boxes joined by lines with arrows to show a process of logical (p. 243) argument (p. 243), much used in computer (p. 42) programming (p. 70).

decision box a box shaped like a rhombus (p. 181) with its diagonals (p. 177) horizontal and vertical, used in a flow chart (↑) to show that a choice of paths must be made.

loop[1] (*n*) a part of a flow chart (↑) which is repeated two or more times. *See also* pp. 189, 240.

critical path the way through a network (p. 239) or flow chart (p. 69) which allows something to be done in the best way, usually in the shortest time.

programming (*n*) writing a mathematical problem in a form which can then be used to obtain a solution, especially by a computer (p. 42). The form *program* is usually used for a plan that is to be put into a computer; the form *programme* is usually used more generally for a plan of work, e.g. Erlangen programme (p. 235).

string (*n*) a finite sequence (p. 91) of symbols used in mathematics and especially in computing (p. 42).

cybernetics (*n*) the study of control systems in animals and machines, including computers (p. 42).

feedback (*n*) the use, e.g. in cybernetics (↑), of the results obtained at one point to change input (p. 289) at another point in the working of a machine.

optimization (*n*) any method which finds a solution which can be regarded in some sense as the best solution to a problem with many solutions. **optimize** (*v*).

linear programming finding solutions (usually the best solution) to sets of simultaneous (p. 69) linear inequations (p. 58), usually applied to real-life examples.

mathematical programming similar to linear programming (↑), but allowing non-linear inequations (p. 58).

integral programming linear programming (↑) in which the variables can only take integer values.

trial and error the solution of a problem by guessing an answer and then improving on the guess. *See also* iteration (p. 44).

false position a value used in a solution by trial and error (↑).

double false position taking two false positions (↑), one too large and one too small, in order to guess which is the correct solution between them.

critical path a network of roads from A to G showing time of travel in minutes. The quickest path is marked in red

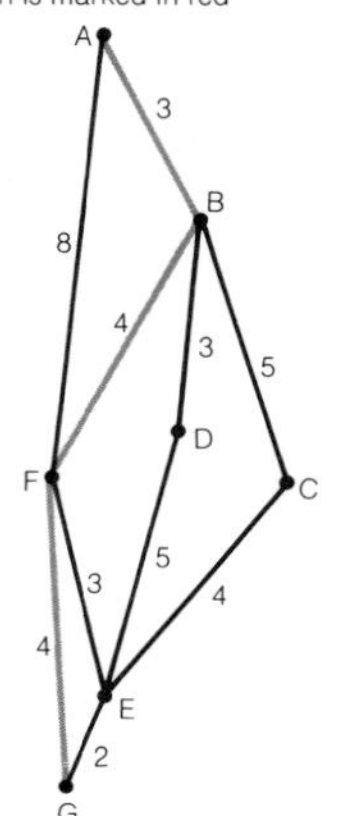

linear programming

if $x + 2y \leqslant 8$
$3x + y \leqslant 9$

find the maximum value of $x + y$

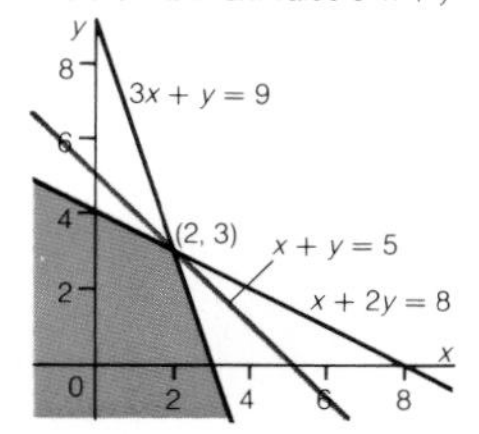

maximum value is the largest value of $x + y$ for the feasible points (shaded). This is given by $x + y = 5$. A false position for the best solution (2, 3) would be (2, 2.5) and (2, 3) is an extreme value of the shaded set of points

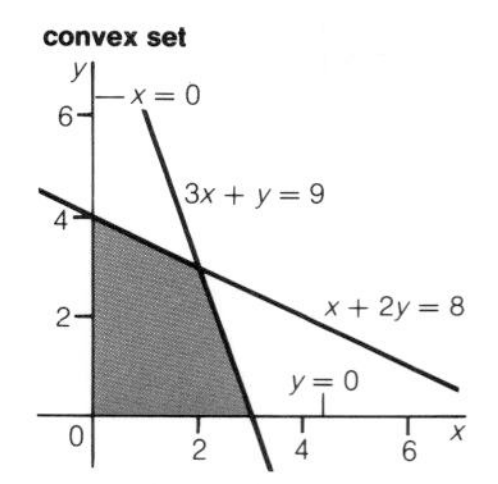

the set of inequalities

$x \geqslant 0$
$y \geqslant 0$
$x + 2y \leqslant 8$
$3x + y \leqslant 9$

form a convex set (in red)

convex set a set of points lying in a region whose sides are a convex polygon (p. 178), the set of equations giving the edges of this polygon used in linear programming (↑).

extreme value theorem if the domain (p. 17) of a function is a convex set (↑), then the extreme values (p. 146) of the function are at the vertices or on the edges of the convex set. It can be extended to more than two dimensions.

slack variable a new variable used to change an inequation (p. 58) into an equation with an additional unknown, e.g. $x + 2y + 6 \leqslant 0$ is an equivalent equation (*see* equivalent expression, p. 59) to $x + 2y + t = 0$ and $t \geqslant 6$ where t is a slack variable.

tableau a table or matrix (p. 83) of the detached coefficients (p. 62) of a set of equations, often used in linear programming (↑). **tableaux** (*pl*).

feasible points the set of solutions in a linear programming (↑) problem which obey the constraints (p. 57), but which need not be the best solution.

ill-conditioned equations a set of equations whose solutions change by a large amount if their coefficients (p. 61) are only changed by a small amount.

stable equations a set of equations which is not ill-conditioned (↑).

minimax method a method of solving linear programming (↑) problems by looking at the vertices of the polygon formed by the inequations (p. 58) of the constraints (p. 57).

simplex method a method of solving linear programming (↑) problems which involves changing the inequations (p. 58) for equations with slack variables (↑) and then setting out the results in a tableau (↑) of coefficients (p. 61).

relaxation method a method of solving linear programming (↑) and similar problems by guessing solutions and changing them by small amounts in a regular way to produce better solutions.

group (*n*) a set of elements which can be combined (p. 14) by a binary operation (p. 14) and where:
- I the set is closed (p. 14);
- II the operation is associative (p. 15);
- III the set has an identity element (p. 15);
- IV every element has an inverse element (p. 16).

commutative group a group (↑) in which the binary operation (p. 14) is also commutative (p. 15), e.g. the group of integers modulo (p. 36) 5 under addition. The transformation group (↓) for the equilateral (p. 177) triangle is an example of a non-commutative group. Also known as **Abelian group**.

group table a table giving the rules for combining (p. 14) all the elements of a finite group (↑). It is an example of a structure table (↓).

structure table *see* group table (↑).

subgroup (*n*) a set of some (or all) elements of a group (↑) which themselves form a group, e.g. the rotation group (p. 74) for the equilateral (p. 177) triangle is a subgroup of the transformation group (↓) for the equilateral triangle.

order of a finite group the number of elements in the group (↑). *See also* pp. 20, 190.

groupoid (*n*) a set of elements which can be combined (p. 14) by a binary operation (p. 14) which is closed. Also written **gruppoid**.

monoid (*n*) a groupoid (↑) for which the binary operation (p. 14) is also associative (p. 15), e.g. the even integers under multiplication, which do not have an identity element (p. 15). Also known as **hemigroup**, **semigroup**.

module (*n*) a commutative group (↑) whose binary operation (p. 14) is addition, e.g. the group of integers modulo (p. 36) 5 under addition. Also known as **additive group**.

submodule (*n*) a set of some (or all) elements of a module (↑) which themselves form a module.

multiplicative group a group (↑) whose binary operation (p. 14) is similar to multiplication.

group
the group of integers under addition modulo 5

+	0	1	2	3	4
0	0	1	2	3	4
1	1	2	3	4	0
2	2	3	4	0	1
3	3	4	0	1	2
4	4	0	1	2	3

here the identity element is 0 and the inverses of 0, 1, 2, 3, 4 are 0, 4, 3, 2, 1. This particular group is commutative and the table is a group table

group table
transformation group for an equilateral triangle

→ second element; ↓ first element

	e	*f*	*g*	*p*	*q*	*r*
e	*e*	*f*	*g*	*p*	*q*	*r*
f	*f*	*g*	*e*	*q*	*r*	*p*
g	*g*	*e*	*f*	*r*	*p*	*q*
p	*p*	*r*	*q*	*e*	*g*	*f*
q	*q*	*p*	*r*	*f*	*e*	*g*
r	*r*	*q*	*p*	*g*	*f*	*e*

permutation group

the set of permutations of *ABCD* which gives the above 24 orders is the symmetric group of degree 4. Those in red and brown form a permutation group which is the cyclic group of order 4. Those in blue and brown form an alternating group which is a permutation group of order 12

A B C D	B A C D	C A B D	D A B C
A B D C	B A D C	C A D B	D A C B
A C B D	B C A D	C B A D	D B A C
A C D B	B C D A	C B D A	D B C A
A D B C	B D A C	C D A B	D C A B
A D C B	B D C A	C D B A	D C B A

symmetric group a group (↑) which is formed by the set of all permutations (p. 97) of *n* objects which has order *n*! (*n* factorial, p. 97), e.g. the symmetric group of degree 3 has order 6 and the same structure table (↑) as the transformation group (↓) for the equilateral (p. 177) triangle.

degree of a symmetric group the number of objects *n* which are permuted (p. 97) to form a symmetric group (↑). *See also* pp. 60, 173.

permutation group either a symmetric group (↑) or one of its subgroups (↑) in which only a subset (p. 12) of all the possible permutations (p. 97) is used, e.g. the cyclic group (p. 74) of order 3 is a subgroup of the symmetric group (↑) of degree 3; these are respectively isomorphic (p. 54) to the rotation group (p. 74) and the transformation group (↓) for the equilateral (p. 177) triangle.

alternating group a subgroup (↑) of a symmetric group (↑) consisting of only the even permutations (p. 98).

Cayley's theorem every finite group (↑) is isomorphic (p. 54) (i.e. has an equivalent, p. 247, structure table, ↑) to a permutation group (↑), e.g. the transformation group (↓) for the equilateral (p. 177) triangle is isomorphic to the permutation group (↑) of degree 3.

transformation group a group (↑) whose elements are transformations (p. 54), e.g. the transformation group of the equilateral (p. 177) triangle. Particular groups may be named from the geometric shape whose transformations they describe, e.g. octahedral group from octahedron (p. 201).

transformation group

an equilateral triangle *PQR* has the six operations *e, f, g, p, q, r*

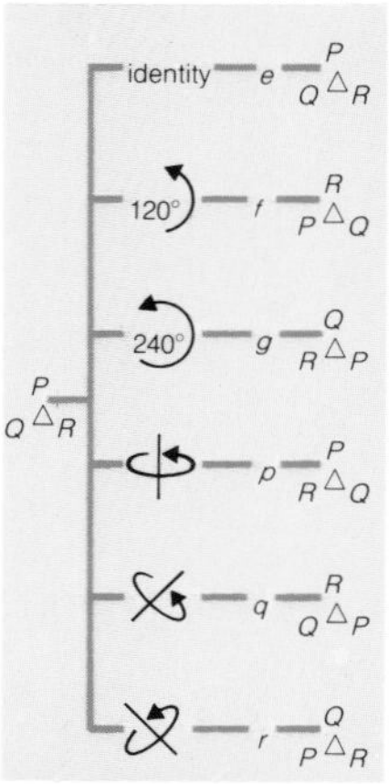

identity — e
120° — f
240° — g
p
q
r

rotation group

	e	f	g
e	e	f	g
f	f	g	e
g	g	e	f

the rotation group for an equilateral triangle is a subgroup of the transformation group

cyclic group a group (p. 72) whose structure table (p. 72) is isomorphic (p. 54) to that for the rotations of a regular polygon, especially note the **cyclic four-group** which describes the rotations of the square and has four elements. Known in geometry as **rotation group**. *See also* Klein's four-group (↓).

Klein's four-group a group (p. 72) with four elements whose structure table (p. 72) is the same as that for the transformations (p. 54) of a rectangle using the identity (p. 57) transformation and rotations about one or both of the lines of symmetry in its plane. The only other four-group is the cyclic four-group (↑).

generator[1] (*n*) of a cyclic group (↑): any element whose powers form all the elements of that group, e.g. *b* and *d* in the structure table (p. 72) of the cyclic four-group (↑). *See also* p. 209.

Lagrange's theorem if a group (p. 72) of order *n* has a subgroup (p. 72) of order *p*, then *p* is a factor of *n*.

coset (*n*) if *G* is a group (p. 72) with an element *g*, and *H* is a subgroup (p. 72) of it, the set formed by operating (p. 14) with *g* on the elements of *H* is the coset of *H* with respect to *g*. *See also* left coset (↓).

left coset if, in the definition of coset (↑), *g* is the left-hand element in the operation (p. 14), it is called a left coset. Similarly **right coset**.

normal subgroup a subgroup (p. 72) whose left and right cosets (↑) are the same. Also known as **invariant subgroup**, **self-conjugate subgroup**.

factor group the set of all cosets (↑) of a normal subgroup (↑), which forms a group (p. 72) under the same operation (p. 14). Also known as **quotient group**.

ring (*n*) a set of elements which can be combined (p. 14) by two binary operations (p. 14) usually called addition and multiplication. Under the additive operation, the elements form a commutative group (p. 72). The second operation need, however, be only closed and associative (p. 15), though it must be distributive (p. 15) over the first operation. *See also* torus (p. 210).

cyclic four group

	a	*b*	*c*	*d*
a	*a*	*b*	*c*	*d*
b	*b*	*c*	*d*	*a*
c	*c*	*d*	*a*	*b*
d	*d*	*a*	*b*	*c*

Klein's four group

	a	*b*	*c*	*d*
a	*a*	*b*	*c*	*d*
b	*b*	*a*	*d*	*c*
c	*c*	*d*	*a*	*b*
d	*d*	*c*	*b*	*a*

coset

if G is the group of real numbers under addition, H is the set of integers under addition, then the set

$$\{\ldots, -2\tfrac{1}{2}, -1\tfrac{1}{2}, -\tfrac{1}{2}, \tfrac{1}{2}, 1\tfrac{1}{2}, 2\tfrac{1}{2}, \ldots\}$$

forms a coset relative to the subgroup H obtained by adding $\frac{1}{2}$ to each element of H

commutative ring a ring (↑) in which the second or multiplicative operation (p. 14) is also commutative (p. 15), e.g. the even integers form a commutative ring without unity.

ring with unity a ring (↑) in which the second or multiplicative operation (p. 14) has an identity element (p. 15), e.g. the integers form a commutative ring (↑) with unity, as does the set of polynomials in a single variable.

divisors of zero elements which are factors of zero in a ring (↑) or other algebraic structure (p. 57); the ring of integers modulo (p. 36) 10 has $2 \times 5 = 0$, so that 2 and 5 are divisors (p. 26) of zero.

integral norm a function ϕ on the elements of a ring (↑) whose range (p. 113) is the non-negative integers, such that if a, b are elements of the ring, then (1) $\phi(a) = 0$ if and only if $a = 0$ and (2) $\phi(ab) = \phi(a)\,\phi(b)$ for all a and b. An example of such a function is $\phi(a) = 0$ if $a = 0$ and $\phi(a) = 1$ if $a \neq 0$. Often known as **norm**, but then the range of the function can be different or the conditions can be changed.

subring (*n*) a subset (p. 12) of a ring (↑) which is a ring itself.

adjunction (*n*) joining a set to a ring (↑) (or a field, p. 76) to give another ring (or field) which contains both the elements of the first ring (or field) and the set. **adjoin** (*v*).

ideal[1] (*n*) a subring (↑) of a ring (↑) for which for every element a of the ring and every element i of the subring both ai and ia are in the subring. This is a two-sided ideal; if only ai is in the subring, it is a left ideal; if only ia it is a right ideal. *See also* p. 289.

quotient ring if I is an ideal (↑) of a ring (↑) R with an element a then the cosets (↑) $a + I$ form a ring called a quotient ring. Also known as **difference ring**, **factor ring**.

residue class ring a quotient ring (↑) formed by a ring of residues (p. 36) of given modulus (p. 36).

integral domain a commutative ring (↑) with unity which has no divisors (p. 26) of zero.

ordered integral domain an integral domain (p. 75) in which some elements are called positive, for which both operations (p. 14) give elements which are also positive, and in which every element is either positive or zero or has positive additive inverse (p. 250). A very important example is the set of integers.

normed domain an integral domain (p. 75) with a norm (integral norm, p. 75).

field[1] (*n*) a set of elements which can be combined (p. 14) by two operations (p. 14) usually called addition and multiplication. Both operations are closed, associative (p. 15) and commutative (p. 15), have distinct (p. 60) identity elements (p. 15) (called zero and unity respectively) and each element has an inverse (p. 250) under each operation except for zero under multiplication. Multiplication is distributive (p. 15) over addition. E.g. the rational numbers (p. 26); the real numbers (p. 46); the complex numbers (p. 47); all numbers of the form $a + \sqrt{b}$ where a and b are rationals. *See also* p. 274.

subfield (*n*) a subset (p. 12) of a field (↑) which is a field itself, e.g. the rational numbers (p. 26) form a subfield of the real numbers (p. 46).

prime field a field (↑) which has no subfields (↑) other than itself.

ordered field a field (↑) in which any two elements have an ordering relation (p. 16), e.g. the real numbers (p. 46), but not the complex numbers (p. 47).

skew field a set of elements having all the properties of a field (↑) except that multiplication is not commutative (p. 15). Also known as **division ring**.

field extension a field (↑) which contains a given field as a subset (p. 12), e.g. the real numbers (p. 46) are a field extension of the rational numbers (p. 26).

simple field extension a field extension (↑) all of whose elements can be obtained from a single added element, e.g. the field of complex numbers (p. 47) is obtained from the field of real numbers (p. 46) with the added element *i*.

vector
the length of the line gives the magnitude, the arrow the direction

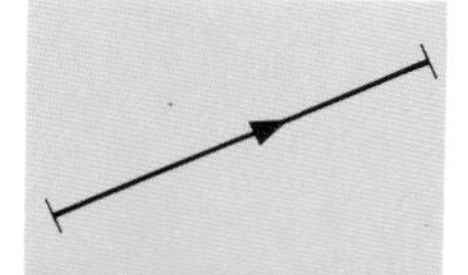

vector (*n*, *adj*) (1) an ordered set (p. 19) of *n* elements obeying the rules of vector algebra (↓). *See also* vector space (p. 78); (2) if *n* is 2 or 3, a directed line segment (p. 167) representing the ordered set; (3) a quantity which has both magnitude (p. 284) and direction and which may be represented by a directed line segment. Examples of vector quantities are velocity (p. 258) and force. Symbols: heavy type or underline, e.g. $\mathbf{x}$ or $\underline{x} = (x_1, x_2, \ldots, x^n)$, where the elements in brackets are separated by commas (,) and are the *n* elements or components (p. 80); if this form is used, care must be taken not to mistake them for coordinates. Vectors are also written in terms of their components as column matrices (p. 83), (often written as row matrices, p.83, transposed, p. 84, to save space) or less often as row matrices, without any commas. The vector represented by *AB* is also written $\overrightarrow{AB}$ or $\underline{AB}$.

dimension of a vector the number *n* in the definition of vector. *See also* p. 170.

fourth dimension the fourth element in a vector of dimension 4, sometimes used for time in applied mathematics (p. 8).

vector analysis the use of infinitesimal calculus (p. 143) in the study of vectors, often taken to include vector algebra (↓) as well.

vector algebra the rules for operating (p. 14) on vectors using vector addition (↓) and scalar multiplication (p. 78).

vector addition
adding the vectors (2, 1) and (1, 3) gives (3, 4)

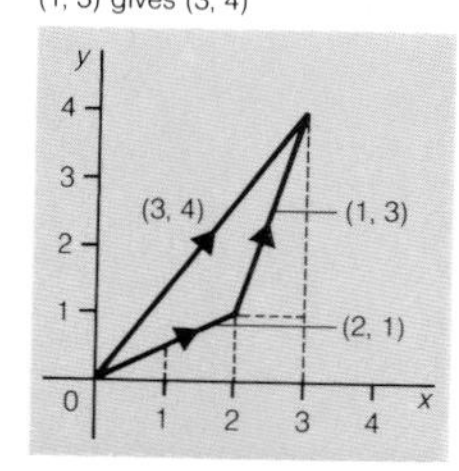

vector addition if $\mathbf{x}$, $\mathbf{y}$ are two vectors of dimension *n*, then

$$\mathbf{x} + \mathbf{y} = (x_1, x_2, \ldots, x_n) + (y_1, y_2, \ldots, y_n)$$
$$= (x_1 + y_1, x_2 + y_2, \ldots, x_n + y_n).$$

scalar (*n*) (1) a number or quantity which has magnitude (p. 284) but no direction, e.g. temperature (p. 284), energy (p. 269); (2) one of the elements of the field (p. 76) over which a vector space (p. 78) is defined. **scalar** (*adj*).

pseudoscalar (*n*) a quantity which is usually regarded as a scalar (↑) but does in fact have direction, e.g. angle in two dimensions where the direction is perpendicular to the plane in which the angle lies such that the direction of measurement of the angle makes a right-hand screw.

scalar multiplication if **x** is a vector of dimension n and a is a number (or scalar, p. 77), then $a\mathbf{x} = (ax_1, ax_2, \ldots, ax_n)$. A similar definition holds for scalar multiplication of matrices (p. 83), each element being multiplied separately by the scalar. *Do not mistake for* scalar product (↓).

scalar multiplication

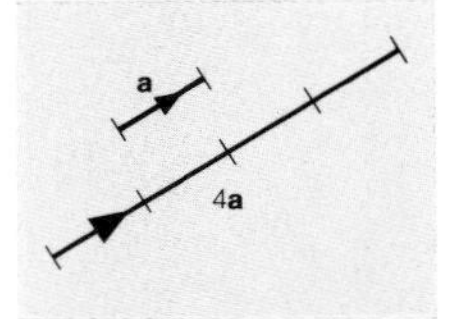

vectors **a** and 4**a**
if **a** is (1, 2), 4**a** is (4, 8)

null vector a vector of n zero elements. Also known as **zero vector**.

vector space a generalization of vector algebra (p. 77). A set of elements forms a vector space over a given field (p. 76) if:

I it is a commutative group (p. 72) under an operation (p. 14) usually called addition;

II for all elements of the field $a, b, \ldots$ and all elements of the vector space $\mathbf{x}, \mathbf{y}, \ldots$ there are elements of the vector space $a\mathbf{x}, a\mathbf{y}, \ldots, b\mathbf{x}, b\mathbf{y}, \ldots$ for which:

(a) $a(\mathbf{x} + \mathbf{y}) = a\mathbf{x} + a\mathbf{y}$

(b) $(a + b)\mathbf{x} = a\mathbf{x} + b\mathbf{x}$

(c) $a(b\mathbf{x}) = (ab)\mathbf{x}$

(d) $1\mathbf{x} = \mathbf{x}$, where 1 is the unity of the field.

Also known as **linear space**, **linear manifold**.

linearly dependent a set of vectors $\mathbf{x}, \mathbf{y}, \mathbf{z}, \ldots$ is linearly dependent if there is a set of scalars (p. 77) $a, b, c, \ldots$, not all zero, for which $a\mathbf{x} + b\mathbf{y} + c\mathbf{z} + \ldots = 0$. Similar definitions apply to other sets such as matrices (p. 83).

linearly independent not linearly dependent (↑).

linear algebra the algebra of vector or linear spaces (↑), usually used with rather wider meaning than vector algebra (p. 77).

phase space a vector space (↑) which gives the set of vectors representing a physical system and which may have components (p. 80) which give quantities such as time, temperature (p. 284), etc. Also known as **parameter space.**

scalar product if $\mathbf{x}, \mathbf{y}$ are vectors of dimension n, then the quantity $\mathbf{x}.\mathbf{y}$ or $\langle \mathbf{x}, \mathbf{y} \rangle$ which is equal to $x_1y_1 + x_2y_2 + \ldots + x_ny_n$ is the scalar product of **x** and **y**. *Do not mistake for* scalar multiplication (↑). Also known as **inner product**, especially when n is greater than 3, and as **dot product**. *See also* angle between two vectors (p. 77).

scalar product

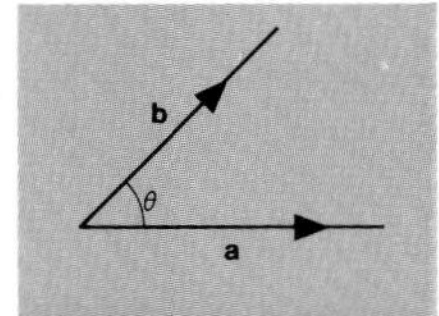

the scalar product **a** . **b** is the number $|\mathbf{a}|\,|\mathbf{b}|\cos\theta$

inner product = scalar product (↑).

unit vector

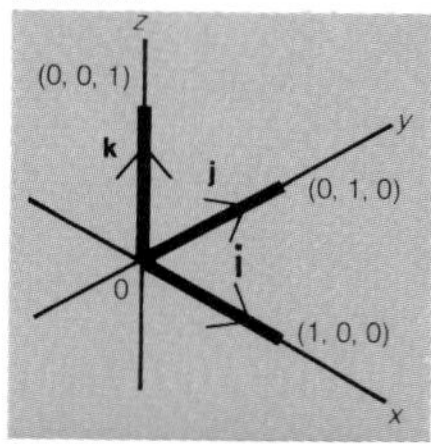

the unit vectors **i**, **j**, **k**

triangle inequality

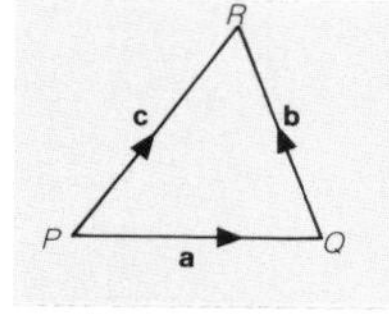

if **c** = **a** + **b**
|**c**| ⩽ |**a**| + |**b**|
PR ⩽ *PQ* + *QR*

Euclidean space a vector space (↑) together with an inner product (↑).

magnitude of a vector the positive square root (p. 47) of the inner product (↑) of a vector **x** with itself, that is $\sqrt{\mathbf{x}.\mathbf{x}}$. It is the mathematical basis (p. 287) of the concept of length. *See also* p. 284. Also known as **length of a vector**. Symbol |**x**|.

separation (*n*) the quantity |**x** − **y**| where **x** and **y** are two vectors.

unit vector a vector of magnitude (p. 284) 1. In three dimensions the important unit vectors (1, 0, 0), (0, 1, 0) and (0, 0, 1) are often called **i**, **j** and **k** respectively.

triangle inequality[1] if **x**, **y** and **z** are three vectors, whose sum is zero, then |**z**| ⩽ |**x**| + |**y**|. *See also* distance (p. 169), p. 177.

Schwartz inequality if **x** and **y** are two vectors then |**x.y**| ⩽ |**x**|.|**y**|, that is the value of the scalar product (↑) is less than or equal to the product of the magnitudes (p. 284) of the two vectors.

position vector a vector from a fixed origin defining the position of a point, usually of dimension 2 or 3. The components (p. 80) of the vector are then the same as the coordinates of the point with respect to that origin.

bound vector a vector, usually of dimension 2 or 3, from a fixed point in space defining a displacement (↓). Also known as **localized vector**.

tied vector a vector, usually of dimension 2 or 3, acting along a fixed line, but not necessarily from a given point, e.g. force. Also known as **line-localized vector**.

free vector a vector, usually of dimension 2 or 3, which does not act along a given line but only has magnitude (p. 284) and direction, e.g. the vector defining a translation (p. 233) in space.

line of action the line along which a tied vector (↑) or a bound vector (↑) acts.

displacement (*n*) generally any movement, particularly the movement or translation (p. 233) of a point defined by a vector, the magnitude (p. 284) of the vector being the size of the displacement. **displace** (*v*).

plane vector a vector whose dimension is 2, used in plane geometry (p. 172).

directional angles the angles between a vector of three dimensions and the directions of the three coordinate axes given by the directions of the vectors (1, 0, 0), (0, 1, 0) and (0, 0, 1).

direction cosines the cosines (p. 131) of the directional angles (↑) of a vector which are the components (↓) of a unit vector (p. 79) in the same direction as the given vector resolved (↓) along the coordinate axes. Also known as **direction ratios**.

component (*n*) (1) one of the *n* elements of the ordered set (p. 19) which makes up a vector **x** of dimension *n*; (2) one of a set of *n* vectors in *n* different given directions (i.e. one of *n* linearly independent, p. 78, vectors) whose vector sum is **x**; (3) the magnitude (p. 284) of one of these *n* vectors. Also known as **resolute**, **projection**.

resolve (*v*) to find a component (↑) of, usually in a given direction. **resolution** (*n*).

resultant (*n*) the sum of two or more vectors.

base vector of a vector space (p. 78) of dimension *n*, one of a set of *n* linearly independent (p. 78) vectors which can be used to define any vector in the space. Also known as **basis vector**, **generator**.

span[2] (*v*) the base vectors (↑) of a vector space (p. 78) are said to span the space. Also known as **generate**. *See also* p. 66

tangential component the component (↑) of a vector at a point on a curve resolved (↑) in the direction of the tangent to that curve at that point. The tangential component of the acceleration (p. 258) is dv/dt where v is the velocity (p. 258) along the curve at that point and t is the time.

normal component the component (↑) of a vector at a point on a curve resolved (↑) in the direction of the positive normal (p. 188) to that curve at that point. The normal component of the acceleration (p. 258) is $v^2/|\rho|$ where v is the velocity (p. 258) along the curve and ρ is the radius of curvature (p. 189) at that point.

directional angles

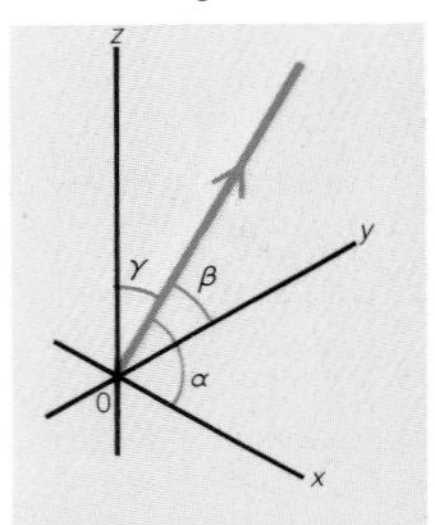

direction cosines
if $\vec{OP}$ is a unit vector and PX, PY, PZ are the perpendiculars to the axes Ox, Oy, Oz, then OX, OY and OZ give the direction cosines of any vector in the direction OP

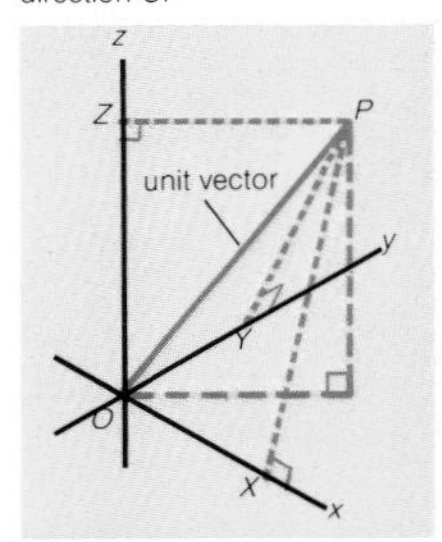

tangential and normal components
the accelerations of a particle moving along a curve with velocity **v** at a point P given along the tangent and the normal

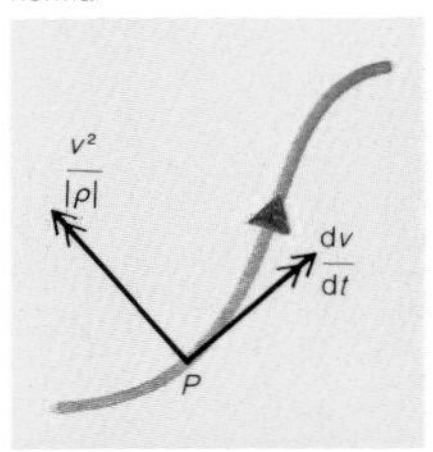

radial and transverse components
velocity and acceleration of particle moving along $r = f(\theta)$ in the radial and transverse directions

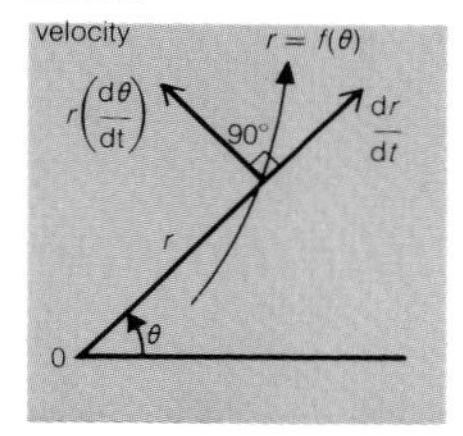

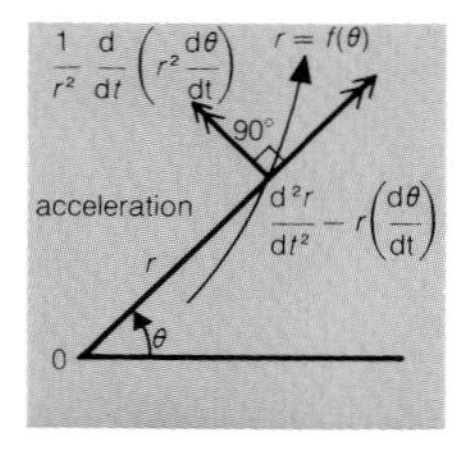

radial component the component (↑) of a vector at a point on a curve resolved (↑) along the positive radius vector (p. 216) in polar coordinates (p. 216). For a particle moving along $r = f(\theta)$ at time t this component of velocity (p. 258) is dr/dt and of acceleration (p. 258) it is $\{d^2r/dt^2 - r(d\theta/dt)^2\}$.

transverse component the component (↑) of a vector at a point on a curve resolved (↑) at an angle of $+90°$ to the positive radius vector (p. 216) in polar coordinates (p. 216). For a particle moving along $r = f(\theta)$ at time t this component of velocity (p. 258) is $r(d\theta/dt)$ and of acceleration (p. 258) it is $(1/r)d\{r^2(d\theta/dt)\}/dt$.

angle between two vectors the angle between **x** and **y** may be defined as the angle between 0 and π radians (p. 133) whose cosine (p. 131) is $(\mathbf{x}\,.\,\mathbf{y})/(|\mathbf{x}||\mathbf{y}|)$, where $\mathbf{x}\,.\,\mathbf{y}$ is the scalar product (p. 78) and $|\mathbf{x}|$ and $|\mathbf{y}|$ are the magnitudes (p. 284) of **x** and **y**. This allows the scalar product to be calculated as $|\mathbf{x}||\mathbf{y}|\cos\theta$ where θ is the angle between **x** and **y**. *See also* p. 173.

orthogonal vectors two vectors whose scalar product (p. 78) is zero. In geometry such vectors are at right angles.

orthogonal complement in a finite dimensional vector space (p. 78) the set of vectors which are all orthogonal (p. 175) to every vector in a given subspace (p. 13).

orthonormal vectors unit orthogonal vectors (↑).

vector triangle a triangle whose sides represent three vector quantities whose sum is zero. Similarly **vector polygon** for more than three vectors.

vector triangle
if $\mathbf{a} + \mathbf{b} + \mathbf{c} = 0$, then the three vectors can be represented by the sides of a triangle

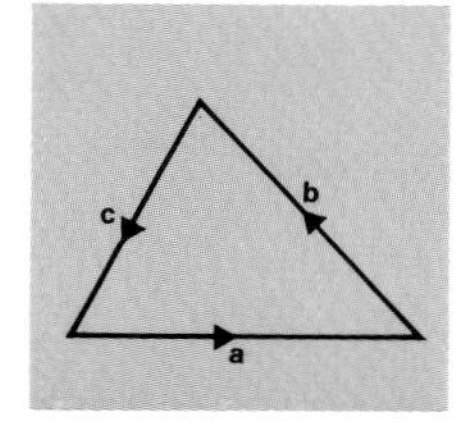

parallelogram of vectors if two vectors are represented by two sides $\overrightarrow{AB}$, $\overrightarrow{AD}$ of a parallelogram $ABCD$ then their sum is represented by the diagonal $\overrightarrow{AC}$.

parallelogram of vectors

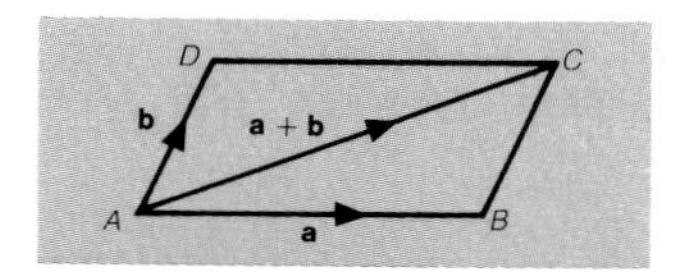

polygon of vectors a generalization of the parallelogram of vectors (p. 81). The sum of n vectors is a polygon of $n + 1$ sides, n of which represent in magnitude (p. 284) and direction the n vectors which have directions in order round the polygon; the last side represents the sum of these vectors, the sense (p. 174) being in the opposite direction to that of the remaining n vectors. If the n vectors themselves form a closed polygon, then their resultant (p. 80) is zero. Similarly **triangle of vectors**.

polygon of vectors

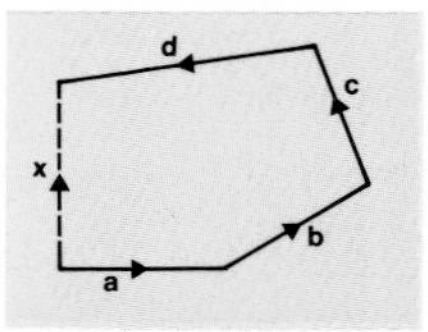

$\mathbf{x} = \mathbf{a} + \mathbf{b} + \mathbf{c} + \mathbf{d}$

vector product if **x**, **y** are the vectors (x_1, x_2, x_3) and (y_1, y_2, y_3) respectively, the vector $\mathbf{x} \times \mathbf{y} = \mathbf{x} \wedge \mathbf{y} = (x_2y_3 - x_3y_2, x_3y_1 - x_1y_3, x_1y_2 - x_2y_1)$, e.g. the moment (p. 264) of a force is calculated by means of a vector product. Also known as **cross product**, e.g. $(3, 1, 2) \times (2, 4, 0) = (1.0 - 2.4, 2.2 - 3.0, 3.4 - 1.2) = (-8, 4, 10)$.

right-handed set if **x** and **y** are orthogonal vectors (p. 81) in three dimensions, then **x**, **y** and the vector product $\mathbf{x} \times \mathbf{y}$ form a right-handed set. A right-handed screw turned from **x** to **y** would move in the direction $\mathbf{x} \times \mathbf{y}$. Also known as **right-handed system**.

right-handed set

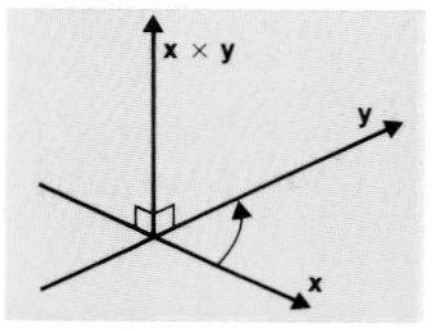

right-handed rule a rule for showing a right-handed set (↑). Also known as **corkscrew rule**.

scalar triple product if **x**, **y** and **z** are three vectors of three dimensions, then the function defined by $[\mathbf{x}, \mathbf{y}, \mathbf{z}] = (\mathbf{x} \times \mathbf{y}) . \mathbf{z}$ is the scalar triple product and gives the volume of the parallelepiped (p. 201) which has the three edges through one vertex defined by the three vectors.

scalar triple product

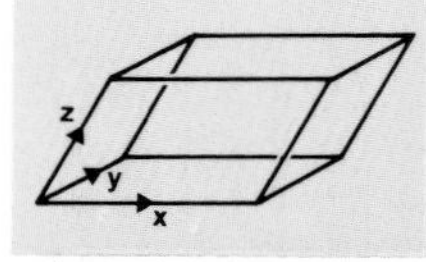

volume $= (\mathbf{x} \times \mathbf{y}) . \mathbf{z}$

vector triple product if **x**, **y** and **z** are three vectors with three dimensions, this is the vector function $\mathbf{x} \times (\mathbf{y} \times \mathbf{z})$.

tensor (n) a generalization of a vector into arrays (↓) of components (p. 80) in two or more dimensions of space.

quaternion (n) an ordered quadruple (p. 19) giving the scalar (p. 77) magnitude (p. 284) and three directions in space of a scalar field (p. 274); not now often used.

matrix
matrices of order 2 × 3 and 3 × 3

$$\begin{pmatrix} 2 & 3 & -1 \\ 7 & -6 & 4.5 \end{pmatrix}$$

$$\begin{pmatrix} a & -b & c \\ b & c & -a \\ -c & a & b \end{pmatrix}$$

row matrix
a 1 × 4 row matrix (or row vector)

$$(7 \quad 6 \quad -3 \quad -2)$$

column matrix

$$\begin{pmatrix} a \\ b \\ c \end{pmatrix}$$
a 3 × 1 column matrix (or column vector)

double suffix notation
in the matrix

$$\mathbf{A} = \begin{pmatrix} 3 & -2 & 4 & 7 \\ 2 & 1 & 6 & -3 \end{pmatrix}$$
the element a_{23} is 6

null matrix of order 2 × 3

$$\begin{pmatrix} 0 & 0 & 0 \\ 0 & 0 & 0 \end{pmatrix}$$

square matrix of order 3

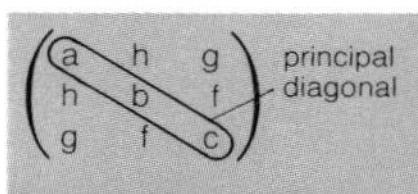

unit matrix of order 4

$$\begin{pmatrix} 1 & 0 & 0 & 0 \\ 0 & 1 & 0 & 0 \\ 0 & 0 & 1 & 0 \\ 0 & 0 & 0 & 1 \end{pmatrix}$$

diagonal matrix of order 3

$$\begin{pmatrix} 3 & 0 & 0 \\ 0 & -1 & 0 \\ 0 & 0 & 7 \end{pmatrix}$$

array (*n*) a regular arrangement of numbers or symbols, often of rectangular shape, sometimes having more than two dimensions, when more than one table is needed to write it on paper.

matrix (*n*) a rectangular array (↑). **matrices** (*pl*). Symbols: the array is given between large brackets (p. 9) or parentheses, by a capital letter **A** or in hand-writing as $\underline{A}$.

row (*n*) a set of numbers or symbols in a line across, often in a matrix (↑) or determinant (p. 86).

column (*n*) a set of numbers or symbols in a line down, often in a matrix (↑) or determinant (p. 86).

element[2] (*n*) any one of the set of numbers or symbols in a matrix (↑) or determinant (p. 86). *See also* p. 10.

order of a matrix a matrix with *m* rows and *n* columns has order $m \times n$. For a square matrix (↓) (or a determinant, p. 86) with *n* rows and *n* columns, the order is more often given as just *n*. Also known as **dimension**, **size**.

row matrix a matrix (↑) with only one row. Also known as **row vector**.

column matrix a matrix (↑) with only one column. Also known as **column vector**.

double suffix notation the element of a matrix (↑) **A** in the *i*th row and the *j*th column is symbolized by a_{ij}.

null matrix any matrix (↑) all of whose elements are zero. Also known as **zero matrix**. Symbol: **0**, the order is usually clear.

square matrix a matrix (↑) with an equal number of rows and columns.

principal diagonal the set of elements of a square matrix (↑) in a line from the upper left-hand corner to the lower right-hand corner. Also known as **leading diagonal**.

unit matrix any square matrix (↑) in which the elements in the principal diagonal (↑) are each unity, and all the other elements are zero. Also known as **identity matrix**. Symbol: **I**, the order is usually clear.

diagonal matrix a square matrix in which non-zero elements are found only in the principal diagonal (↑).

scalar matrix a diagonal matrix (p. 83) in which all the elements in the principal diagonal (p. 83) are equal.

trace (*n*) of a square matrix (p. 83): the sum of the elements in the principal diagonal (p. 83). *See also* sketch (p. 9). Also known as **spur**.

triangular matrix a square matrix (p. 83) which has all the elements on one side of the principal diagonal (p. 83) equal to zero. Known as upper or lower depending on whether the non-zero elements are above or below the principal diagonal. Also known as **echelon matrix**, **reduced matrix**.

transpose (*n*) of a matrix (p. 83) or determinant (p. 86) with elements a_{ij}: the one with elements a_{ji} formed by interchanging rows and columns. *See also* p. 59. Symbols (for matrices): $\mathbf{A}^T$ or $\mathbf{A}'$.

symmetric matrix a square matrix (p. 83) which is equal to its transpose (↑).

skew-symmetric used of a square matrix (p. 83) or determinant (p. 86) in which every element is the negative of the corresponding element of its transpose (↑). Each element of the principal diagonal (p. 83) must be zero.

sum of two matrices of order 2 × 3

$$\begin{pmatrix} 3 & 6 & -1 \\ 0 & 7 & 2 \end{pmatrix} + \begin{pmatrix} 1 & 1 & 7 \\ 0 & -4 & -2 \end{pmatrix} = \begin{pmatrix} 4 & 7 & 6 \\ 0 & 3 & 0 \end{pmatrix}$$

similarly difference of two matrices

$$\begin{pmatrix} 3 & 6 & -1 \\ 0 & 7 & 2 \end{pmatrix} - \begin{pmatrix} 1 & 1 & 7 \\ 0 & -4 & -2 \end{pmatrix} = \begin{pmatrix} 2 & 5 & -8 \\ 0 & 11 & 4 \end{pmatrix}$$

sum of two matrices for matrices of the same order, this is formed by adding corresponding elements. Similarly their **difference** is found by subtracting corresponding elements.

product of two matrices if **A** is an $m \times n$ matrix and **B** is an $n \times p$ matrix, then the $m \times p$ matrix **C** for which $c_{ij} = \sum_{\lambda} a_{i\lambda} b_{\lambda j}$ is their product. This is sometimes known as **row by column multiplication**.

scalar matrix of order 2

$$\begin{pmatrix} 3 & 0 \\ 0 & 3 \end{pmatrix}$$

trace
the trace of this matrix is $4 + (-6) + 3 = 1$

$$\begin{pmatrix} 4 & 8 & 3 \\ 7 & -6 & 1 \\ 4 & 0 & 3 \end{pmatrix}$$

triangular matrix
an upper triangular matrix of order 4

$$\begin{pmatrix} 4 & -6 & 3 & 4 \\ 0 & 3 & -8 & 1 \\ 0 & 0 & 7 & -7 \\ 0 & 0 & 0 & 2 \end{pmatrix}$$

transpose
a matrix of order 2 × 3 and its transpose of order 3 × 2

$$\text{if } \mathbf{A} = \begin{pmatrix} 7 & 4 & -6 \\ 3 & 8 & 2 \end{pmatrix}$$

$$\text{then } \mathbf{A}^T = \begin{pmatrix} 7 & 3 \\ 4 & 8 \\ -6 & 2 \end{pmatrix}$$

symmetric matrix of order 4

$$\begin{pmatrix} 3 & 4 & 7 & -2 \\ 4 & -5 & 8 & -1 \\ 7 & 8 & 2 & 4 \\ -2 & -1 & 4 & 6 \end{pmatrix}$$

skew-symmetric
matrix of order 3

$$\begin{pmatrix} 0 & p & -q \\ -p & 0 & r \\ q & -r & 0 \end{pmatrix}$$

product of two matrices
a 3 × 2 matrix and a 2 × 3 matrix give a 3 × 3 matrix. The element in the 2nd row and 3rd column is obtained from the 2nd row of the first matrix and the 3rd column of the second
(3 × 4) + (1 × 2) = 12 + 2 = 14

$$\begin{pmatrix} 7 & 4 \\ 3 & 1 \\ 4 & 0 \end{pmatrix} \begin{pmatrix} 5 & 2 & 4 \\ 1 & 3 & 2 \end{pmatrix} = \begin{pmatrix} 39 & 26 & 36 \\ 16 & 9 & 14 \\ 20 & 8 & 16 \end{pmatrix}$$

conformable matrices an ordered pair (p. 19) of matrices (p. 83) in which the number of columns (p. 83) in the first equals the number of rows (p. 83) in the second, so that they have a product, are conformable matrices under multiplication. Under addition they are a pair of matrices of the same order (p. 83), which can thus be added.

orthogonal matrix

$$\begin{pmatrix} \cos\alpha & -\sin\alpha \\ \sin\alpha & \cos\alpha \end{pmatrix}$$

orthogonal matrix a square matrix (p. 83) **A** which when multiplied by its transpose (↑) $\mathbf{A}^T$ gives a unit matrix $\mathbf{I} = \mathbf{A}.\mathbf{A}^T$.

orthogonal group the orthogonal group of order n is the group of orthogonal matrices (↑) of order n under multiplication.

cross product the cross product of two square matrices (p. 83) **A** and **B** of the same order is the matrix $\mathbf{A}.\mathbf{B} - \mathbf{B}.\mathbf{A}$. Also known as **Lie product**, **commutation**. Symbol: $\mathbf{A} \times \mathbf{B}$.

Hermitian matrix

$$\begin{pmatrix} a & p+iq & -i \\ p-iq & b & r \\ i & r & c \end{pmatrix}$$

Hermitian matrix a square matrix (p. 83) **A** whose elements may be complex numbers (p. 47) and whose transpose (↑) $\mathbf{A}^T$ is equal to the matrix $\bar{\mathbf{A}}$ which is **A** with its elements replaced by the corresponding conjugate complex numbers (p. 48), so that $\mathbf{A}^T = \bar{\mathbf{A}}$. If the elements are real (p. 49), it is a symmetric matrix (↑).

skew-Hermitian matrix as for Hermitian matrix (↑) except that $\mathbf{A}^T = -\bar{\mathbf{A}}$. If the elements are real (p. 49) it is a skew-symmetric matrix (↑).

unitary matrix a square matrix (p. 83) **U** whose elements are complex numbers (p. 47) and for which $\bar{\mathbf{U}}^T.\mathbf{U} = \mathbf{I}$ where $\bar{\mathbf{U}}^T$ is the transpose (↑) of the matrix whose elements are the conjugate complex numbers (p. 48) of the elements of **U**, and **I** is a unit matrix (p. 83). If the elements are real (p. 49), it is an orthogonal matrix (↑).

unitary group the unitary group of order n is the group (p. 72) of unitary matrices (p. 85) of order n under multiplication.

normal matrix a square matrix (p. 83) whose elements may be complex numbers (p. 47) and for which $\mathbf{A}.\bar{\mathbf{A}}^T = \bar{\mathbf{A}}^T.\mathbf{A}$, where $\bar{\mathbf{A}}^T$ is the matrix whose elements are the conjugate complex numbers (p. 48) of the elements of the transpose (p. 84) of **A**. Diagonal (p. 83), unitary (p. 85), Hermitian (p. 85) and skew-Hermitian matrices (p. 85) are all normal matrices.

gram matrix if **A** is a matrix (p. 83) (usually not a square matrix, p. 83), then the matrix $\bar{\mathbf{A}}^T.\mathbf{A}$ is the gram matrix of **A**, where $\bar{\mathbf{A}}^T$ is the transpose (p. 84) of the matrix **A** with each element replaced by its complex conjugate (p. 48).

determinant (n) a number obtained from a square matrix (p. 83) **A**. If **A** is of order 2,

$$\det \mathbf{A} = |\mathbf{A}| = \begin{vmatrix} a_{11} & a_{12} \\ a_{21} & a_{22} \end{vmatrix} = a_{11}a_{22} - a_{12}a_{21}.$$

For order 3, $\det \mathbf{A} = |\mathbf{A}| = \begin{vmatrix} a_{11} & a_{12} & a_{13} \\ a_{21} & a_{22} & a_{23} \\ a_{31} & a_{32} & a_{33} \end{vmatrix}$

$$= a_{11}\begin{vmatrix} a_{22} & _{23} \\ a_{32} & a_{33} \end{vmatrix} - a_{12}\begin{vmatrix} a_{21} & a_{23} \\ a_{31} & a_{33} \end{vmatrix} + a_{13}\begin{vmatrix} a_{21} & a_{22} \\ a_{31} & a_{32} \end{vmatrix}$$

and similarly for higher orders, with alternate (p. 287) positive and negative signs.

determinant

$$\begin{vmatrix} 3 & 1 & 7 \\ 4 & -2 & 6 \\ 2 & 0 & 5 \end{vmatrix}$$

$$= 3\begin{vmatrix} -2 & 6 \\ 0 & 5 \end{vmatrix} - 1\begin{vmatrix} 4 & 6 \\ 2 & 5 \end{vmatrix} + 7\begin{vmatrix} 4 & -2 \\ 2 & 0 \end{vmatrix}$$

$$= 3 \times \{-10 - 0\} - 1\,\{20 - 12\} + 7\,\{0 - (-4)\}$$

$$= -30 - 8 + 28 = -10$$

minor[1] (n) a determinant (↑) of order $n - 1$ obtained from one of order n by striking out one row and one column. Thus the minor of the element a_{12} in a determinant **A** of order 3 is

$$\begin{vmatrix} a_{21} & a_{23} \\ a_{31} & a_{33} \end{vmatrix}$$

Sometimes used similarly of matrices (p. 83).

minor and cofactor

in $\begin{vmatrix} 3 & 1 & 7 \\ 4 & -2 & 6 \\ 2 & 0 & 5 \end{vmatrix}$

the minor of the element 1 is

$$\begin{vmatrix} 4 & 6 \\ 2 & 5 \end{vmatrix} = 8$$

the cofactor is

$$-\begin{vmatrix} 4 & 6 \\ 2 & 5 \end{vmatrix} = -8$$

principal minor a principal minor of a matrix (p. 83) **A** is a minor (↑) whose principal diagonal (p. 83) is part of the principal diagonal of **A**.

cofactor (*n*) the minor (↑) of the element a_{ij} of a determinant (↑) multiplied by -1 if $i + j$ is odd. Thus the cofactor of a_{12} in a determinant of order 3 is $-\begin{vmatrix} a_{21} & a_{23} \\ a_{31} & a_{33} \end{vmatrix}$

alien cofactor a cofactor (↑) of an element of a determinant (↑) other than the element being talked about.

true cofactor a cofactor (↑) which is not alien (↑).

eliminant (*n*) a determinant (↑), usually one used in solving a set of simultaneous equations (p. 69). Also known as **characteristic**.

Cramer's rule a method of solving linear simultaneous equations (p. 69) using determinants (↑).

Cramer's rule

$$\text{if } \begin{cases} 3x + 2y = 9 \\ 5x - y = 2 \end{cases}$$

$$\text{then } x = \frac{\begin{vmatrix} 9 & 2 \\ 2 & -1 \end{vmatrix}}{\begin{vmatrix} 3 & 2 \\ 5 & -1 \end{vmatrix}}$$

$$y = \frac{\begin{vmatrix} 3 & 9 \\ 5 & 2 \end{vmatrix}}{\begin{vmatrix} 3 & 2 \\ 5 & -1 \end{vmatrix}}$$

$$x = \frac{-13}{-13} = 1$$

$$y = \frac{-39}{-13} = 3$$

Gaussian condensation a method of finding the value of a determinant (↑) by lowering the order one step at a time. Also known as **pivotal condensation**.

Gaussian elimination a method of solving linear simultaneous equations (p. 69) using Gaussian condensation (↑); the same idea as transforming (p. 55) the corresponding matrix (p. 83) of coefficients (p. 61) to a triangular matrix (p. 84) by elementary row operations (p. 88).

augmented matrix a matrix (p. 83) to which additional elements have been added, usually the matrix of the coefficients (p. 61) of a set of linear simultaneous equations (p. 69) with the constant terms added.

alternant (*n*) a determinant (↑) whose form is similar to $\begin{vmatrix} 1 & 1 & 1 \\ a & b & c \\ a^2 & b^2 & c^2 \end{vmatrix}$

circulant (*n*) a determinant (↑) whose rows or columns are cyclic permutations (p. 97) of one another such as $\begin{vmatrix} a & b & c \\ c & a & b \\ b & c & a \end{vmatrix}$

gram determinant the determinant (↑) of a gram matrix (↑).

Pfaffian (*n*) a skew-symmetric (p. 84) determinant (p. 86) of even order, which must be a perfect square (p. 35).

Jacobian determinant a determinant (p. 86) of partial derivatives (p. 163) arranged similarly to

$$\begin{vmatrix} \partial x/\partial u & \partial y/\partial u & \partial z/\partial u \\ \partial x/\partial v & \partial y/\partial v & \partial z/\partial v \\ \partial x/\partial w & \partial y/\partial w & \partial z/\partial w \end{vmatrix}$$

Also known as **functional determinant**.

Wronskian (*adj*) used of square matrices (p. 83) or determinants (p. 86) having elements which are functions and their derivatives (p. 143) of the form

$$\begin{vmatrix} x & y & z \\ \mathrm{d}x/\mathrm{d}t & \mathrm{d}y/\mathrm{d}t & \mathrm{d}z/\mathrm{d}t \\ \mathrm{d}^2x/\mathrm{d}t^2 & \mathrm{d}^2y/\mathrm{d}t^2 & \mathrm{d}^2z/\mathrm{d}t^2 \end{vmatrix}$$

where *x*, *y* and *z* are functions of *t*. Similarly for higher orders.

Pfaffian

$$\begin{vmatrix} 0 & 2 & -1 & 3 \\ -2 & 0 & 4 & -2 \\ 1 & -4 & 0 & 1 \\ -3 & 2 & -1 & 0 \end{vmatrix}$$

determinant has value $144 = 12^2$

elementary row operation

1 $$\begin{pmatrix} 3 & 0 & 7 \\ 2 & 1 & 6 \\ 1 & -1 & 4 \end{pmatrix} \longrightarrow \begin{pmatrix} 2 & 1 & 6 \\ 3 & 0 & 7 \\ 1 & -1 & 4 \end{pmatrix}$$

2 $$\begin{pmatrix} 3 & 0 & 7 \\ 2 & 1 & 6 \\ 1 & -1 & 4 \end{pmatrix} \longrightarrow \begin{pmatrix} 3 & 0 & 7 \\ 6 & 3 & 18 \\ 1 & -1 & 4 \end{pmatrix}$$

3 $$\begin{pmatrix} 3 & 0 & 7 \\ 2 & 1 & 6 \\ 1 & -1 & 4 \end{pmatrix} \longrightarrow \begin{pmatrix} 3 & 0 & 7 \\ 2 & 1 & 6 \\ 7 & 2 & 22 \end{pmatrix}$$

elementary row operation an operation (p. 14) on the rows of a matrix (p. 83) or determinant (p. 86) which either:

I interchanges two rows of elements, or
II multiplies a row of elements by a non-zero constant, or
III adds to the elements of any one row a constant multiple of the corresponding elements of another row.

Also known as **row operation**, or sometimes (**elementary**) **row transformation**. Similarly for **elementary column operation** etc.

elementary matrix a matrix (p. 83) which carries out an elementary row or column operation (p. 14) when another matrix is multiplied by it.

elementary matrix

3 above is obtained from the elementary matrix

$$\begin{pmatrix} 1 & 0 & 0 \\ 0 & 1 & 0 \\ 0 & 3 & 1 \end{pmatrix}$$

equivalent matrices matrices (p. 83) which can be obtained from each other by elementary row or column operations (p. 14).

triangular decomposition a method of solving simultaneous linear equations (p. 69) by transforming (p. 55) the square matrix (p. 83) of the coefficients (p. 61) to a triangular matrix (p. 84) by using elementary row or column operations (p. 14).

adjoint matrix the square matrix (p. 83) whose elements are the cofactors (p. 87) of the transpose (p. 84) of a given square matrix. Also known as **adjugate matrix**.

inverse matrix

$$\begin{pmatrix} 2 & 3 \\ 1 & 2 \end{pmatrix} \text{ and } \begin{pmatrix} 2 & -3 \\ -1 & 2 \end{pmatrix}$$

are inverse matrices since

$$\begin{pmatrix} 2 & 3 \\ 1 & 2 \end{pmatrix}\begin{pmatrix} 2 & -3 \\ -1 & 2 \end{pmatrix} = \begin{pmatrix} 2 & -3 \\ -1 & 2 \end{pmatrix}\begin{pmatrix} 2 & 3 \\ 1 & 2 \end{pmatrix} = \begin{pmatrix} 1 & 0 \\ 0 & 1 \end{pmatrix}$$

singular matrix

$\begin{pmatrix} 3 & 6 \\ 1 & 2 \end{pmatrix}$ has determinant

$(3 \times 2) - (6 \times 1) = 0$

rank of a matrix

$$\begin{pmatrix} 3 & 1 & 0 \\ 0 & 0 & 1 \\ 6 & 2 & 0 \end{pmatrix}$$

a matrix of rank two. Its determinant is zero, but it has non-zero determinants of minors of order two

e.g. $\begin{vmatrix} 0 & 1 \\ 2 & 0 \end{vmatrix} = -2$

inverse matrix a square matrix (p. 83) $\mathbf{A}^{-1}$ is the inverse of a given square matrix **A** if $\mathbf{A}.\mathbf{A}^{-1} = \mathbf{A}^{-1}.\mathbf{A} = \mathbf{I}$ where **I** is the unit matrix (p. 83). Also known as **reciprocal matrix**.

singular matrix a square matrix (p. 83) whose determinant (p. 86) is zero and which cannot therefore have an inverse matrix (↑).

non-singular matrix a square matrix (p. 83) which is not a singular matrix (↑).

rank of a square matrix the highest order (p. 83) of the non-zero determinants (p. 86) of minors (p. 86) of the matrix (p. 83). *See also* p. 122.

partitioned matrix a matrix (p. 83) which has been divided into blocks by horizontal and/or vertical lines.

submatrix (*n*) a block of elements in a partitioned matrix (↑).

exponential matrix a square matrix (p. 83) exp **A** (*see* exponential function, p. 138) defined by $\mathbf{I} + \mathbf{A}/(1!) + \mathbf{A}^2/(2!) + \mathbf{A}^3/(3!) + \ldots$

eigenequation

the eigenequation of $\begin{pmatrix} 2 & -3 \\ -2 & 1 \end{pmatrix}$ is $\begin{vmatrix} 2-\lambda & -3 \\ -2 & 1-\lambda \end{vmatrix} = 0$

$\Rightarrow (2-\lambda)(1-\lambda) - 6 = 0$

$\Rightarrow \lambda = 4$ or -1

these are the eigenvalues.

Since $\begin{pmatrix} 2 & -3 \\ -2 & 1 \end{pmatrix}\begin{pmatrix} 1 \\ 1 \end{pmatrix} = -1\begin{pmatrix} 1 \\ 1 \end{pmatrix}$

$\begin{pmatrix} 1 \\ 1 \end{pmatrix}$ is an eigenvector corresponding to the eigenvalue -1

eigenequation (*n*) the equation in λ obtained by putting the determinant (p. 86) of $\mathbf{A} - \lambda\mathbf{I}$ equal to zero where **A** is a square matrix (p. 83) and **I** is the unit matrix (p. 83) of the same order; in symbols $|\mathbf{A} - \lambda\mathbf{I}| = 0$. Also known as **secular equation**, **characteristic equation**.

eigenvalue (*n*) one of the values of λ in the eigenequation (↑). Also known as **eigenroot**.

eigenvector (*n*) a vector **x** corresponding to a given eigenvalue (↑) λ of a square matrix (p. 83) **A** and satisfying the equation $\mathbf{Ax} = \lambda\mathbf{x}$.

characteristic[1] (*adj*) (of a square matrix, p. 83) the same as eigen- (↑). Also known as **latent**. *See also* p. 139.

Cayley-Hamilton theorem

from the above eigenequation $\begin{pmatrix} 2 & -3 \\ -2 & 1 \end{pmatrix}$ satisfies $\begin{pmatrix} 2 & -3 \\ -2 & 1 \end{pmatrix}^2 - 3\begin{pmatrix} 2 & -3 \\ -2 & 1 \end{pmatrix} - 4\begin{pmatrix} 1 & 0 \\ 0 & 1 \end{pmatrix} = \mathbf{0}$

Cayley-Hamilton theorem every square matrix (p. 83) satisfies its own eigenequation (↑).

signature (*n*) if **A** is a symmetric matrix (p. 84), then the number of positive roots (p. 47) minus the number of negative roots of its eigenequation (↑) is the signature of **A**.

transformation matrix a square matrix (p. 83) which describes a transformation (p. 54), especially one in transformation geometry (p. 233).

linear operator a linear transformation (p. 55) which maps (p. 51) a vector space (p. 78) onto itself and which can be represented by a square matrix (p. 83).

sequence
infinite sequences

1	1, 2, 4, 8, 16, ...
2	$a, 2a^2, 3a^3, 4a^4, \ldots$

series

1	$1 + 2 + 4$
2	$1 + 2 + 4 + 8 + 16 + \ldots$
3	$a + 2a^2 + 3a^3 + 4a^4$

2 is an infinite series

sequence (*n*) an ordered set (p. 19) of members which is determined by a rule. *Members* is a better word than *elements* or *terms* (p. 59) which can then be used for sets and series (↓) respectively. Also known as **progression**.

series (*n*) a sum of successive (p. 292) members of a sequence (↑); each member is then best called a *term* (p. 59) of the series, e.g. the sequence 1, 2, 3, 4, ... gives the series $1 + 2 + 3 + 4 + , \ldots,$ the sum to n terms being the series $1 + 2 + 3 + \ldots + n$. Also used generally of things in line one after the other (e.g. *see* switching algebra, p. 244).

partial sum

$1 + \frac{1}{2} + \frac{1}{4} + \frac{1}{8}$ is a partial sum
of the series $1 + \frac{1}{2} + \frac{1}{4} + \frac{1}{8} + \ldots$

partial sum the sum of some of the members of a sequence (↑), often an infinite sequence, and usually those members at the beginning.

sigma notation a notation which uses the Greek letter Σ (capital sigma) to show that the members of a sequence (↑) must be added, e.g. $\sum_{i=1}^{4} i^2 = 1^2 + 2^2 + 3^2 + 4^2$.

continued product a product shown by the symbol Π (capital pi), e.g.

$\prod_{i=1}^{4} i^2 = 1^2 . 2^2 . 3^2 . 4^2$. Also known as **pi notation**.

	arithmetic sequence	common differences
1	3, 5, 7, 9, ...	2
2	8, 6.5, 5, 3.5 ...	−1.5

arithmetic sequence a sequence (p. 91) in which each element is formed by adding a constant amount to the one before. Also known as **arithmetic progression**. Similarly **arithmetic series**.

common difference the difference between successive (p. 292) members of an arithmetic sequence (↑).

geometric sequence a sequence (p. 91) in which each member is formed by multiplying the one before it by a constant amount. Also known as **geometric progression**, **geometric series**.

common ratio the amount by which each member in a geometric sequence (↑) is multiplied to give the next member.

hypergeometric sequence a sequence (p. 91) whose members are of the form: 1, $(abx)/c$,

$$\frac{a(a+1)b(b+1)x^2}{c(c+1)(2!)},$$

$$\frac{a(a+1)(a+2)b(b+1)(b+2)x^3}{c(c+1)(c+2)(3!)}, \ldots$$

where a, b and c are greater than zero. Similarly **hypergeometric series**.

harmonic sequence a sequence (p. 91) in which each member is the reciprocal (p. 27) of a member in an arithmetic sequence (↑), e.g. 1, 1/2, 1/3, 1/4, ... Also known as **harmonic progression**. Similarly **harmonic series**.

alternating sequence a sequence (p. 91) in which each member is the opposite sign (positive or negative) to the one before it, e.g. 1, −2, 4, −8, 16, −32, ... Similarly **alternating series**.

Fibonacci sequence a sequence (p. 91) in which each member is the sum of the two members immediately before it, most often used for the sequence 0, 1, 1, 2, 3, 5, 8, 13, ...

Fibonacci number a member of the Fibonacci sequence (↑) 0, 1, 1, 2, 3, 5, 8, 13, ...

	geometric sequence	common ratios
1	1, 2, 4, 8, 16, ...	2
2	$1, \frac{1}{2}, \frac{1}{4}, \frac{1}{8}, \frac{1}{16}, \ldots$	$\frac{1}{2}$
3	1, −1, 1, −1, 1, ...	−1
4	$a, \frac{1}{2}a^2, \frac{1}{4}a^3, \frac{1}{8}a^4, \ldots$	$\frac{1}{2}a$

Farey sequence

sequences of order 4, 5 and 6

$n = 4 \quad \frac{0}{1}, \frac{1}{4}, \frac{1}{3}, \frac{1}{2}, \frac{2}{3}, \frac{3}{4}, 1$

$n = 5 \quad \frac{0}{1}, \frac{1}{5}, \frac{1}{4}, \frac{1}{3}, \frac{2}{5}, \frac{1}{2}, \frac{3}{5}, \frac{2}{3}, \frac{3}{4}, \frac{4}{5}, 1$

$n = 6 \quad \frac{0}{1}, \frac{1}{6}, \frac{1}{5}, \frac{1}{4}, \frac{1}{3}, \frac{2}{5}, \frac{1}{2}, \frac{3}{5}, \frac{2}{3}, \frac{3}{4}, \frac{4}{5}, \frac{5}{6}, 1$

Farey sequence a sequence (p. 91) of fractions (a/b) such that $0 \leqslant (a/b) \leqslant 1$, $0 < b \leqslant n$ and $0 \leqslant a \leqslant b$ where a, b and n are positive integers. The members of the sequence are arranged in increasing order and equivalent fractions (p. 31) are each given once only in their simplest form.

power sequence a sequence (p. 91) whose members are powers of the form a_0, a_1x, a_2x^2, a^3x^3, ... *See also* power series (p. 149).

arithmetico-geometric sequence a power sequence (↑) whose members have coefficients (p. 61) which form an arithmetic sequence (↑). Similarly **arithmetico-geometric series**.

monotonic sequence a sequence (p. 91) in which either every member is greater than or equal to the one before it, or every member is less than or equal to the one before it.

strictly monotonic sequence as for a monotonic sequence (↑) except that no member may equal the one before it.

increasing sequence a sequence (p. 91) in which every member is greater than or equal to the one before it. Often used with the words monotonic (↑) or strictly monotonic (↑). Similarly **decreasing sequence**.

recurrence relation an equation which gives a relation (p. 16) between consecutive members of a sequence (p. 91), often used to define the members of the sequence one by one from the first few members, as in the Fibonacci sequence (↑). E.g. $a_{i+2} = a_i + 2a_{i+1}$ together with $a_1 = 1$ and $a_2 = 2$ define the sequence $a_3 = 5$, $a_4 = 12$, $a_5 = 29$, ... Also known as **difference equation**.

recurring sequence a power sequence (p. 93) in which the coefficients (p. 61) of consecutive members satisfy a linear equation (p. 62), e.g. using the example under recurrence relation (p. 93) with $a_0 = 0$, the sequence x, $2x^2$, $5x^3$, $12x^4$, $29x^5$, ...

recursive definition[2] a definition of a function as a sequence (p. 91) using a recurrence relation (p. 93) for the natural numbers (p. 21), e.g. $a^i = a \,.\, a^{i-1}$, $a^0 = 1$ which can be used to define powers of a positive integer a. *See also* p.10.

recursion (*n*) the use of recursive definitions (↑). **recursive** (*adj*).

functional equation similar to recursive definition (↑) except that the variable is continuous, rather than just being the discrete set of natural numbers (p. 21), e.g. the gamma function (p. 157) can be defined by $\Gamma(x + 1) = x\Gamma(x)$.

converge (*v*) a sequence (p. 91) or series (p. 91), whose nth member is p_n (which may be a function $f_n(x)$), is said to converge to a limit L if the difference between p_n and L can be made as small as we wish by taking a large enough value of n. Also known as **tend**. *See also* p. 64. **convergence** (*n*), **convergent** (*adj*). Symbols: if for any positive number ε we can find a positive integer N such that $| p_n - L | < \varepsilon$ when $n \geqslant N$ then $p_n \rightarrow L$.

limit[2] (*n*) the number L in the definition of convergence (↑). *See also* p. 64. **limiting** (*adj*).

fundamental limit theorem if every member of a sequence (p. 91) is positive, then its sum tends either to infinity or to a finite number.

null sequence a sequence (p. 91) whose limit is zero, e.g. $1, \frac{1}{2}, \frac{1}{4}, \frac{1}{8}, \frac{1}{16}, \ldots$.

diverge (*v*) a sequence (p. 91) or series (p. 91) which does not converge (↑) is said to diverge. **divergence** (*n*), **divergent** (*adj*).

Cauchy sequence a convergent (↑) sequence (p. 91).

convergent series a series (p. 91) is convergent (↑) if the sequence (p. 91) of successive partial sums (p. 91) is convergent.

converge

consider the sequence

$\frac{1}{2}, \frac{3}{4}, \frac{7}{8}, \frac{15}{16}, \cdots$

whose nth term is $1 - (\frac{1}{2})^n$

the limit L is 1

if ϵ is $\frac{1}{10}$, $N = 4$ and
$|p_N - L| = |\frac{15}{16} - 1| = \frac{1}{16} < \frac{1}{10}$

if ϵ is $\frac{1}{100}$, $N = 7$ and
$|p_N - L| = |\frac{127}{128} - 1| = \frac{1}{128} < \frac{1}{100}$

if ϵ is $\frac{1}{1000}$, $N = 10$ and
$|p_N - L| = |\frac{1023}{1024} - 1| = \frac{1}{1024} < \frac{1}{1000}$

in each case $|p_N - L| < \epsilon$ whenever $n \geqslant N$. For any ϵ, no matter how small, we can find a suitable N, so the sequence is convergent

completeness (*n*) a set of numbers for which the limit of every convergent (↑) sequence (p. 91) in the set also lies in the set is said to have the property of completeness, e.g. the real numbers (p. 46) have completeness, but not the rational numbers (p. 26), since sequences of rational numbers can have irrational numbers as their limit. **complete** (*adj*).

absolute convergence a sequence (p. 91) or series (p. 91) of real (p. 46) or complex numbers (p. 47) whose *n*th member or term is p_n is absolutely convergent if the sequence of moduli (pp. 36, 49) $|p_n|$ is convergent (↑).

conditional convergence convergence (↑) which is not absolute convergence (↑). Also known as **semi-convergence** (*n*).

oscillating (*adj*) infinite sequences (p. 91), series (p. 91) or functions which neither have a limit nor tend to infinity are said to be oscillating, e.g. the sequence 1, −1, 1, −1, 1, −1, . . . ; the sine (p. 131) function. Also known as **non-convergent**.

upper limit the greatest cluster point (p. 65) of a sequence (p. 91) of members if one exists. Also known as **superior limit**. **lim sup** (*abbr*).

lower limit the least cluster point (p. 65) of a sequence (p. 91) of members if one exists. Also known as **inferior limit**. **lim inf** (*abbr*).

limiting sum the limit of the sequence (p. 91) of partial sums (p. 91) of a convergent series (↑). Also known as **sum to infinity**.

neighbourhood (*n*) the neighbourhood of a point is the set of points within an arbitrarily (p. 287) small distance of a given point; in topology (p. 238) any subset (p. 12) of a topological space (p. 238) which contains an open set (p. 10) of which the point is an element.

circle of convergence a circle within which a complex power series (p. 149) is absolutely convergent (↑) for all values of the variable inside it, and divergent (↑) for all values outside it. Similarly **radius of convergence** is the radius of this circle.

uniform convergence convergence (p. 94) of a sequence (p. 91) of functions $f_n(x)$ defined in an interval such that the integer N given in the definition of converge (p. 94) depends on ε, but not on the value of x in the interval.

comparison test if $0 \leqslant u_n \leqslant v_n$ for all n, then the series (p. 91) $\Sigma\, u_n$ converges (p. 94) if $\Sigma\, v_n$ converges. If $0 \leqslant v_n \leqslant u_n$, then $\Sigma\, u_n$ diverges (p. 94) if $\Sigma\, v_n$ diverges.

d'Alembert's ratio test when $\Sigma\, u_n$ is a series (p. 91) of positive terms, then

I if u_n / u_{n+1} tends to a limit greater than 1, $\Sigma\, u_n$ is convergent (p. 94);

II if u_n / u_{n+1} tends to a limit less than 1, it is divergent (p. 94).

Raabe's ratio test when $\Sigma\, u_n$ is a series (p. 91) of positive terms, then

I if $n\{(u_n / u_{n+1}) - 1\}$ tends to a limit greater than 1, $\Sigma\, u_n$ is convergent (p. 94);

II if $n\{(u_n / u_{n+1}) - 1\}$ tends to a limit less than 1, it is divergent (p. 94).

general ratio test a generalized form of the above two tests.

Leibnitz's theorem[1] if $u_1 - u_2 + u_3 - u_4 + \ldots$ is an alternating series (p. 92) of positive terms and $u_{n+1} < u_n$, and if the limit of u_n tends to zero as n tends to infinity, then the series is convergent (p. 94). *Do not mistake for* Leibnitz's theorem (p. 146).

Dirichlet's test if $\Sigma\, a_n$ converges (p. 94) or oscillates (p. 275) between finite limits, and if the sequence (p. 91) v_n has uniform convergence (↑) to zero as n tends to infinity, then $\Sigma\, a_n v_n$ is also uniformly convergent.

Abel's test if $\Sigma\, a_n$ is convergent (p. 94) and if the sequence (p. 91) v_n is monotonic (p. 93) and tends to a finite limit as n tends to infinity, then $\Sigma\, a_n v_n$ is also convergent.

integral test if a function $f(x)$ is greater than zero when x is greater than zero and if $f(x)$ is a decreasing function (p. 67), then

$s_n = f(1) + f(2) + \ldots + f(n)$ and

$I_n = \int_1^n f(x)\,dx$ (*see* integral, p. 152) are either both convergent (p. 94) or both divergent (p. 94). Also known as **Maclaurin-Cauchy test**.

factorial

0! = 1
1! = 1
2! = 2
3! = 6
4! = 24
5! = 120
6! = 720
7! = 5040
8! = 40320
9! = 362880

factorial (*n*) the factorial of a positive integer *n* is the product of all the positive integers up to and including *n*. Symbols: $n!$ or $\lfloor n$, e.g. $5! = 1 \times 2 \times 3 \times 4 \times 5 = 120$.

Stirling's formula an approximate formula for the values of large factorials. $n!$ is approximately equal to $n^n\sqrt{(2\pi n)}e^{-n}$, where *e* is the exponential constant (p. 138).

permutation (*n*) a way in which a set of *r* elements may be chosen in order from a set of *n* elements is called a permutation of *r* from *n*. This can be done in $n!/(n-r)!$ ways, e.g. choosing the first, second and third from a set of ten can be done in $10!/(10-3)! = 10 \times 9 \times 8 = 720$ ways. Also used for a rearrangement (↓) of the *n* elements of an ordered set (p. 19). **permute** (*v*). Symbol: ^nP_r or $_nP_r$.

permutation

choosing two elements in order from *A*, *B*, *C*, *D*, *E* (permuting two from five)

AB	*BA*	*CA*	*DA*	*EA*
AC	*BC*	*CB*	*DB*	*EB*
AD	*BD*	*CD*	*DC*	*EC*
AE	*BE*	*CE*	*DE*	*ED*

$$^5p_2 = \frac{5!}{(5-2)!} = 20$$

subfactorial (*n*) the subfactorial of a positive integer *n* is $n!\{1 - (1/1!) + (1/2!) - (1/3!) + \ldots + (-1)^n(1/n!)\}$ (*see* factorial, ↑) and arises from the problem of placing *n* objects in *n* boxes so that every object is in the wrong box, e.g. subfactorial 4 is $4!\{1 - (1/1!) + (1/2!) - (1/3!) + (1/4!)\} = 24\{1 - 1/1 + 1/2 - 1/6 + 1/24\} = 9$. Symbol: $n¡$

rearrangement (*n*) a permutation (↑) of all the elements of a set in which they are placed in a different order. Also known as **shuffle**, especially when cards are used.

cyclic permutation a permutation (↑) in which, if the elements are written in order around a circle, the order after permuting remains the same, but the starting element may change.

cyclic order taken as if written in order round a circle, e.g. *abcd*, *bcda*, *cdab* and *dabc* are all in the same cyclic order.

cyclic permutation

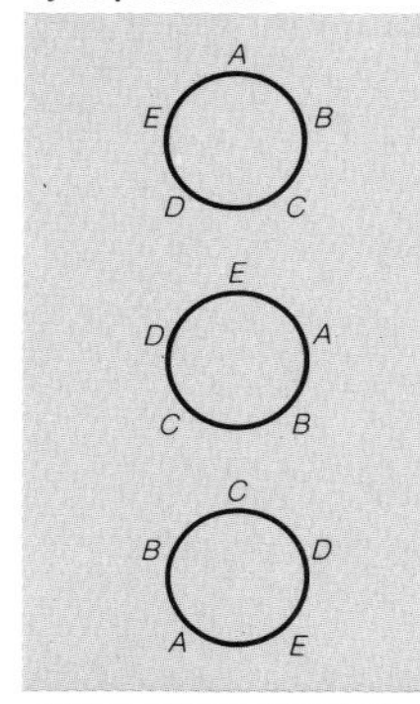

ABCDE, *EABCD*, *CDEAB* are three ways of writing *A*, *B*, *C*, *D*, *E* in the same cyclic order

cyclic function a function which remains the same function if all the variables are changed in cyclic order (p. 97), e.g. $(x + y - z)(y + z - x)(z + x - y)$.

even permutation a rearrangement (p. 97) of the elements of an ordered set (p. 19) which can be obtained by interchanging pairs of elements for each other an even number of times. Similarly **odd permutation**.

combination (*n*) a way in which *r* elements can be chosen without ordering from a set of *n* elements is called a combination of *r* from *n*. This can be done in $n!/\{r!(n-r)!\}$ ways, e.g. choosing 3 from a set of 10 can be done in $10!/\{3! \times (10-3)!\}$ $= (10 \times 9 \times 8) \div (3 \times 2 \times 1) = 120$ ways. **combinatorial** (*adj*). Symbols: ${}^nC_r$, ${}_nC_r$ or $\binom{n}{r}$.

Vandermonde's theorem the theorem on combinations (↑) ${}^nC_r = {}^nC_{r-1} + {}^{n-1}C_r$.

binomial theorem
some simple cases

$(1 + x)^2 = 1 + 2x + x^2$
$(1 + x)^3 = 1 + 3x + 3x^2 + x^3$
$(1 + x)^4 = 1 + 4x + 6x^2 + 4x^3 + x^4$
$(1 + x)^{\frac{1}{2}} = 1 + \frac{1}{2}x - \frac{1}{8}x^2 + \frac{1}{16}x^3 - \frac{5}{128}x^4$
$+ \ldots (-1 < x < 1)$

binomial theorem the expansion (p. 61) of $(x + a)^n$. If *n* is a positive integer, then this is $x^n + {}^nC_1x^{n-1}a + {}^nC_2x^{n-2}a^2 + \ldots + {}^nC_{r-1}xa^{n-1} + a^n$, where the coefficient (p. 61) of $x^{n-r}a^r$ is the combination (↑) ${}^nC_r$. If *n* is not a positive integer, then only if $-1 < h < 1$ can this be generalized to the formula $(1 + h)^n = 1 + nh + n(n-1)h^2/2! + n(n-1)(n-2)h^3/3! + \ldots$ where *n* must now be a rational number (p. 26), the coefficient of h^r is $\{n(n-1)(n-2)\ldots(n-r+1)\}/r!$ and the series (p. 91) is now convergent (p. 94).

binomial coefficients the coefficients (p. 61) ${}^nC_r$ of $x^{n-r}a^r$ in the binomial theorem (↑) when *n* is a positive integer, or the coefficients of h^r in the general case.

binomial series the power series (p. 149) given by the generalized binomial theorem (↑).

even permutation
for four elements *A, B, C, D*

identity

ABCD	*BDCA*	*CDAB*
ACDB	*BCAD*	*DCBA*
ADBC	*CABD*	*DBAC*
BADC	*CBDA*	*DACB*

Similarly the odd permutations of *A, B, C, D* are

ABDC	*BCDA*	*CDBA*
ADCB	*BDAC*	*DBCA*
ACBD	*CBAD*	*DCAB*
BACD	*CADB*	*DABC*

combination
choosing two elements (regardless of order) from *A, B, C, D, E*

AB	*AD*	*BC*	*BE*	*CE*
AC	*AE*	*BD*	*CD*	*DE*

$${}^5C_2 = \frac{5!}{(5-2)!\,3!} = 10$$

Pascal's triangle

```
                1
              1   1
            1   2   1
          1   3   3   1
        1   4   6   4   1
      1   5   10  10  5   1
    1   6   15  20  15  6   1
  1   7   21  35  35  21  7   1
1   8   28  56  70  56  28  8   1
.   .   .   .   .   .   .   .   .
```

Pascal's triangle the triangular array (p. 83) of binomial coefficients (↑) for positive integers *n* (*see* figure) which has the property (among many others) that each number is the sum of the two numbers immediately above it.

Bernoulli's inequality if $x > 0$, and $n > 0$, then $(1 + x)^n > 1 + nx$.

multinomial theorem a generalization of the binomial theorem (↑) where there are more than two terms in the brackets (p. 9).

population (*n*) a set of elements, especially in statistics, from which some are to be chosen. Also known as **universe**.

sample (*n*) a set of elements which is chosen from a population, especially in statistics. **sample** (*v*), **sampling** (*adj*).

replacement (*n*) a type of sampling in which, when any element of a population has been chosen it is returned to the population and may be chosen again with the same probability as the other elements of the population. Because the theory is easier, it is the usual method of sampling, e.g. a sample of six letters chosen with replacement might be *P, S, U, S, T, A*.

sample space the set of all possible samples from a population, including the null set (p. 11) and the whole population. Also known as **event space**, **outcome space**, **possibility space**.

random sample a sample chosen in such a way that any other sample of the same size from the same sample space (↑) would be equally likely to have been chosen.

random (*adj*) used of anything chosen in the same way as a random sample (↑), so that any particular choice can only be given as a probability (p. 101). **randomness** (*n*), **randomize** (*v*), **randomization** (*n*).

random number
a set of 100 random digits which can be used to make random numbers

59141	75097	89234	80419
11269	69920	57893	74913
83523	14425	02325	29385
06767	88831	14507	10469
88753	70829	12560	52062

random number a number whose digits are chosen at random with replacement (p. 99) so that it is one of a set of numbers which can be chosen, all of which are equally likely. Often used in the choice of samples.

random number table a list of digits 0 to 9, given in a random way and which can be used to choose samples.

quasi-random number a number which is not truly random, but which may be used as if it were random.

pseudo-random number similar to quasi-random number (↑), but used especially for numbers given by a computer (p. 42) or calculator (p. 42) which can be treated as random.

stochastic (*adj*) random, descriptive of anything produced by random methods.

experiment (*n*) something done to try out a theory or discover something unknown in science. In statistics often based on a sample chosen from a population which is too large to be completely examined. **experiment** (*v*).

trial (*n*) a single part of an experiment which is repeated many times with other samples to complete the whole experiment.

event (*n*) an element of a sample space (p. 99). Also known as **trial**.

outcome (*n*) the result of an event (↑).

systematic sampling sampling from a population by means of a system, e.g. taking every sixth member from a list.

stratified random sampling dividing a population into parts and then sampling randomly from each part.

whole group sampling sampling from a population by taking a set of elements already chosen for another purpose, e.g. taking one class from a year group of children at a large school. Also known as **cluster sampling**.

double sampling sampling with a first sample and from the result deciding whether a further sample needs to be taken.

sequential sampling an extension of double sampling (↑) to more than two steps. Also known as **multistage sampling**.

small sample usually taken to be a sample of less than 25 or 30 for which more difficult statistical tests are needed.

large sample usually taken to be a sample of more than 25 or 30, when simpler approximate statistical tests may be used.

probability (*n*) if all the elements in a sample space (p. 99) of given size are equally likely to happen, the probability of any set of different samples of that size occurring is that number of samples divided by the number of possible samples. Thus the probability of obtaining an even number when throwing a die (p. 105) is $3 \div 6 = \frac{1}{2}$, since there are six numbers of which 3 are even. A complete definition is difficult. *See also* chance (↓), likelihood (↓), odds (↓). **probable** (*adj*), **probablistic** (*adj*). Symbol: the probability of an event (↑) A is $p(A)$.

equiprobable (*adj*) having the same probability.

likelihood (*n*) similar to probability, but used where calculations are based not on a known ratio but on a value obtained by sampling, and therefore only an estimate (p. 109).

chance (*n*) used loosely instead of probability, likelihood (↑).

odds (*n.pl.*) the ratio of the probability of an event happening to that of it not happening, e.g. if the probability of it happening is 0.6 (and of it not happening is thus 0.4), this is 'odds on' 6 to 4; if the probability of it happening is 0.1 this is 'odds against' 9 to 1. *See also* odd (p. 29).

Bernoulli trial an event (↑) with only two results whose probabilities (↑) sum to unity and remain constant throughout an experiment.

success (*n*) one outcome (↑) of a Bernoulli trial (↑), sometimes *but not necessarily* one which is in some way desired.

failure (*n*) the outcome (↑) of a Bernoulli trial (↑) which is not a success (↑).

run (*n*) a repetition of a single result in a set of trials (p. 100). The length of the run is often of interest.

run
in the set of random numbers

06767 88831 14507

there is a run of three 8s

Bernoulli's theorem in a set of n Bernoulli trials (p. 101) the probability of r successes (p. 101) is ${}^nC_r\, p^r\, q^{n-r}$ (*see* combinations, p. 98, binomial theorem, p. 98) where p is the probability of success in each trial and $q = 1 - p$.

occurrence ratio if an outcome (p. 100) occurs r times in n trials (p. 100), this is the ratio r/n. Also known as **success ratio**, **frequency ratio**.

Buffon's needle a famous experiment in which a needle of length l was dropped at random many times onto paper with parallel lines drawn on it a distance a apart where $l < a$. The probability of a needle lying across a line is $2l/\pi a$ and can be used to estimate (p. 109) the value of π (pi, p. 192).

Buffon's needle

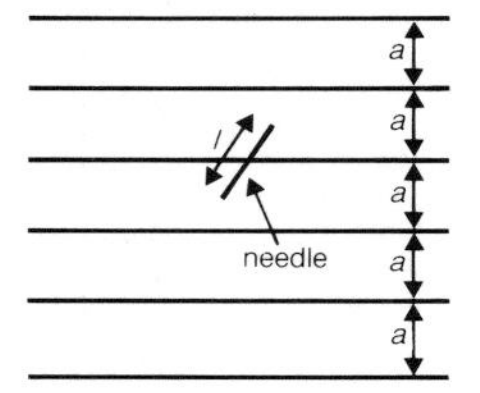

mutually exclusive used of events (p. 100) of which only one can take place, e.g. a coin (p. 105) may land only one way (head, p. 105) or the other (tail, p. 105). The probabilities of all such events must add up to 1.

exhaustive (*adj*) giving all possible choices. Used of a set of events (p. 100) when there is no other possible event which can occur.

independent (*adj*) not depending on, not related; used of a set of events (p. 100) where if one occurs, it makes no difference to the probability of the others. The probability of all the events occurring is the product of all the separate probabilities. **independence** (*n*).

dependent (*adj*) not independent. **dependence** (*n*).

joint probability the probability of two or more events both occurring.

conditional probability the probability of an event (p. 100) A occurring if some other event B on which it is dependent (↑) has already occurred. Symbol: $p(A \mid B)$. This is equal to $p(A \cap B)/p(B)$. *See* intersection (p. 13).

total probability law the law $p(A) = \Sigma_i\, p(A \mid B_i)p(B_i)$, which gives the probability of A in terms of the conditional probabilities (↑) of A on all the events (p. 100) B_i on which A depends, and the probabilities of the events B_i.

tree diagram

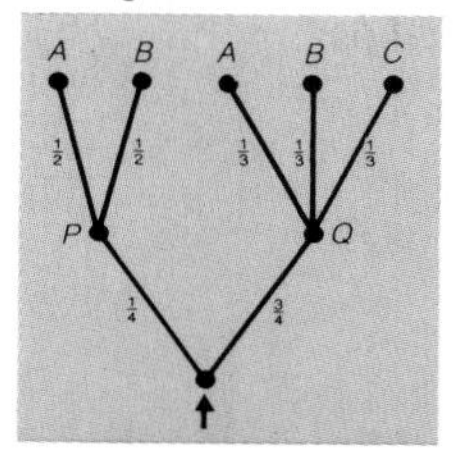

P occurs with probability $\frac{1}{4}$ and *Q* with probability $\frac{3}{4}$. If P occurs *A* and *B* are equiprobable. If *Q* occurs *A*, *B* and *C* are equiprobable

conditional probability
the conditional probability of *B* on *Q* is $p(B|Q) = \frac{1}{3}$

joint probability
the joint probability of *B* and *Q* is $p(B \cap Q) = \frac{1}{3} \times \frac{3}{4} = \frac{1}{4}$

total probability law
the probability of *B* is $(\frac{1}{4} \times \frac{1}{2}) + (\frac{3}{4} \times \frac{1}{3}) = \frac{3}{8}$

tree diagram a diagram with branches like a tree showing sequences (p. 91) of events (p. 100). Against each branch is written the conditional probability (↑) of that event. By multiplying the probabilities along any sequence of events their total probability (↑) may be found.

prior probability an estimate (p. 109) of the probability made before sampling begins and based on earlier results. *See also a priori* (p. 255).

posterior probability the new estimate (p. 109) of probability after sampling based on the prior probability (↑) and the results of the further sampling. *See also a posteriori* (p. 255).

Bayes' theorem a theorem on conditional probability. (↑):

$$p(A|B) = \frac{p(B|A)\,p(A)}{p(B|A)\,p(A) + p(B|\sim A)\,p(\sim A)},$$

where $p(\sim A)$ is the probability of *A* not occurring. This allows the calculation of the conditional probability of *A* on *B* given that the conditional probabilities of *B* on *A* and *B* on $\sim A$ are known.

Cauchy's inequality if $a_1, a_2, \ldots, a_n$ and $b_1, b_2, \ldots, b_n$ are two sets of real numbers (p. 46), then $(a_1b_1 + a_2b_2 + \ldots + a_nb_n)^2 < (a_1^2 + a_2^2 + \ldots + a_n^2)(b_1^2 + b_2^2 + \ldots + b_n^2)$ unless $a_1/b_1 = a_2/b_2 = \ldots = a_n/b_n$, when they are equal.

Tchebycheff's inequality if x is an element of a population of mean μ and standard deviation (p. 118) σ, then the probability that $|x - \mu| \geqslant \lambda\sigma$ is less than $1/\lambda^2$ for all values of λ. It can be used to prove the law of large numbers (↓). Also written **Chebyshev** and many other ways.

law of large numbers if an event (p. 100) occurs r times in n trials (p. 100), then r/n is a measure of the probability p of the event, and the difference between p and r/n will probably be small (**weak law**). By choosing $n > N$ where N is finite and can be found, the probabilities of the differences can be made small (**strong law**).

Khintchine's law an application of the law of large numbers (↑) which shows that the sample mean (p. 119) tends to the population mean as the size of the sample increases.

random variable a quantity which can take any one of a number of values which cannot be known beforehand except as a probability of their occurrence. Also known as **stochastic variable**, **random variate**. *See also* variate (↓).

variate (*n*) a random variable (↑) or similar quantity. A variate in probability and statistics is similar to a variable in algebra and analysis. Often known as **variable**. *See also* p. 56.

response variate a dependent variate (↑) whose value depends on other variates. Also known as **response variable**.

information theory that part of probability theory which deals with sending messages and the possibility of errors (p. 43) in doing so. Also known as **communication theory**.

Huffman code a method of writing messages so that they are as short as possible. This depends on the relative frequency of the letters in the language being used.

Markov process an ordered set (p. 19) of discrete random variates (↑), each of which has at any time a given state or value dependent only on the state of the variate (↑) immediately before it, is said to be subject to a Markov process.

Markov chain a set of states or values in a Markov process (↑).

transition probability the conditional probability (p. 102) of one random variate (↑) upon another in a Markov process (↑).

transition matrix a matrix (p. 83) whose elements give the complete transition probabilities (↑) which change one state or value to the next in a Markov process (↑).

ergodic (*adj*) used of a Markov process (↑) in which any one state or value can always in time produce any other.

absorbing (*adj*) used of a Markov process (↑) in which one state or value once obtained can never be left, though it can be reached from all the other states or values.

gambler's ruin a game of chance which is really an absorbing (↑) Markov process (↑) and in which one player will eventually lose everything.

Markov process

a transition matrix. The transition probability of moving from state X_4 to state X_2 is $\frac{1}{6}$. This example is an ergodic Markov process

$$\begin{array}{cc} & \text{to } X_1 \quad X_2 \quad X_3 \quad X_4 \\ \text{from} & \\ \begin{array}{c} X_1 \\ X_2 \\ X_3 \\ X_4 \end{array} & \begin{pmatrix} 0 & \frac{1}{2} & 0 & \frac{1}{2} \\ \frac{1}{3} & 0 & \frac{1}{3} & \frac{1}{3} \\ \frac{1}{4} & \frac{1}{4} & \frac{1}{4} & \frac{1}{4} \\ \frac{1}{6} & \frac{1}{6} & \frac{1}{6} & \frac{1}{2} \end{pmatrix} \end{array}$$

absorbing

a transition matrix for an absorbing Markov process. The state X_2 once reached is never left

$$\begin{array}{cc} & \text{to } X_1 \quad X_2 \quad X_3 \\ \text{from} & \\ \begin{array}{c} X_1 \\ X_2 \\ X_3 \end{array} & \begin{pmatrix} 0 & \frac{1}{10} & \frac{9}{10} \\ 0 & 1 & 0 \\ \frac{9}{10} & \frac{1}{10} & 0 \end{pmatrix} \end{array}$$

random walk
a walk on a rectangular cartesian grid which starts at (0, 0)

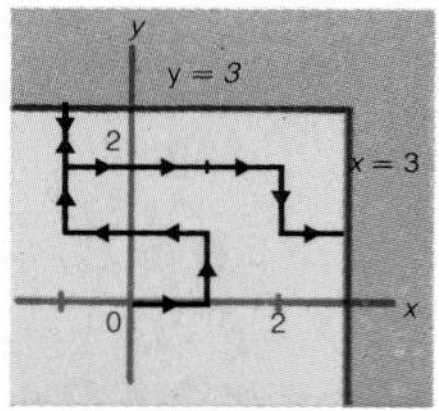

$y = 3$ is a reflecting barrier
$x = 3$ is an absorbing barrier

random walk a stochastic (p. 100) process which is a sequence (p. 91) of steps, each of which is independent of the size and direction of the one before it but follows on from it.

absorbing barrier a point in a random walk (↑) from which no further movement can take place.

reflecting barrier a point in a random walk (↑) where the only possible movement is to return in the direction of arrival.

games theory the study of games based on the probability of different outcomes (p. 100), whose results are useful in studying other areas such as trade or war.

strategy (*n*) a method of playing in games theory (↑), the best strategy being the one which gives the highest probability of winning.

maximum likelihood a strategy (↑) in games theory (↑) which is based on the most likely event (p. 100) to happen.

dice

dice (*n.pl.*) small cubes used in games, usually with the numbers 1 to 6 on the faces. (The correct singular word is **die**, but *dice* is often, though incorrectly, used.)

urn model a method of understanding probability by imagining coloured, but otherwise similar, balls to be taken, without looking, from a large urn or container, each ball having the same probability of being chosen.

coin (*n*) a piece of money which can be thrown (called *tossing* or *spinning*) and has equal probabilities of landing on each of its two sides.

head (*n*) the name of one side of a coin (↑), usually having the picture of a head on it. Also known as **obverse**.

tail (*n*) the side of a coin (↑) which is not a head (↑). Also known as **reverse**.

unbiased (*adj*) fair, having the probabilities expected, e.g. of dice (↑) which have 1/6 probability for each result, used of estimators (p. 109) whose expected values (p. 109) are equal to the quantity being estimated (p. 109). Also written **unbiassed**. Such estimators are also sometimes known as **efficient**.

biased (*adj*) unfair, not unbiased (↑). Also written **biassed**. bias (*n*).

frequency
pay of 155 workers

pay	number of workers	
	frequency	cumulative frequency
£ 4000 – 6000	15	15
£ 6000 – 8000	40	55
£ 8000 – 10000	65	120
£10000 – 15000	20	140
£15000 – 20000	10	150
£20000 – 30000	5	155

frequency[1] (*n*) the number of times an event (p. 100) or set of events occurs. If events are equiprobable (p. 101), the frequency of an event divided by the total frequency of all possible events gives the probability of that event.

distribution (*n*) a graph or table showing the frequency of the probability of the events (p. 100) forming a sample or population.

discrete distribution a distribution where the variable can only take a finite number of values in any interval.

continuous distribution a distribution where the variable can take any value within certain intervals.

frequency density the quantity in the vertical scale of a histogram (p. 112) or frequency distribution (↓) which is the frequency of a given class divided by the class interval (p. 111) for that frequency.

frequency distribution function a function $f(x)$ for each value x of a continuous distribution (↑) which gives the frequency of the values of x for $a \leqslant x \leqslant b$ as the integral $\int_a^b f(x)dx$.

sampling distribution a distribution of a sampling statistic (p. 110).

probability function a function $p(x)$ for each value x of a discrete distribution (↑) which gives the probability of that value x.

probability density function a function $f(x)$ for each value x of a continuous distribution (↑) which gives the probability that $a \leqslant x \leqslant b$ as the area under $f(x)$ between $x = a$ and $x = b$, that is as the definite integral $\int_a^b f(x)dx$.
p. d. f. (*abbr*).

discrete distribution
discrete frequency distribution showing result of throwing a die 60 times

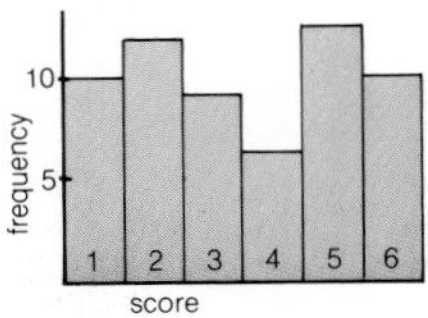

continuous distribution

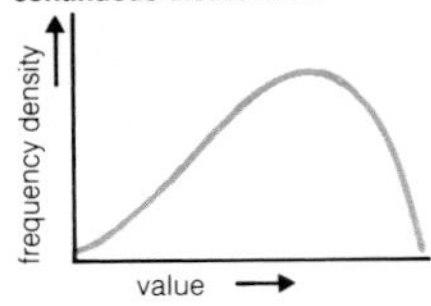

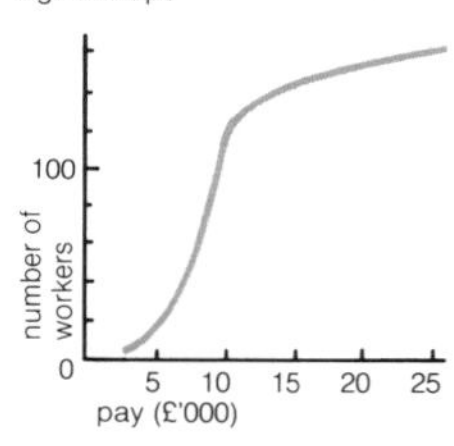

cumulative distribution
pay of 155 workers showing ogive shape

cumulative (*adj*) formed by adding on successive (p. 292) values.

cumulative frequency the total frequency up to a given value.

grade (*n*) the cumulative frequency (↑) up to a given value divided by the total frequency. *See also* p. 173. Also known as **relative cumulative frequency**.

cumulative distribution a distribution which gives probabilities or frequencies of events (p. 100) less than or equal to a given value of the variable *x*.

cumulative distribution function the probability density function (↑) or probability function (↑) of a cumulative distribution (↑). **c. d. f.** (*abbr*).

cumulative sum analysis analysis based on cumulative (↑) sums and the statistics obtained from them, e.g. trend lines are examined to control production in machines. Also known as **cusum analysis**.

ogive (*n*) a curve whose shape is similar to the shape of a cumulative distribution (↑), especially to that of the cumulative normal distribution (p. 120).

generating function *see* probability generating function (↓), moment generating function (p. 121).

probability generating function a function $G(t)$ which gives a power series (p. 149) $p_0 + p_1t + p_2t^2 + \ldots$. The coefficients (p. 61) $p_0, p_1, p_2, \ldots$ of the power series give the probabilities of the values of a given variate (p. 104) x being equal to 0, 1, 2, ... respectively.

bivariate distribution a distribution which depends on two variables. *See also* joint distribution (↓).

joint distribution a distribution which depends on two or more variables; for two variables it can be represented by a surface.

uniform distribution a distribution in which all the values of the variable in a given interval have equal probability. Also known as **rectangular distribution**.

uniform distribution
over the interval from *a* to *b*

y 0 a b x

binomial distribution a discrete probability distribution where the probability of r successes (p. 101) in n Bernoulli trials (p. 101) is given by the quantity ${}^{n}C_{r}p^{r}q^{n-r}$ in the binomial theorem (p. 98) expansion (p. 61) of $(p+q)^{r}$ where p is the constant probability of success and $q = 1 - p$. Its mean is np and its variance (p. 118) is npq, e.g. the probability of throwing exactly three 6s in ten throws of a die (p. 105) is ${}^{10}C_{3}(1/6)^{3}(5/6)^{7}$. The mean number of 6s in ten throws is 10/6 and its variance is 50/6. Also known as **Bernoulli distribution**.

quincunx (n) a machine in which balls are dropped against pins and are equally likely to fall on either side of each pin. It demonstrates (p. 288) the binomial distribution (↑).

multinomial distribution the distribution obtained from the multinomial theorem (p. 99) which replaces the binomial theorem (p. 98) when there are more than two outcomes (p. 100) of an event (p. 100).

geometric distribution the discrete distribution (p. 106) of the number of Bernoulli trials (p. 101) up to and including the first success (p. 101) in which successive (p. 292) frequencies become smaller in geometric progression (p. 92). Its mean is $1/p$, and its variance (p. 118) is q/p^2 where p is the probability of a success and $q = 1 - p$.

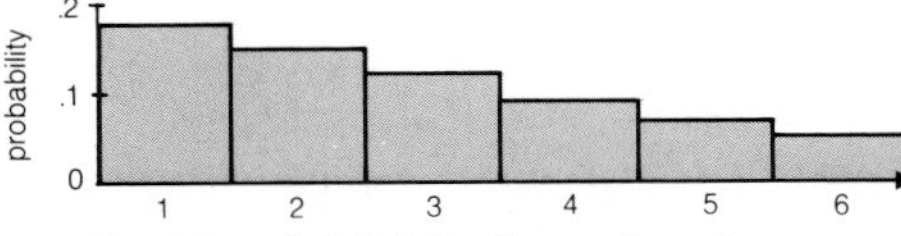

negative binomial distribution a discrete distribution (p. 106) which is that of the number of Bernoulli trials (p. 101) up to and including the rth success (p. 101). Its mean is r/p and its variance (p. 118) is rq/p^2. Also known as **Pascal's distribution**.

exponential distribution the continuous form of the geometric distribution (↑) whose shape is like that of the negative exponential function (p. 138) $y = e^{-x}$.

binomial distribution
the probability of the given number of sixes when throwing four dice

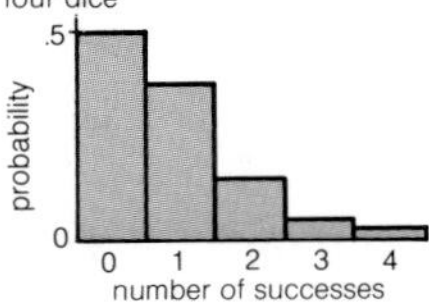

Here $n = 4$ and $p = \frac{1}{6}$. The probabilities are $\frac{625}{1296}, \frac{500}{1296}, \frac{150}{1296}, \frac{20}{1296}$ and $\frac{1}{1296}$ respectively

quincunx

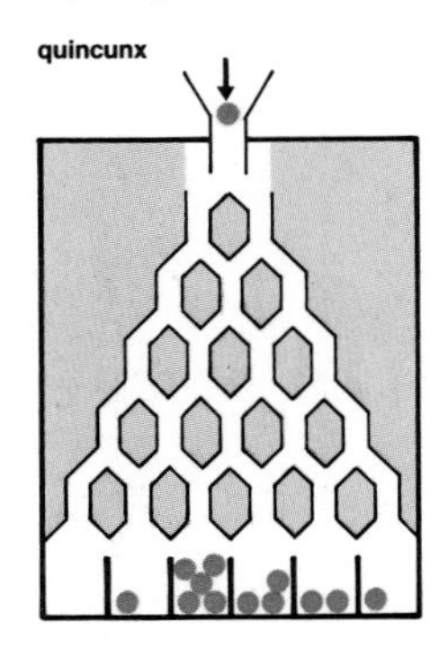

geometric distribution
the probability of throwing the first six with a die on the 1st, 2nd, 3rd, ... throw. The probabilities are $\frac{1}{6}$, $\frac{1}{6} \times \frac{5}{6}$, $\frac{1}{6} \times (\frac{5}{6})^2$, $\frac{1}{6} \times (\frac{5}{6})^3$, ...

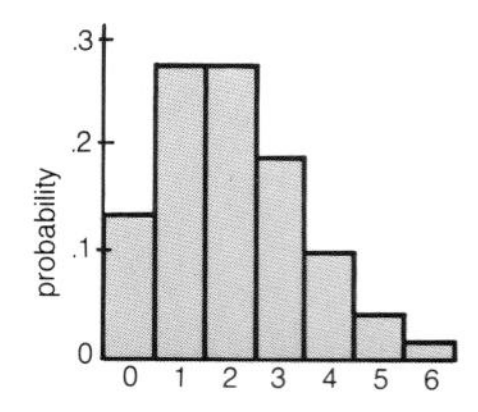

Poisson distribution for $p = 2$ if a mean number of 2 events happen in a given period, the probabilities of 0, 1, 2, 3, ... events happening are 0.14, 0.27, 0.27, 0.18, ... (or e^{-2}, $2e^{-2}$, $2^2e^{-2}/2!$, $2^3e^{-2}/3!$, ...)

Poisson distribution a discrete probability distribution which is the limiting case of the binomial distribution (↑) with a small probability of success (p. 101) p in a large number of trials (p. 100) n. Both the mean and the variance (p. 118) are np and the probability of r successes is $(e^{-\mu}\mu^r)/r!$ where $\mu = np$ is the mean number of successes. It is useful for dealing with random events which do not happen often, or which happen at discrete points in a continuous variable such as time or space.

hypergeometric distribution a discrete probability distribution for Bernoulli trials (p. 101) where sampling is *without* replacement (p. 99).

probabilistic model a theoretical probability distribution such as the binomial distribution (↑) or normal distribution (p. 120) against which a random variable may be tested.

estimate (*n*) an approximate value of some quantity, e.g. the mean, for a whole population obtained from the value for a sample. Also used outside statistics for a rough calculation which gives an approximate answer. *See also* estimator (↓).

estimator (*n*) a quantity of which an estimate (↑) is a particular example, e.g. the mean of a sample is in general an estimator of the mean of a population, the mean of a particular sample is an estimate. The best estimator of a population parameter (p. 110) is shown by placing ˆ over its symbol, e.g. $\hat{\sigma}$ for the best estimator of the standard deviation (p. 118).

estimation (*n*) the process of finding an estimate (↑) or estimator (↑).

point estimation estimation (↑) of a quantity as a single value.

interval estimation estimation (↑) of a quantity as a probability of it lying in an interval.

expectation (*n*) the mean value of an estimator (↑) in repeated sampling, e.g. with a die (p. 105) the expectation in any throw is $(1 + 2 + 3 + 4 + 5 + 6)/6 = 3.5$. Also known as **expected value**. Symbol: the expectation of a variable x is $\mathscr{E}(x)$ or $E(x)$.

statistics (1) (*n.sing.*) the part of mathematics which studies the classification (↓), tabulation (p. 112) and analysis of numerical data, and the inferences (p. 255) which can be made from these; (2) (*pl*) *see* statistic (↓). **statistical** (*adj*).

statistic (*n*) a quantity which describes some property of a sample, e.g. its arithmetic mean (p. 113).

parameter[2] (*n*) a quantity which describes some property of a whole population, e.g. its arithmetic mean (p. 113). *See also* p. 67, phase space (p. 78). Also known as **population parameter**. **parametric** (*adj*).

data (*n.pl.*) a set of quantities obtained from a sample of a population. **datum** (*sing.*) is not often used.

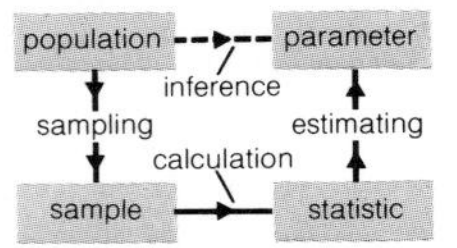

statistic and parameter
estimating a population parameter from a sample statistic

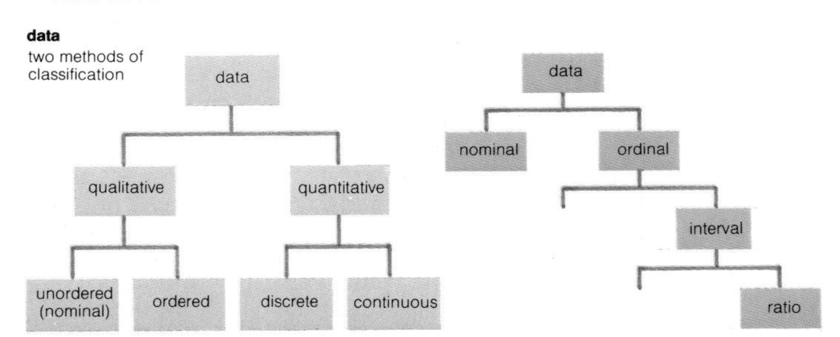

data
two methods of classification

attribute (*n*) any property possessed by a set of data, e.g. sex, age, place of birth.

class[2] (*n*) a subset (p. 12) of a set of data which is grouped into ranges (p. 113) of values, or in other ways, for analysis. *See also* pp. 13, 187. **classify** (*v*), **classification** (*n*).

grouped data data given in classes (↑) having consecutive ranges (p. 113).

class limits the greatest and least values of a class (↑) which can be obtained. For a continuous distribution (p. 106) these are the same as the class boundaries (↓).

class boundaries (1) for continuous distributions (p. 106) the same as class limits (↑); (2) for discrete distributions (p. 106), the mid-point values between the class limits of consecutive classes (↑).

grouped data
note that with ages the class limits of the class 15–24 are 15 and 25, since '24' means 'up to 25'. With other data the class limits would be 15 and 24 for discrete data and 14.5 and 24.5 for continuous data

age	number
0–4	7
5–14	18
15–24	16
25–34	13

class limit   class   frequency

class interval the difference between the class boundaries (↑) of a class (↑); the range (p. 113) of a class.

class mark the value of the mid-point of a class interval (↑). Also known as **class mid-point**, **mid-interval value**.

intraclass (*adj*) within a class (↑).

interclass (*adj*) between classes (↑).

qualitative (*adj*) having differences depending on some quality other than number, used especially of such data as colour, nationality, type of employment, sex. Some types of qualitative data can be put in order, e.g. beauty.

quantitative (*adj*) having differences depending on number or quantity, used especially of data in which each observation (p. 290) is given a number.

nominal (*adj*) used of qualitative (↑) data which can only be put into unordered groups such as eye colour, type of shop, etc.

ordinal (*adj*) able to be ordered, as of all quantitative (↑) data and some types of qualitative (↑) data. *See also* p. 24.

interval^2 (*adj*) used of data or scales of measurement where differences between values on the scale have the same meaning anywhere along the scale, but the ratio of two values need not have meaning. All interval data is also ordinal (↑). E.g. in degrees Celsius 30° is not twice as hot as 15°. *See also* p. 65.

ratio^2 (*adj*) used of data or scales of measurement which are interval but where ratios of quantities also have meaning, e.g. temperature (p. 284) in degrees Kelvin, most scales of measurement of length and weight. *See also* p. 28.

pictogram (*n*) a diagram showing quantities by drawing numbers of small pictures, e.g. of men, the numbers being proportional to the quantities. Also known as **ideograph**.

bar chart a graph having a set of bars whose lengths show the frequencies of different quantities. Also known as **bar graph**.

block graph similar to bar chart (↑) but with blocks or rectangles touching each other rather than bars. The data can be discrete as well as continuous. Also known as **block diagram**.

pictogram

number of people working for a company; each figure represents 10 people so that the figures are 20, 30 and 50

bar chart

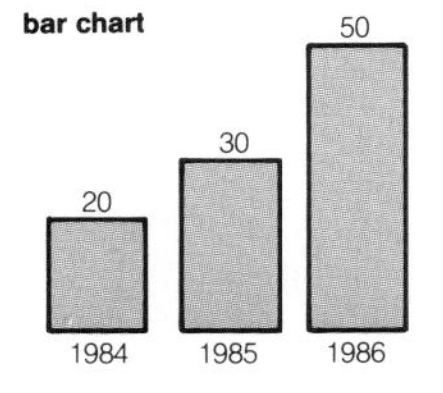

block graph

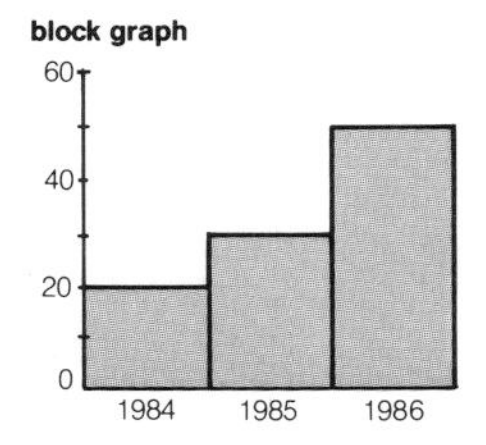

histogram (*n*) a block graph (p. 111) in which the area of each rectangle represents the frequency of each class.

histogram
ages of workers in a company

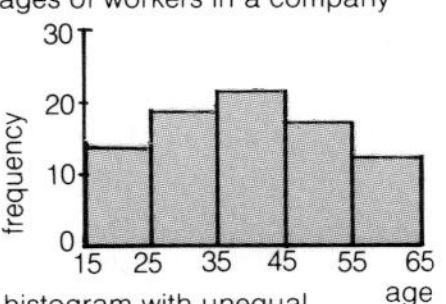

histogram with unequal class intervals showing pay of workers in a company.
NB 24 workers (not 12) receive between £0 and £10 000

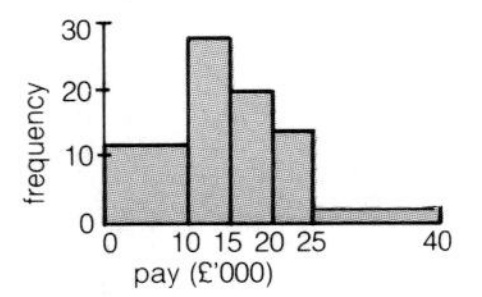

frequency polygon a polygon-shaped graph obtained by joining the mid-points of the tops of consecutive rectangles in a histogram (↑). The ends of the polygon are the class marks (p. 111) of classes (p. 110) of zero frequency at each end of the histogram.

frequency curve when data are continuous, a smooth curve through the vertices of the frequency polygon (↑).

frequency polygon
polygon of age histogram above and its associated frequency curve

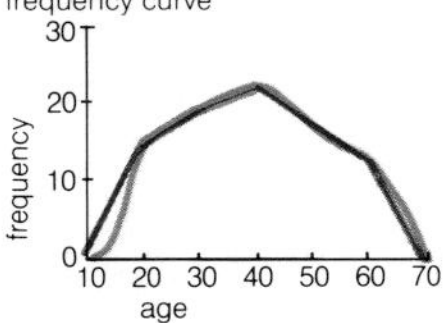

pie chart a diagram showing proportions of a whole as sectors (p. 193) of a circle. When more than one pie chart is used, the total amounts of the quantities can be compared by making the areas of the circles proportional to them. Also known as **circular diagram**.

pie chart
age distribution of workers in a company

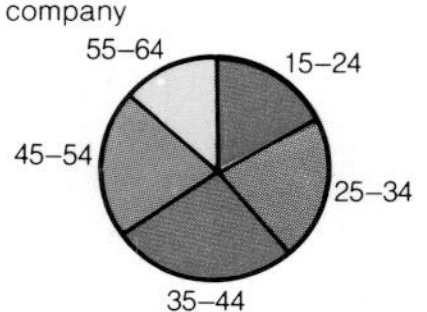

scatter diagram a diagram in which the points of a bivariate distribution (p. 107) are plotted as the points on a cartesian graph (p. 215).

hierarchy
branches of mathematics

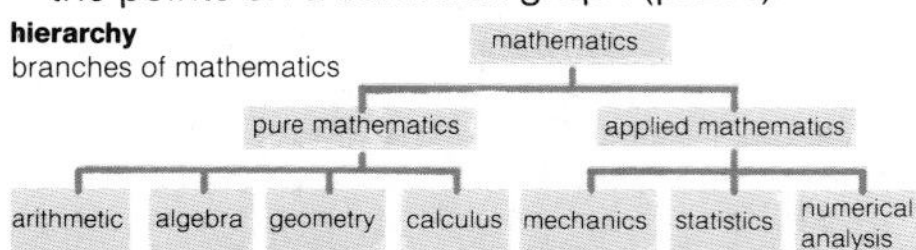

hierarchy (*n*) a classification (p. 110) at different levels, each class (p. 110) at a lower level being a dichotomy (p. 13) or polytomy (p. 13) of one of the classes at the next higher level.

table (*n*) a method of giving data by arranging frequencies of other quantities in a systematic way, usually with rows and columns of figures. **tabular** (*adj*), **tabulate** (*v*), **tabulation** (*n*).

margin (*n*) the edge, especially the rows or columns at the edge of a table which show sums or other quantities obtained from the other rows or columns. **marginal** (*adj*).

age	male	female	total
15–24	4	10	14
25–34	14	5 (cell)	19
35–44	11	10	21
45–54	6	12	18
55–64	11	2	13
total	46	39	85

(margins: total row and total column)

table
workers in a company by age and sex. Modal classes are 25–34 for males, 45–54 for females and 35–44 for all workers

cell (*n*) a place in a table in which a particular value is written.

working zero a value of a variable which is equated to zero as a base for other values, usually for ease of calculation. Also known as **false origin**. *See* working mean (p.115).

range[2] (*n*) the difference between the greatest and least values of a set of data. *See also* pp. 17, 167, 268.

measure of central tendency a quantity which gives an idea of the commonest or most central member of a set of data, e.g. mode (↓), arithmetic mean (↓). Also known as **measure of location**.

average (*n*) any measure of central tendency (↑) of a set of data, often used in everyday life for arithmetic mean (↓).

location (*n*) position generally, that which is measured by any of the quantities called *average* (↑).

mode (*n*) the most frequent measure in any sample; the maximum value in a probability distribution. **modal** (*adj*).

modal class the most frequent class (p. 110) in a frequency table.

crude mode the point which divides the modal class (↑) in the inverse ratio (p. 28) of the frequencies of the classes (p. 110) next to it.

modal point the point on a probability or frequency distribution at which the probability or frequency is greatest.

unimodal (*adj*) having only one mode (↑).

bimodal (*adj*) having two separate modes or maximum points.

antimode (*n*) a minimum value in a probability or frequency distribution, usually one between two maximum values.

arithmetic mean of *n* quantities $x_1, x_2, x_3, \ldots, x_n$: the sum of the quantities divided by the number of quantities, i.e. $\Sigma x_i / n$. If the quantities have frequencies $f_1, f_2, f_3, \ldots, f_n$, respectively, so that $\Sigma f_i = n$, then it is $\Sigma f_i x_i / n$. E.g. the arithmetic mean of 6, 16, and 18 is $(6 + 16 + 18) \div 3 = 13.33$. Often known as just **mean** (p. 114) and in everyday life as **average** (↑). Symbol: $\overline{x}$.

mode
frequency of scores in a game, with mode of 3

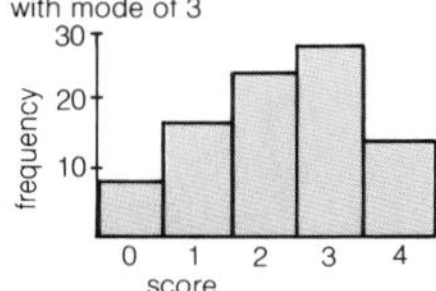

crude mode
ages of male workers in company

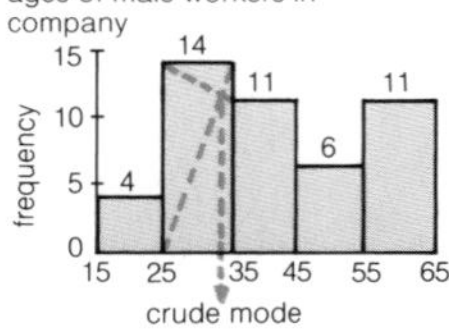

$$\frac{(4 \times 25) + (11 \times 35)}{4 + 11} = 32.3 \text{ years}$$

modal point
for a frequency distribution

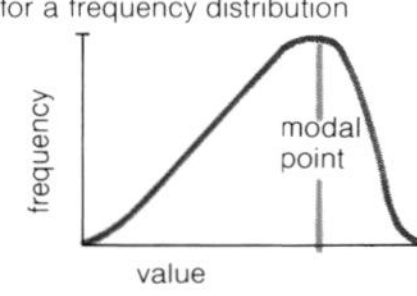

bimodal
a bimodal distribution

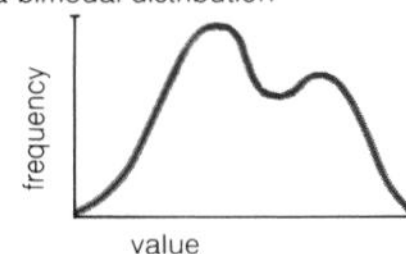

arithmetic mean
calculation for scores in a game from a frequency distribution

score (*x*)	frequency (*f*)	*fx*
0	8	0
1	16	16
2	22	44
3	27	81
4	14	56
total	87	197

the average score is
197 ÷ 87 = 2.26

geometric mean of n quantities, the nth root (p. 47) of the product of the quantities, e.g. the geometric mean of 6, 16 and 18 is $\sqrt[3]{(6 \times 16 \times 18)} = 12$.

harmonic mean of n quantities $x_1, x_2, x_3, \ldots, x_n$: the quantity H given by $(n/H) = (1/x_1) + (1/x_2) + (1/x_3) + \ldots + (1/x_n)$, e.g. the harmonic mean of 3, 4 and 8 is given by $(3/H) = (1/3) + (1/4) + (1/8)$, so that $H = 72/17 = 4.24$.

inequality of means if A is the arithmetic mean (p. 113), G is the geometric mean (↑) and H is the harmonic mean (↑) of n positive numbers, then $A \geqslant G \geqslant H$, equality (p. 10) occurring only when the n numbers are all equal, e.g. for 6, 16 and 18, $A = 13.33$, $G = 12$ and $H = 10.54$.

mean (n) a measure of central tendency (p. 113), the arithmetic (p. 113), geometric (↑) or harmonic means (↑), the median (↓), the mode (p. 113) or the mid-range (↓). Often used for arithmetic mean particularly.

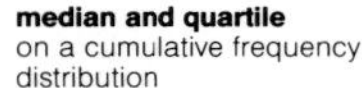

median and quartile
on a cumulative frequency distribution

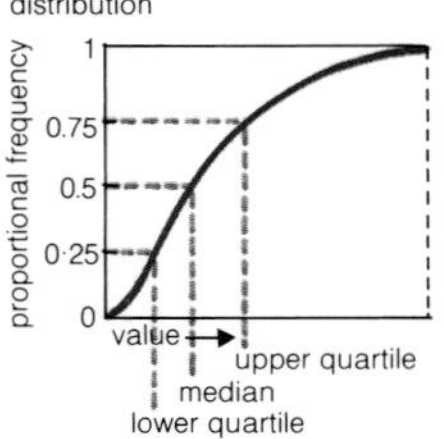

median[1] (n) the value of the central number in an ordered set (p. 19) of data. If there is an even number of data, it is taken to be the arithmetic mean (p. 113) of the two central ones. It is the fiftieth percentile (↓), e.g. the median of 7, 8, 4, 9, 10, 15, 2 and 7 is $\frac{1}{2}(7 + 8) = 7\frac{1}{2}$. *See also* p. 196.

mid-range (n) the mean of the end points of the range (p. 113) of a set of data, e.g. the mid-range of 7, 8, 4, 9, 10, 15, 2 and 7 is $\frac{1}{2}(2 + 15) = 8\frac{1}{2}$. It is a very simple measure, but not accurate since it depends only on the largest and smallest values.

percentile (n) if a set of data is arranged in order of size, the nth percentile is the value which separates the lower n percent of the data from the rest. For a discrete distribution (p. 106) this might not be an exact value; percentiles should therefore be used only when the number of data is large so that a good approximation can be obtained. Also known as **centile**.

decile (n) one of the percentiles (↑) where n is a multiple of 10, e.g. the 30th percentile is the 3rd decile.

quartile (*n*) the 25th percentile (↑) (the *lower quartile*) or the 75th percentile (the *upper quartile*).

quantile (*n*) a general word for percentiles (↑), deciles (↑), quartiles (↑) and the median (↑).

working mean an approximate guessed value of a mean used to make calculation easier, and which is later corrected, e.g. the mean of 98, 103 and 105 is easily found by using a working mean of 100 and calculating the mean of -2, 3 and 5 as $(-2 + 3 + 5) \div 3 = 2$ to give $100 + 2 = 102$. Also known as **assumed mean**.

weighted mean if n quantities $x_1, x_2, x_3, \ldots, x_n$ each have weights $w_1, w_2, w_3, \ldots, w_n$ respectively, then their weighted mean is $\Sigma w_i x_i / \Sigma w_i$. This is similar to finding the arithmetic mean (p. 113) when the quantities have frequencies corresponding to the weights.

index[3] (*n*) a quantity which gives the relation of a variable to a base measure (usually taken to be equal to 100) and can be used to compare values over time or space, e.g. price index, cost of living index. *See also* pp. 8, 35.

weighted index an index (↑) based (as most are) on the weighted mean (↑) of a set of values relative to the weighted mean of a base set of values.

Laspeyres' index the usual form of weighted index (↑) where the weights used are those found at the time that the base value was fixed.

Paasche's index a form of weighted index (↑) in which the weights used are those found at the time for which the value of the index is being calculated. In, e.g. the cost of living index, this allows for changes in buying custom.

absolute measure a measure which is a pure number, e.g. the coefficient of variation (p. 119). Often such measures are given as percentages.

time series a series (p. 91) of variates (p. 104) taken at regular intervals of time, and often given as a graph.

time series
with four point moving average

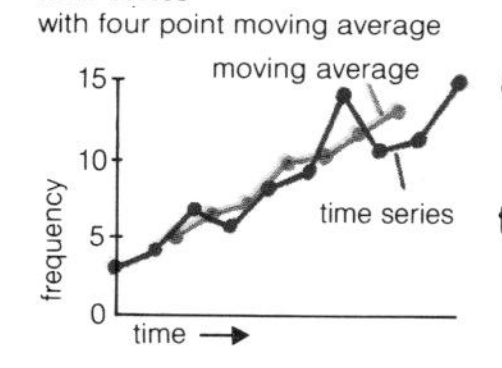

moving average an average (p. 113) used mostly with time series (p. 115) where cyclic and random variations can be smoothed by taking the arithmetic means (p. 113) of a given number of consecutive observations (p. 290). The next value of the moving average is then found by removing the first observation and adding a new one at the end, the process being repeated as each new value is found.

moving total similar to moving average (↑) but using the total values rather than average ones.

progressive average an average which changes as each new value is found, e.g. the progressive averages of 3, 7, 2 and 8 are 3, $(3 + 7)/2$, $(3 + 7 + 2)/3$ and $(3 + 7 + 2 + 8)/4$.

trend (*n*) the change in a moving average (↑) or other graph which approximately follows a straight line or other smooth curve and lasts over a long time. Also known as **secular trend**. **trend** (*v*), **trend** (*adj*).

trend line a straight line which fits a time series (p. 115) or other series best and gives an idea of how the series is changing. *See also* line of best fit (p. 125).

seasonal variation the difference between a time series (p. 115) and its moving average (↑) or trend line (↑) which depends on the particular time, e.g. the day of the week, the month of the year.

ratio to trend a ratio of observations (p. 290) to a trend line (↑), usually given as a percentage.

ratio to moving average a ratio of observations (p. 290) to a moving average (↑), usually given as a percentage. It is useful for showing seasonal changes.

autocovariance (*n*) the autocovariance of two terms x_i and x_j in a time series (p. 115) is $(x_i - \mu)(x_j - \mu)$ where μ is the arithmetic mean (p. 113) of the values of the series.

lag[1] (*n*) the value $j - i$ in autocovariance (↑). *See also* p. 276.

serial correlation correlation (p. 122) between two time series (p. 115) with a fixed lag (↑) between corresponding values.

autocorrelation (*n*) the sum of the autocovariances (↑) for a given lag (↑) in a time series (p. 115) divided by the product of their standard deviations (p. 118) which gives a measure of the smoothness of the series. It is the serial correlation (↑) between values of the same time series.

stationary time series a time series (p. 115) showing no change in time so that its trend line (↑) is horizontal.

control chart
the last reading has passed the lower warning limit

control chart a graph for plotting a time series (p. 115) used to control quality in a product by noting if the actual value falls outside certain limits which are marked on the graph. Also known as **quality control chart**.

control limit a limit in a control chart (↑). Types of control limit are known as **action limit**, **warning limit**.

lattice diagram a diagram made by joining the points of a lattice (p. 172), especially one used as a control chart (↑).

dispersion (*n*) spread, the way in which a set of values differ from one another, measured by quantities like range (p. 113), mean deviation (p. 118), standard deviation (p. 118). Also known as **spread**.

deviate (*n*) the difference between a particular value of a variate (p. 104) and some constant measure, often the mean.

relative deviate a deviate (↑) divided by the constant measure.

deviation (*n*) the amount that one quantity differs from another, often the difference between a value in a distribution and its arithmetic mean (p. 113), median (p. 114) or other average (p. 113).

mean deviation a measure of dispersion (p. 117), the arithmetic mean (p. 113) of the deviations (p. 117) of a set of values x_i from some quantity, usually their arithmetic mean $\bar{x}$ or median (p. 114), no notice being taken of whether the signs of the deviations are positive or negative. In symbols, the mean deviation from the arithmetic mean is $\{\Sigma|x_i - \bar{x}|\}/n$, where n is the number of values in the set.

mean square deviation the arithmetic mean (p. 113) of the squares of the deviations (p. 117) of a set of values from some quantity; the variance (↓) is a particular example.

variance (*n*) a measure of dispersion (p. 117) of a set of values x_i; their mean square deviation (↑) from their arithmetic mean x. In symbols $\{\Sigma(x_i - \bar{x})^2\}/n$. *See also* standard deviation (↓).

root mean square deviation the square root (p. 47) of the mean square deviation (↑).

standard deviation a measure of dispersion (p. 117), the square root (p. 47) of the variance (↑) of a set of values x_i, their root mean square deviation (↑) from their arithmetic mean (p. 113) $\bar{x}$. If the standard deviation of a sample is used as an estimator (p. 109) of the standard deviation of the population, the formula should have the sample size n replaced by $n - 1$ in order to be an unbiased (p. 105) estimator. Symbols: $\sqrt{\{\Sigma(x_i - \bar{x})^2/n\}}$ where n is the total number of values.

Bessel's correction the quantity $n/(n - 1)$ which when used as a multiplier (p. 23) turns the variance (↑) of a set of values into an estimator (p. 109) of the variance of the population from which the values are drawn.

interquartile range a measure of dispersion (p. 117), the difference between the lower and upper quartiles (p. 115).

semi-interquartile range a measure of dispersion (p. 117), half the interquartile range (↑). Also known as **quartile deviation**, **interquartile deviation**.

interdecile range a measure of dispersion (p. 117), the difference between the first and ninth deciles (p. 114).

mean deviation

for 6, 7, 11, 13, 13

1 the mean is

$$\frac{(6 + 7 + 11 + 13 + 13)}{5} = 10$$

2 deviations from the mean are

−4, −3, 1, 3, 3

3 moduli of deviations are

4, 3, 1, 3, 3

4 their mean is

$$\frac{(4 + 3 + 1 + 3 + 3)}{5} = 2{\cdot}8$$

variance and standard deviation

for 6, 7, 11, 13, 13

1 the mean is 10
2 deviations from the mean are

−4, −3, 1, 3, 3

3 squared deviations are

16, 9, 1, 9, 9

4 their mean is

$$\frac{(16 + 9 + 1 + 9 + 9)}{5} = 8.8$$

this is the variance

5 the standard deviation is

$\sqrt{8.8} = 2.97$

there are ways of making this calculation easier

Bessel's correction

when the values above are used for an estimate of a population variance, then **4** becomes

$$\frac{(16 + 9 + 1 + 9 + 9)}{4} = 11$$

and the standard deviation is

$\sqrt{11} = 3.32$

interquartile range

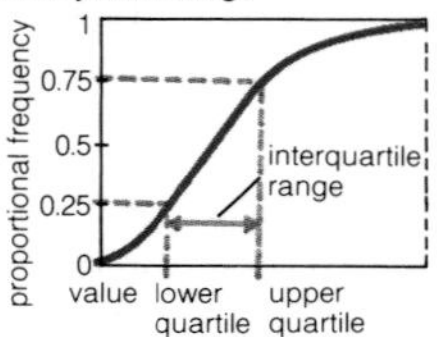

coefficient of variation an absolute measure (p. 115) of dispersion (p. 117) which is the standard deviation (↑) divided by the mean, often given as a percentage. It is sometimes useful in comparing (p. 287) the dispersions of two distributions with different means.

relative variance the square of the coefficient of variation (↑). Also known as **relvariance**.

Snedecor's check a rough check for the calculation of standard deviation (↑) which should be between one half and one fifth of the range (p. 113), depending on the size of the sample.

Charlier's check the formula

$$\Sigma f(x + 1)^2 = \Sigma fx^2 + 2\Sigma fx + \Sigma f$$

used as a check in the calculation of standard deviation (↑) and variance (↑).

continuity correction (1) a correction (p. 288) made to a statistic (p. 110) which is calculated from grouped data (p. 110) to decrease the error (p. 43) caused by the grouping; (2) a correction used when a continuous distribution (p. 106), e.g. normal distribution (p. 120) is used as an approximation to a discrete distribution (p. 106), e.g. binomial distribution (p. 108).

Sheppard's correction a continuity correction (↑) made in calculating variance (↑) and standard deviation (↑) from grouped data (p. 110) to reduce (p. 291) the error (p. 43) caused by grouping into classes (p. 110). The variance is decreased by $h^2/12$ where h is the range (p. 113) of each class interval (p. 111).

sample mean the arithmetic mean (p. 113) of a sample.

sample variance the variance (↑) of a sample calculated from the sample mean (↑).

sample mean distribution the distribution of sample means (↑) when many samples of the same size n are taken from a population of mean μ and variance (↑) σ^2. It also has mean μ and has variance (↑) σ^2/n when all possible samples of this size are taken.

sampling error the difference between a sample mean (↑) and the population mean. Its mean value is zero and its variance (↑) the same as for the sample mean distribution (↑).

mean square error the variance (p. 118) of a set of sample means (p. 119) when more than one sample is being considered.

standard error the standard deviation (p. 118) of the difference between a sample statistic (p. 110) and the corresponding population parameter (p. 110). The standard error of the sample mean (p. 119) is σ^2/n where σ^2 is the population variance (p. 118) and n is the sample size. The standard error of the sample mean is usually meant unless another statistic is named.

normal distribution the continuous distribution (p. 106) arising in many practical cases with a bell-shaped equation $y = \exp\{-(x - \mu)^2/2\sigma^2\}$ where μ is its arithmetic mean (p. 113), σ^2 is its variance (p. 118) and exp stands for the exponential function (p. 138). Also known as **Gaussian distribution**, **normal curve of error**, **$N(\mu,\sigma^2)$ distribution**.

abnormal (*adj*) not usual, sometimes used of non-normal distributions (↑).

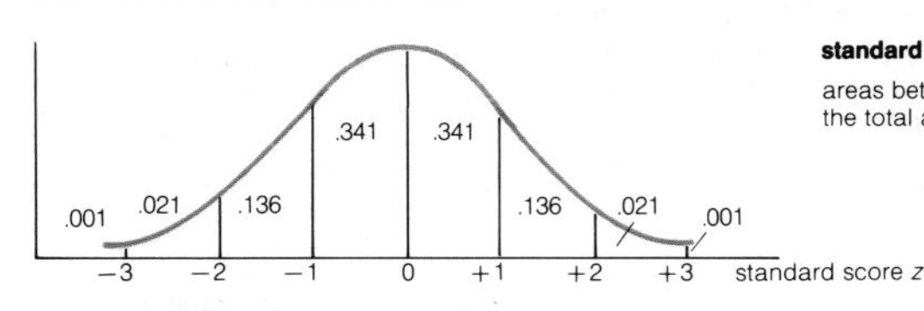

standard normal distribution
areas between the ordinates
the total area is 1

standard normal distribution the normal distribution (↑) with mean 0 and variance (p. 118) 1, which is the form given in tables and gives standard scores (↓).

standard score the value z of a variate (p. 104) x given as its number standard deviations (p. 118) σ from the arithmetic mean μ of the distribution (p. 106) from which it is obtained. It is given by $z = (x - \mu)/\sigma$. Also known as **standardized score**, **z-score**, **normal deviation**.

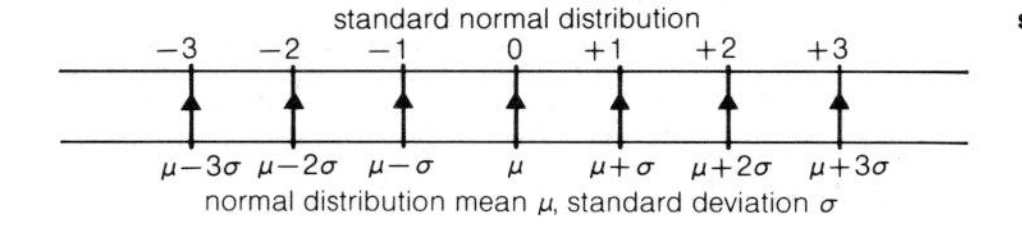

standard score

standardization (*n*) obtaining a standard score (↑) and other similar processes. **standardize** (*v*).

sigma scale a scale which changes the standard score (↑) z from -3 to $+3$ (values of z which are not between -3 and 3 are highly unusual) onto a scale from 0 to 100, so that the score on the sigma scale is given by $\Sigma = (50/3)z + 50$.

Hull scale a scale which changes the standard score (↑) z to a scale whose arithmetic mean (p. 113) is 50 by finding $14.28z + 50$.

normal approximation the use of the normal distribution (↑) as an approximation to a discrete distribution (p. 106) of similar shape such as the binomial (p. 108) or Poisson distributions (p. 109).

central limit theorem if samples are drawn from any population, as their number becomes large the distribution of their arithmetic means (p. 113) tends to become a normal distribution (↑). It is used, e.g., to obtain the arithmetic mean and variance (p. 118) of the distribution of sample means (p. 119).

reproductive (*adj*) used of types of distribution for which when two or more random sets of variates (p. 104) are added a distribution of the same type results, as in the normal distribution (↑).

moment[1] (*n*) the arithmetic mean (p. 113) of the deviations (p. 117) of powers of a variate (p. 104) from some value. The first moment about the arithmetic mean is zero, the second moment about the arithmetic mean is the variance (p. 118) and the rth moment μ_r about the arithmetic mean is $\{\Sigma(x_i - \bar{x})^r\}/n$, where the letters stand for the same quantities as under variance. The use is similar to that in mechanics.

moment generating function a function which may be expanded in a power series (p. 149) and which may be used to generate (p. 209) the moments (↑) about the origin of a probability density function (p. 106).

characteristic function the expectation (p. 109) of e^{ixt} where i is the square root (p. 47) of -1, and x is a variate (p. 104). It can be expanded in a power series (p. 149) and is a moment generating function (↑).

skewness (*n*) lack of symmetry in the graph of a distribution. One measure is the moment (p. 121) of the third powers about the arithmetic mean (p. 113) μ_3 divided by the variance (p. 118) μ_2 to the power 3/2, i.e. $\mu_3/\{\mu_2^{(3/2)}\}$. If the maximum point is to the left of the distribution it has *positive skewness* (since μ_3 is positive), if it is to the right it has *negative skewness* (since μ_3 is negative). **skew** (*v*), **skew** (*adj*). *See also* p. 200.

kurtosis (*n*) a measure of the shape of a distribution, usually given as $(\mu_4/\mu_2^2) - 3$, where μ_4 is the fourth moment (p. 121) about the arithmetic mean (p. 113) and μ_2 is the second moment or variance (p. 118). The normal distribution (p. 120) has kurtosis zero.

leptokurtic (*adj*) used of a distribution which has positive kurtosis (↑), and whose maximum is 'sharper' than that of the normal distribution (p. 120).

mesokurtic (*adj*) used of a distribution (p. 106) which has zero kurtosis (↑) like the normal distribution (p. 120).

platykurtic (*adj*) used of a distribution (p. 106) which has negative kurtosis (↑) and whose maximum is 'flatter' than that of the normal distribution (p. 120).

rank (*n*) the position of any given measure, when a set of measures is placed in order. In everyday life the highest measure is usually given rank 1, and two equal measures are given the higher of the two ranks which would otherwise be given. In statistics, the lowest measure is sometimes given rank 1, since this makes correlations (↓) easier to follow, and equal measures must always be given the arithmetic mean (p. 113) of the ranks which would otherwise be given. *See also* p. 89. **rank** (*v*), **ranked** (*adj*).

correlation (*n*) the amount of agreement between two sets of variates (p. 104). If the two variates are plotted on a graph, the closeness of the points to a straight line or curve. **correlate** (*v*).

linear correlation correlation (↑) to a straight line rather than a curve.

skewness

positive skewness

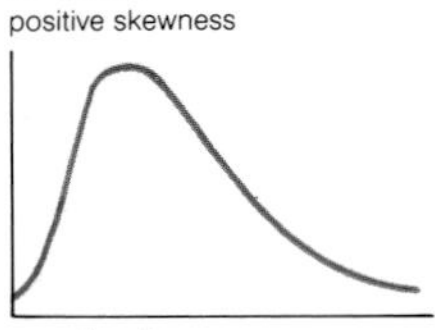

negative skewness

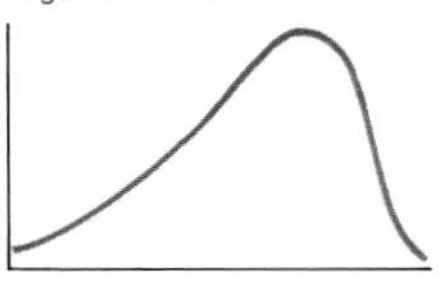

kurtosis

the mesokurtic curve shown is in fact the normal curve

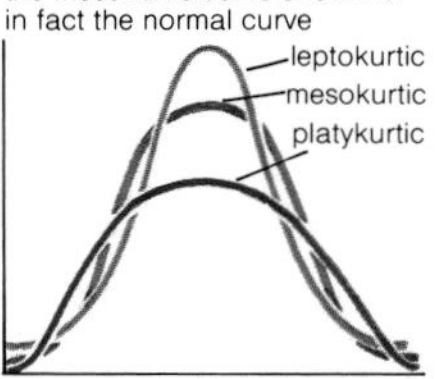

Pearson's product-moment correlation coefficient (r)

for the values (1, 4), (2, 1), (3, 8), (4, 7), (5, 10)

	x	y	x^2	y^2	xy
	1	4	1	16	4
	2	1	4	1	2
	3	8	9	64	24
	4	7	16	49	28
	5	10	25	100	50
sum	15	30	55	230	108

$$r = \frac{108 - (15 \times 30) \div 5}{\sqrt{\{[55 - 15^2/5][230 - 30^2/5]\}}} = \frac{18}{\sqrt{\{10 \times 50\}}} = 0.805$$

Pearson's product-moment correlation coefficient the commonest measure of linear correlation (↑) obtained from the arithmetic mean (p. 113) of the products of the standard scores (p. 120) z_1 and z_2 of the two sets of n variates (p. 104) $\Sigma z_1 z_2/n$. If the actual variates are x and y the formula is

$$r = \frac{\Sigma xy - (\Sigma x \Sigma y)/n}{\sqrt{\{[\Sigma x^2 - (\Sigma x)^2/n][\Sigma y^2 - (\Sigma y)^2/n]\}}}.$$

Also known as **linear correlation coefficient**, **product-moment correlation coefficient**.

Spearman's rank-order correlation coefficient (ρ)

for the rank pairs (1, 2), (2, 3), (3, 5), (4, 1), (5, 4)

x	y	D	D^2
1	2	−1	1
2	3	−1	1
3	5	−2	4
4	1	3	9
5	4	1	1

$\sum D^2 = 16$

so $\rho = 1 - \left\{\frac{6 \times 16}{5 \times 24}\right\} = 0.2$

Spearman's rank-order correlation coefficient a measure of correlation (↑) of ranked (↑) data corresponding to Pearson's coefficient (↑). The formula simplifies (p. 192) to $\rho = 1 - \{(6\Sigma D^2)/[n(n^2 - 1)]\}$ where D is the difference between pairs of corresponding ranks, and n is the number of pairs. Also known as **Spearman's rho**.

Kendall's rank-order correlation coefficient (τ)

for rank pairs (1, 2), (2, 3), (3, 5), (4, 1), (5, 4)

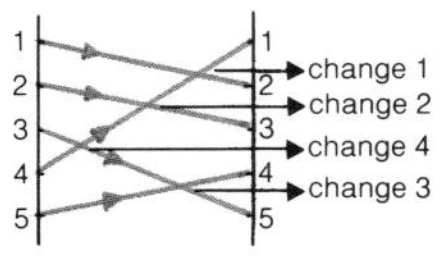

four changes are needed to put the second set in order, so $S = 4$ and since $n = 5$

$$\tau = 1 - \left\{\frac{2 \times 4}{\frac{1}{2} \times 5 \times 4}\right\} = 0.2$$

in this case the value is the same as for Spearman's but this is not always so

Kendall's rank-order correlation coefficient a measure of correlation (↑) of ranked (↑) data given by $\tau = 1 - \{2S/[\frac{1}{2}n(n - 1)]\}$ where S is the number of changes of rank needed to put the second set of ranks in order when the first are already in order, and n is the number of pairs of ranks. Also known as **Kendall's tau**.

biserial correlation the correlation (↑) between two or more sets of dichotomous (p. 13) data.

point biserial correlation the correlation (↑) between a dichotomous (p. 13) variable and a continuous variable.

positive correlation correlation (↑) with positive values where high values of one variate (p. 104) correspond to high values of the other.

negative correlation correlation (p. 122) with negative values where low values of one variate (p. 104) correspond to high values of the other.

uncorrelated (*adj*) having a correlation (p. 122) not significantly (p. 126) different from zero.

alienation (*n*) lack of correlation (p. 122) rather than negative correlation (↑), measured by the coefficient (p. 61) $k = \sqrt{(1 - r^2)}$ where r is Pearson's product-moment correlation coefficient (p. 123).

coefficient of non-determination the quantity k^2 where k is as given under alienation (↑).

concordance (*n*) the amount of agreement between two or more sets of ranked (p. 122) variates (p. 104), usually measured by **Kendall's coefficient of concordance**.

partial correlation the correlation (p. 122) between two variates (p. 104) when other variates are held constant.

multiple correlation coefficient a correlation (p. 122) coefficient (p. 61) in which one variate (p. 104) is correlated with two or more independent variates.

spurious correlation correlation (p. 122) which is not caused by one of the variates (p. 104) acting on the other, but by some third variate acting on the given variates separately.

covariance (*n*) a measure of the agreement between two sets of variates (p. 104) which, unlike correlation (p. 122), depends on the scale used to measure them. If x_i, y_i are values of the two sets of variates whose arithmetic means (p. 113) are $\bar{x}$, $\bar{y}$ and there are n pairs, the covariance is $\{\Sigma(x_i - \bar{x})(y_i - \bar{y})\}/n$. If the $n - 1$ form of standard deviation (p. 118) (*see* Bessel's correction, p. 118) is used, then n must be replaced by $n - 1$. Also known as **first product-moment**.

regression (*n*) the use of the agreement between two sets of variates (p. 104) to predict (p. 290) values of one when values of the other are given. **regress** (*v*).

linear regression regression (↑) based on a straight line graph. Similarly **curvilinear regression** based on a curve.

regression

linear regression of y on x

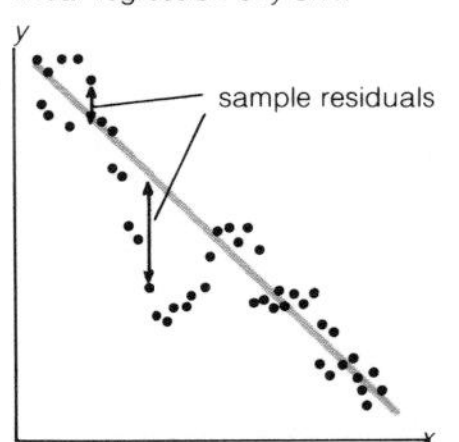

the vertical distances of the points from the line are the residuals

regression line a straight line which gives the best prediction (p. 290) of one variate (p. 104) from another. The regression line of y on x which predicts values of y from those of x is given by $y = a + bx$,
where $a = (\Sigma y_i - b\Sigma x_i)/n$
and $b = \{\Sigma x_i y_i - (\Sigma x_i \Sigma y_i)/n\} / \{\Sigma x_i^2 - (\Sigma x_i)^2/n\}$.
The regression line of x on y is found by interchanging x and y and is usually different from this.

regression curve a curve having a similar purpose to a regression line (↑).

linear regression coefficients the quantities a and b defined under regression line (↑).

regressor (*n*) the independent variable (p. 59) in a regression (↑) equation.

regressand (*n*) the dependent variable (p. 59) in a regression (↑) equation.

residual (*n*) (1) the difference between the value of an ordinate (p. 215) of a y-variate (p. 104) and the regression line (↑) of y on x; (2) the difference between the approximate and the true value of an unknown; similar quantities elsewhere.

homoscedastic (*adj*) having the same variance (p. 118) for all values of another variable, especially used of deviations (p. 117) from the regression line (↑). Similarly **heteroscedastic** having different variances.

least squares the method used in finding regression lines (↑) in which the sum of the squares of the residuals (↑) is made a minimum. The method can be extended to regression curves (↑).

line of best fit often a loose term for a regression line (↑). It should really be used for a different line for which the sums of squares of perpendicular distances of the points from the line is a minimum.

partial regression the regression (↑) of one variate (p. 104) upon another when other variates are held constant.

multiple regression regression (↑) in which one variate (p. 104) is predicted (p. 290) in terms of two or more other variates.

null hypothesis a hypothesis (p. 251) which acts as a statement (p. 243) against which results may be tested. Usually the intention is that it will be disproved. Symbol: H_0.

alternative hypothesis the hypothesis (p. 251) which is true when the null hypothesis (↑) is false and false when it is true. Symbol: H_1.

null distribution a theoretical distribution used as the basis for a null hypothesis (↑).

hypothesis test a test of the truth of a null hypothesis (↑). Often used with the same meaning as significance test (↓),

significance level the probability in a test of rejecting (p. 291) the null hypothesis (↑) when it is in fact true.

significance test a test using the significance level (↑) of a statistic (p. 110) which may be used to test whether a null hypothesis (↑) may be rejected (p. 291) or not at the given significance level (↑). Often used with the same meaning as hypothesis test (↑).

significant (*adj*) having a value whose significance level (↑) is less than some fixed amount, usually 0.05, 0.01 or 0.001.

statistical inference the use of significance tests (↑) to decide the probability of the truth or falsity of statements. Also known as **decision theory**.

type I error the rejection (p. 291) in a significance test (↑) of the null hypothesis (↑) when it is in fact true.

type II error failing to reject (p. 291) the null hypothesis (↑) in a significance test (↑) when it is in fact false. Also known as **acceptance error**.

power of a test the probability that the test rejects the null hypothesis (↑) at a given significance level (↑). It is 1 − β where β is the probability of a type II error (↑). *See also* pp. 35, 270.

power function the function giving the power of a test in terms of the values of the parameter (p. 110) against which the test is being made.

operating characteristic function the function giving the probability β of a type II error (↑) in terms of the values of the parameter (p. 110) against which the test is being made.

type I and type II errors

	H_0 is true	H_1 is true
H_1 is accepted	type I error	correct decision
H_0 is accepted	correct decision	type II error

probability = power of the test

two-tailed test
values are significant in either of two critical regions

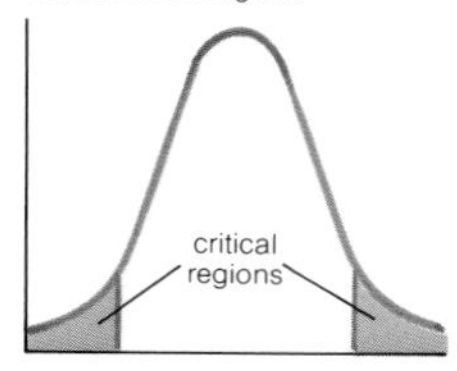

one-tailed test
values are significant in only one critical region (which can be at either end of the distribution)

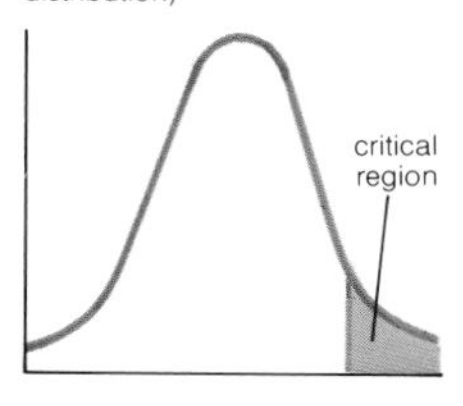

Student's *t*-distribution
shape of normal distribution and two examples of *t*-distribution, with one and ten degrees of freedom

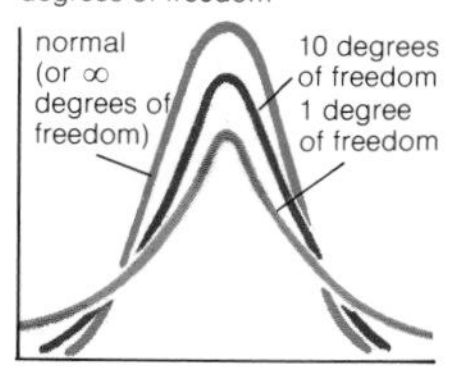

critical value a value at which an important change takes place. In a statistical test the value for which, at a given significance level (↑), the null hypothesis (↑) ceases to be accepted (p. 287) and the alternative hypothesis (↑) is accepted. *See also* stationary value (p. 145).

critical region a region beyond a critical value (↑); in a statistical test where the null hypothesis (↑) is rejected (p. 291) and the alternative hypothesis (↑) accepted (p. 287).

one-sample test any test which uses only one sample from a population.

two-tailed test a test where the alternative hypothesis (↑) has values given on both sides of the null hypothesis (↑) and there are two critical regions (↑), one on each side.

one-tailed test a test where the alternative hypothesis (↑) has values given in one direction only from that of the null hypothesis (↑) and there is only one critical region (↑). Also known as **one-sided test**.

confidence interval an interval determined from a sample where there is a given probability that the population mean (or other population parameter, p. 110) lies within the interval.

confidence limits the bounds (p. 66) of a confidence interval (↑).

Student's *t*-distribution if a random sample of size n is taken from a normally distributed (p. 120) population, the resulting distribution is approximately normal for $n > 30$. For smaller samples the t-distribution corrects this error (p. 43). Also known as ***t*-distribution**.

Cauchy distribution Student's t-distribution (↑) with one degree of freedom (p. 69).

Student's *t*-test the significance test (↑) based on Student's t-distribution (↑), in which the values of t obtained can be tested against the significance level (↑) for a given number of degrees of freedom (p. 69), (usually one less than the sample size). These values are usually obtained from tables. Also known as ***t*-test**.

***F*-distribution** the distribution of the ratio of the variances (p. 118) of two sets of random variates (p. 104).

***F*-test** a test based on the *F*-distribution (p. 127) and used to decide whether two samples are drawn from populations with the same variance (p. 118).

homogeneity of variance used of samples which may be taken to have been drawn from populations with the same variance (or from the same population).

analysis of variance a method of separating the factors which cause variance (p. 118) by using the *F*-test (↑) several times. When there are two or more factors it also allows for the testing of interaction (↓) between the factors.

factorial design a type of experiment in which analysis of variance (↑) is used to find the particular variates (p. 104) or factors which cause variation.

interaction[1] (*n*) the effect of changes in one variate (p. 104) on another variate. *See also* p. 260.

Fisher's *z* an alternative (p. 287) to the *F*-test (↑); *z* is half the natural logarithm (p. 139) of *F*. *Do not mistake for* *z*-score (p. 120) *or for* Fisher's exact probability test (p. 129).

chi-squared distribution the distribution of sums of squares of random variates (p. 104) with standard normal distribution (p. 120). The shape changes with the number of degrees of freedom (p. 69) ν, but always has mean ν and variance (p. 118) 2ν. Symbol: χ^2.

chi-squared test a test based on the chi-squared distribution (↑) and used to test how well a sample distribution fits a theoretical distribution, as well as in the testing of contingency tables (↓).

contingency table a table in which nominal (p. 111) data is classified (p. 110), often in two (or more) ways. An $m \times n$ table has *m* rows and *n* columns. *See also* contingent (p. 249).

Yates' correction a correction (p. 288) for continuity (p. 119) when using the chi-squared test with a 2×2 contingency table (↑) which reduces (p. 291) the differences between expected (p. 109) and observed (p. 290) values by $\frac{1}{2}$ before squaring.

chi-squared distribution
shapes for certain degrees of freedom ν

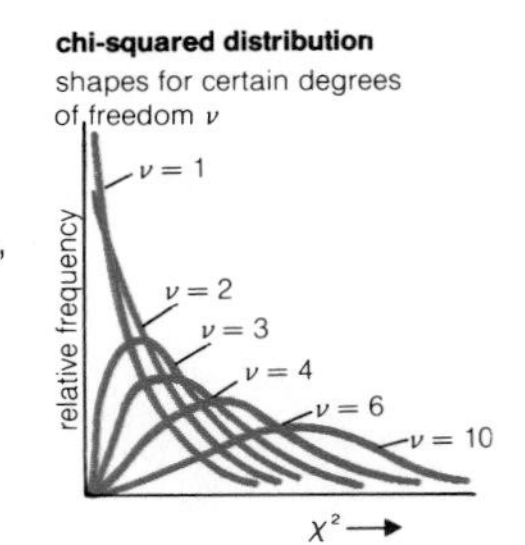

Kolmogorov-Smirnov test
depends on the maximum difference between cumulative distributions

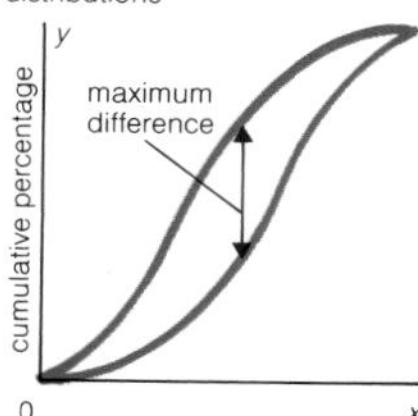

Fisher's exact probability test a test for 2 × 2 contingency tables (↑) with small total sample sizes, <50, where the chi-squared test (↑) is not accurate (p. 43) enough, even with Yates' correction (↑).

association (*n*) similar to correlation (p. 122), but between two sets of nominal (p. 111) data; measured in a number of ways, including the use of the chi-squared test (↑). Also known as **colligation**.

goodness of fit the agreement between a sample distribution and some theoretical distribution; it can be tested by the chi-squared test (↑).

Kolmogorov-Smirnov test a test of goodness of fit (↑) which uses the maximum difference between two cumulative distribution functions (p. 107).

McNemar test a simple test for the significance (p. 126) of the difference of two proportions.

distribution-free (*adj*) not having any theoretical distribution assumed (p. 251), especially used when the assumption of normal distribution (p. 120) is not made. Often taken to mean the same as non-parametric (↓).

non-parametric (*adj*) not depending on any population parameter (p. 110) for a theoretical distribution. Often taken to mean the same as distribution-free. (↑)

robust (*adj*) having few conditions upon which correct (p. 288) use depends, as is usually the case with distribution-free (↑) tests.

Tukey quick test a simple distribution-free (↑) test to decide whether two independent samples could have been drawn from the same population which can be used with small samples of approximately equal size.

Mann-Whitney *U*-test a distribution-free (↑) test which tests the difference between the medians (p. 114) of two distributions.

Wilcoxon rank-sum test a distribution-free (↑) test which is really another way of performing the Mann-Whitney *U*-test (↑).

Kruskal-Wallis test a distribution-free (↑) test similar in idea to analysis of variance (↑) for one factor.

Friedmann test a distribution-free (p. 129) test similar in idea to analysis of variance (p. 128) for two factors.
rank test a test based on the ranks (p. 122) of variates (p. 104).
paired-sample test a test which compares (p. 287) two sets of samples which are related in pairs.
Wilcoxon matched-pairs signed-ranks test a paired-sample test (↑) based on ranking (p. 122) the differences between the corresponding pairs of values.
sign test a distribution-free (p. 129) test which tests the difference of a sample median (p. 114) from a population median by counting the numbers in the sample above and below it.
runs test a test for randomness based on the number of runs (p. 102) of values falling either above or below the median (p. 114). Also known as **randomness test**.
Cox and Stuart test for trend a test based on the sign test (↑) which tests for trends (p. 116).
Monte Carlo method method using a model (p. 290) based on simulation (p. 292) with random numbers.
queue (*n*) a set of elements (often people) together with times of arrival. Each element takes a given time to deal with, the earliest to arrive is dealt with first and then leaves.
stack (*n*) similar to a queue (↑), but the last to arrive is dealt with first.
queueing theory the theory of queues (↑) or stacks (↑) in which Monte Carlo methods (↑) are often used.
magic square a square array (p. 83) of numbers such that the sum of the numbers in any row, column or diagonal (p. 177) is a fixed number called the **magic number**.
Latin square a square arrangement of *n* elements in *n* rows and *n* columns so that each element occurs exactly once in each row and each column; often used in farming experiments.
Graeco-Latin square two different Latin squares (↑) on top of each other so that the same two pairs of elements never occur together in more than one position.

magic square

4 × 4 magic square; the rows, columns and diagonals all add up to 34

10	1	16	7
15	8	9	2
3	12	5	14
6	13	4	11

Latin square

4 × 4 Latin square

A	*B*	*C*	*D*
B	*C*	*D*	*A*
C	*D*	*A*	*B*
D	*A*	*B*	*C*

Graeco-Latin square

3 × 3 Graeco-Latin square

Aa	*Bb*	*Cc*
Bc	*Ca*	*Ab*
Cb	*Ac*	*Ba*

trigonometry (*n*) the study of the ratios of the pairs of sides of a right angled triangle (p. 179) and of the generalizations of these ratios. **trigonometric** (*adj*).

sine (*n*) if a point $P(x_1, y_1)$ lies on a unit circle with centre at the origin *O*, and θ is the angle between *OP* and the positive direction of the *x*-axis (p. 215), then y_1 is the sine of the angle θ for all values of θ. If $0° < \theta < 90°$, the sine of θ may also be defined in a right angled triangle (p. 179) *ABC* in which the angle *ABC* is a right angle and angle $CAB = \theta$, by BC/AC, often remembered by 'opposite (p. 171) over hypotenuse (p. 179)'. *See also* power series (p. 149). **sin θ** (*abbr*).

sine and cosine graphs

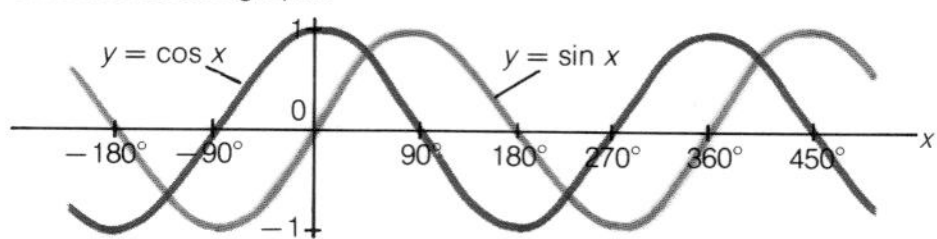

cosine (*n*) with the same definitions as for sine (↑), the cosine of θ is x_1, or if $0° < \theta < 90°$ it is AB/AC, often remembered by 'adjacent (p. 171) over hypotenuse (p. 179)'. *See also* power series (p. 149). **cos θ** (*abbr*).

tangent[1] (*n*) with the same definitions as for sine (↑), the tangent of θ is y_1/x_1, or if $0° < \theta < 90°$ it is BC/AB, often remembered as 'opposite (p. 171) over adjacent (p. 171)'. It may also be defined as $\sin\theta/\cos\theta$. The tangent is not defined when $\cos\theta = x_1 = 0$, i.e. for angles $\pm 90°$, $\pm 270°$, $\pm 450°$, ... *See also* p. 187. **tan θ** (*abbr*).

tangent and cotangent graphs

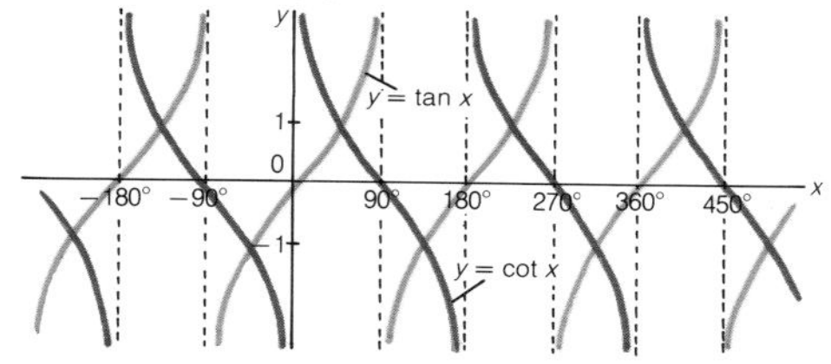

sine and cosine
in each case $\sin\theta = y_1$, $\cos\theta = x_1$, in two cases of each these are negative

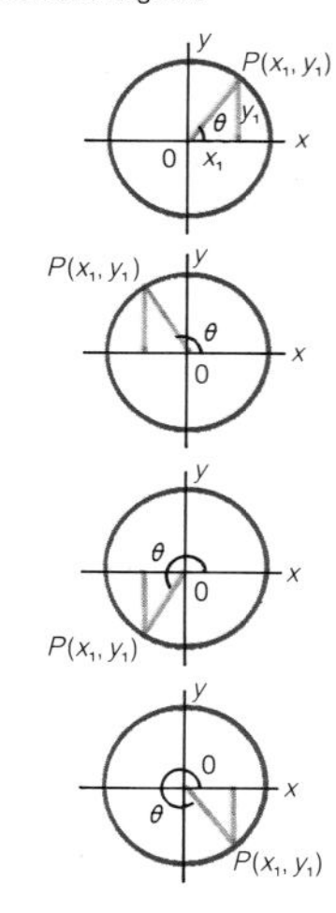

sine and cosine
for $0 < \theta < 90°$
$\sin\theta = BC/AC$
$\cos\theta = AB/AC$

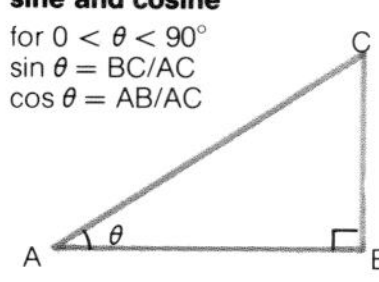

trigonometry
signs of functions

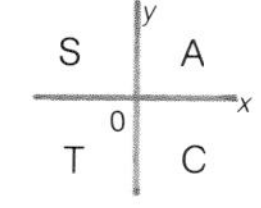

in the four quadrants
quadrant 1 – A (all trig functions are positive)
quadrant 2 – S (sin is positive but not cos, tan)
quadrant 3 – T (tan is positive but not sin, cos)
quadrant 4 – C (cos is positive but not sin, tan)

secant and cosecant graphs

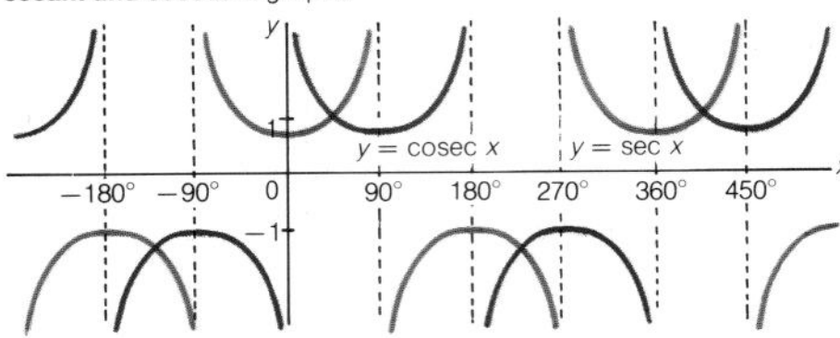

secant[1] (*n*) the secant of the angle θ is unity divided by the cosine (p. 131) of the angle θ, i.e. $1/\cos\theta$. As for $\tan\theta$ it is not defined when $\cos\theta = 0$. *See also* p. 187. **sec θ** (*abbr*).

cosecant (*n*) the cosecant of the angle θ is unity divided by the sine (p. 131) of the angle θ, that is $1/\sin\theta$. It is not defined when $\sin\theta = 0$, i.e. for angles 0°, ±180°, ±360°, ±540°, ... **cosec θ** (*abbr*) or **csc θ** (*abbr*).

cotangent (*n*) the cotangent of the angle θ is unity divided by the tangent of the angle θ. It may also be defined as $\cos\theta/\sin\theta$. As for cosec θ it is not defined when $\sin\theta = 0$. **cot θ** (*abbr*).

trigonometric functions the six functions, sine (p. 131), cosine (p. 131), tangent (p. 131), secant (↑), cosecant (↑), cotangent (↑) which repeat their values every 360° (or for tangent and cotangent every 180°). Also known as **circular functions**. Symbols: powers of such functions like $(\cos\theta)^3$ are written $\cos^3\theta$, and *never* as $\cos\theta^3$ which means $\cos(\theta^3)$.

general solutions of trigonometric equations usually given as follows:

if $\sin x = \sin\alpha$, $x = n\pi + (-1)^n\alpha$,
if $\cos x = \cos\alpha$, $x = 2n\pi \pm \alpha$ and
if $\tan x = \tan\alpha$, $x = n\pi + \alpha$,

where α is a given angle and n is any integer.

complementary trigonometric functions for each trigonometric function (↑) there is a corresponding function of the complementary angle (p. 175) which has the same value:
$\sin\theta = \cos(90° - \theta)$, $\cos\theta = \sin(90° - \theta)$,
$\tan\theta = \cot(90° - \theta)$, $\cot\theta = \tan(90° - \theta)$,
$\sec\theta = \mathrm{cosec}(90° - \theta)$, $\mathrm{cosec}\,\theta = \sec(90° - \theta)$.
These are true for all values of θ for which the functions are defined.

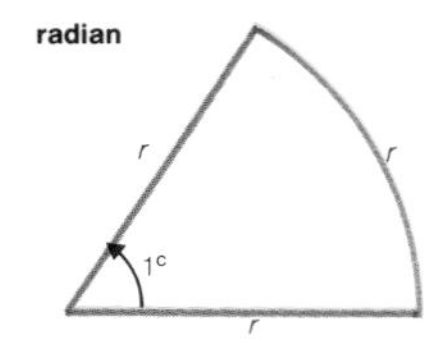

radian (*n*) the angle subtended (p. 175) by an arc of a circle of length equal to its radius, so that π (pi, p. 192) radians is equal to 180° and 1 radian is about 57.29°. The anticlockwise (p. 174) direction of turning is taken to be positive. Used widely in analysis. Symbol ^{c}.

Euler's exponential forms the equations
$\exp(ix) = \cos x + i \sin x$ and
$\exp(-ix) = \cos x - i \sin x$,
which give the exponential function (p. 138) in terms of the trigonometric functions (↑) and the square root (p. 47) of -1.

period[2] (*n*) the time or angle at which something repeats its position or values, e.g. a pendulum (p. 275) of a clock often has period 2 seconds, sine (p. 131) and cosine (p. 131) have periods 360°, tangent (p. 131) and cotangent (↑) have periods 180°. *See also* pp. 39, 276.

compound angles angles which are the sum or difference of two or more separate angles.

addition formulae (1) for sine (p. 131) and cosine (p. 131), for all angles A and B:

$$\sin(A + B) = \sin A \cos B + \cos A \sin B,$$
$$\sin(A - B) = \sin A \cos B - \cos A \sin B,$$
$$\cos(A + B) = \cos A \cos B - \sin A \sin B,$$
$$\cos(A - B) = \cos A \cos B + \sin A \sin B.$$

(2) for tangent (p. 131), for all values for which the angles are defined:

$$\tan(A + B) = \frac{\tan A + \tan B}{(1 - \tan A \tan B)},$$
$$\tan(A - B) = \frac{\tan A - \tan B}{(1 + \tan A \tan B)}.$$

double angle formulae for all angles A for which the functions are defined:

$$\sin 2A = 2 \sin A \cos A,$$
$$\cos 2A = \cos^2 A - \sin^2 A$$
$$= 1 - 2\sin^2 A = 2\cos^2 A - 1,$$
$$\tan 2A = 2 \tan A/(1 - \tan^2 A).$$

multiple angle formulae for all angles A:

$$\sin 3A = 3 \sin A - 4 \sin^3 A,$$
$$\cos 3A = 4 \cos^3 A - 3 \cos A.$$

Formulae for higher multiples may be obtained from the addition formulae (↑) or from Demoivre's theorem (p. 50).

Tchebycheff's polynomials polynomials giving multiple angles of trigonometric functions (p. 132) in terms of power series (p. 149) of those functions as in the multiple angle formulae (p. 133). Also written **Chebyshev** and many other ways.

product formulae for all angles A and B:

$2 \sin A \cos B = \sin (A + B) + \sin (A - B)$,
$2 \cos A \sin B = \sin (A + B) - \sin (A - B)$,
$2 \cos A \cos B = \cos (A + B) + \cos (A - B)$,
$-2 \sin A \sin B = \cos (A + B) - \sin (A - B)$.

These formulae are useful in integration.

sum formulae for all angles x and y:

$\sin x + \sin y = 2 \sin \frac{1}{2}(x + y) \cos \frac{1}{2}(x - y)$,
$\sin x - \sin y = 2 \cos \frac{1}{2}(x + y) \sin \frac{1}{2}(x - y)$,
$\cos x + \cos y = 2 \cos \frac{1}{2}(x + y) \cos \frac{1}{2}(x - y)$,
$\cos x - \cos y = -2 \sin \frac{1}{2}(x + y) \sin \frac{1}{2}(x - y)$.

Also known as **factorization formulae**.

half-angle formulae for all values of the angle x:

$\cos^2 \frac{1}{2}x = \frac{1}{2}(1 + \cos x)$,
$\sin^2 \frac{1}{2}x = \frac{1}{2}(1 - \cos x)$.

Further, for all values of x for which $t = \tan \frac{1}{2}x$ and $\tan x$ are defined:

$\sin x = 2t/(1 + t^2)$,
$\cos x = (1 - t^2)/(1 + t^2)$,
$\tan x = 2t/(1 - t^2)$.

These are useful in integration, when:

$dx/dt = 2/(1 + t^2)$.

(Care should be taken that the integration does not include undefined values.)

small angles the limits of $\sin \theta$ and $\cos \theta$ as θ tends to zero are respectively θ and $1 - \frac{1}{2}\theta^2$ when the angles are measured in radians (p. 133).

trigonometric function inequalities for all angles x measured in radians (p. 133) such that $0 < x < \frac{1}{2}\pi$, $2x/\pi < \sin x < x < \tan x$.

sine formula in any triangle ABC with sides $a = BC$, $b = CA$, $c = AB$,
$a/\sin A = b/\sin B = c/\sin C = 2R$,
where R is the radius of the circumcircle (p. 197) of the triangle. If two angles and a side of a triangle are given, this may always be used to calculate the other two sides. Also known as **sine rule**. *See also* ambiguous case (↓).

trigonometric function inequalities

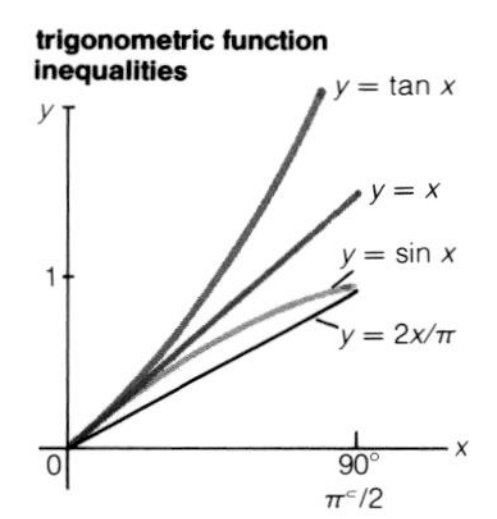

sine formula

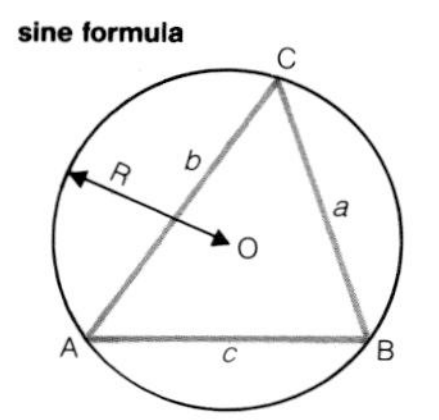

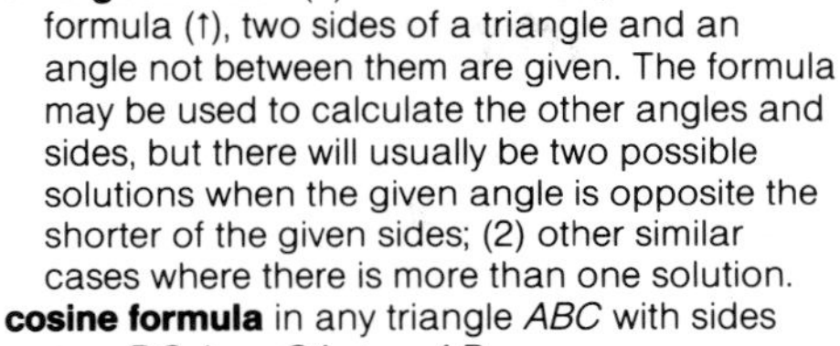

ambiguous case (1) the case when, in the sine formula (↑), two sides of a triangle and an angle not between them are given. The formula may be used to calculate the other angles and sides, but there will usually be two possible solutions when the given angle is opposite the shorter of the given sides; (2) other similar cases where there is more than one solution.

ambiguous case
two alternative triangles with the same dimensions

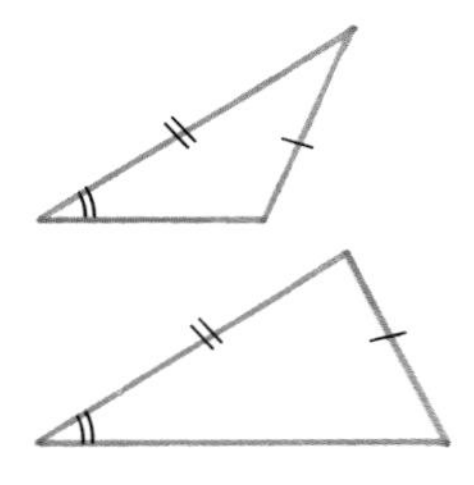

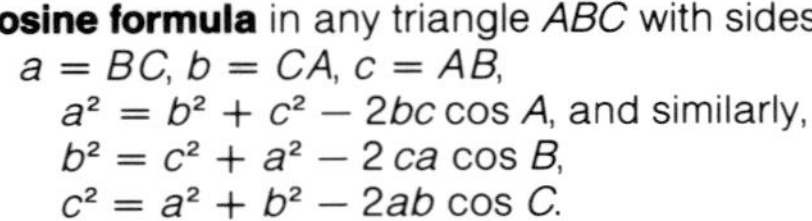

cosine formula in any triangle ABC with sides $a = BC$, $b = CA$, $c = AB$,

$a^2 = b^2 + c^2 - 2bc \cos A$, and similarly,
$b^2 = c^2 + a^2 - 2\,ca \cos B$,
$c^2 = a^2 + b^2 - 2ab \cos C$.

If two sides and the angle between them are given, the third side may be calculated; if three sides are given the angles may be found. Also known as **cosine rule**.

cosine formula

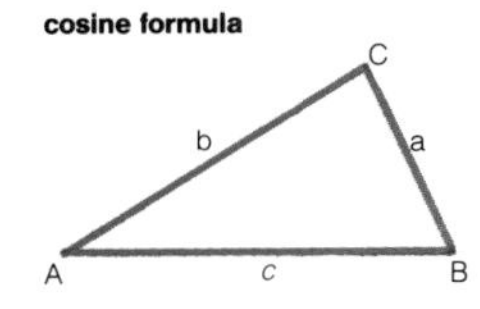

half-angle formulae for triangles if in a triangle ABC, $a = BC$, $b = CA$, $c = AB$ and $s = \frac{1}{2}(a + b + c)$:

$\sin \frac{1}{2}A = \sqrt{\{(s - b)(s - c)/bc\}}$,
$\cos \frac{1}{2}A = \sqrt{\{s(s - a)/bc\}}$,
$\tan \frac{1}{2}A = \sqrt{\{(s - b)(s - c)/s(s - a)\}}$.

The last of these is the best formula for calculating the angle A of a triangle given the three sides. *See also* cosine formula (↑).

area of triangle the usual formula is $\frac{1}{2}bh$ where b is the length of the base and h is the perpendicular height. Also useful are Heron's formula (↓) and $\triangle = \frac{1}{2}bc \sin A$, where $\triangle$ is the area (p. 191), b and c are the lengths of two sides, and A is the angle between them.

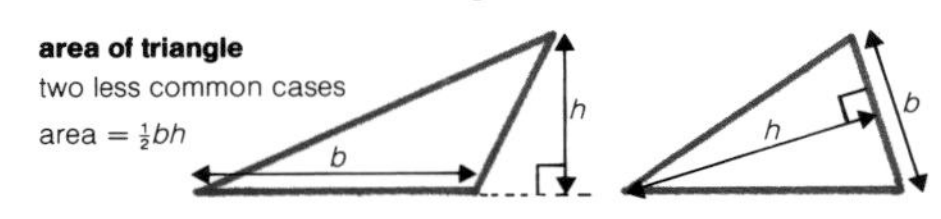

area of triangle
two less common cases
area = $\frac{1}{2}bh$

Heron's formula in any triangle with sides length a, b and c and where $s = \frac{1}{2}(a + b + c)$, the area (p. 191) $\triangle$ is given by
$\triangle = \sqrt{\{s(s - a)(s - b)(s - c)\}}$.
Also known as **Hero's formula**.

Heronian triangle a triangle whose sides and area (p. 191) are all rational numbers (p. 26). Also known as **arithmetical triangle**.

cotangent formula in any triangle ABC where D divides AB in the ratio $m:n$, angle $ACD = \alpha$, angle $BCD = \beta$ and angle $CDB = \theta$, then:
$(m + n) \cot \theta = m \cot \alpha - n \cot \beta$, and
$(m + n) \cot \theta = n \cot A - m \cot B$.
Also known as **cotangent rule**.

cotangent formula

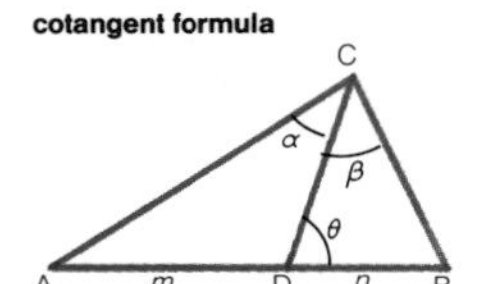

quadrilateral formulae if a, b, c, d, x and y are the sides and diagonals (p. 177) AB, BC, CD, DA, AC and BD respectively of a quadrilateral (p. 177) whose diagonals intersect at X, $s = \frac{1}{2}(a + b + c + d)$ and S is its area (p. 191) then:
$S^2 = (s - a)(s - b)(s - c)(s - d) - abcd \cos^2\alpha$, $S = \frac{1}{2}xy \sin \theta$ and
$16S^2 = 4x^2y^2 - (a^2 - b^2 + c^2 - d^2)^2$,
where 2α = angle A + angle C and θ = angle AXB. (When the quadrilateral is cyclic, then $\cos \alpha = 0$.)

quadrilateral formulae

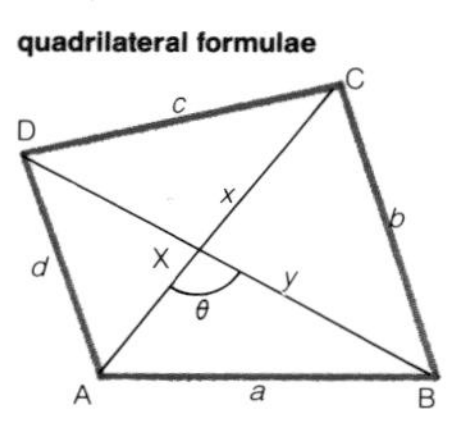

versine (n) the function $1 - \cos \theta$ where θ is any angle.

haversine (n) the function $\frac{1}{2}(1 - \cos \theta)$ where θ is any angle. Symbol: hav θ.

haversine formula the formula in haversines (↑) for spherical triangles (p. 204) corresponding to the cosine formula (p. 135) for plane triangles. If N is the pole (p. 205) of a sphere, A, B are two other points on it with colatitudes (p. 205) χ_a, χ_b, the difference in longitudes (p. 205) between A and B is λ and the angle $z = AOB$ where O is the centre of the sphere then:
$\text{hav}\, z = \text{hav}\, |\chi_a - \chi_b| + \sin \chi_a \sin \chi_b \,\text{hav}\, \lambda$.
It is used in finding the courses of ships.

haversine formula

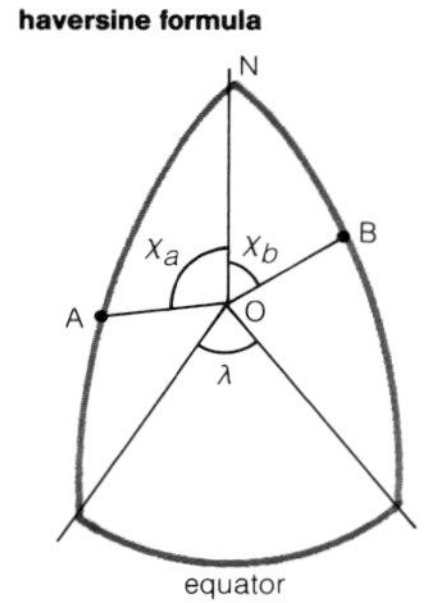

the angle between AO and OB is λ

Napier's formulae in a spherical triangle (p. 204) ABC whose sides BC, CA, AB subtend (p. 175) angles a, b, c respectively at the centre of the sphere:

$$\frac{\sin \frac{1}{2}(a - b)}{\sin \frac{1}{2}(a + b)} = \frac{\tan \frac{1}{2}(A - B)}{\tan \frac{1}{2}C},$$

$$\frac{\cos \frac{1}{2}(a - b)}{\cos \frac{1}{2}(a + b)} = \frac{\tan \frac{1}{2}(A + B)}{\tan \frac{1}{2}C}$$

$$\frac{\sin \frac{1}{2}(A - B)}{\sin \frac{1}{2}(A + B)} = \frac{\tan \frac{1}{2}(a - b)}{\cot \frac{1}{2}c}$$

$$\frac{\cos \frac{1}{2}(A - B)}{\cos \frac{1}{2}(A + B)} = \frac{\tan \frac{1}{2}(a - b)}{\cot \frac{1}{2}c}$$

inverse trigonometric functions and relations graphs

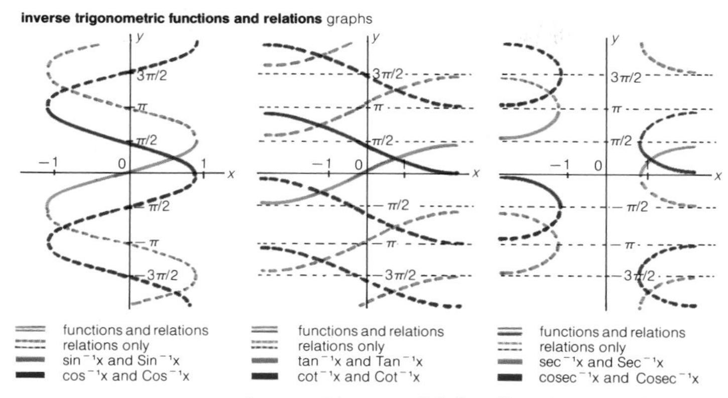

inverse trigonometric functions because the trigonometric functions (p. 132) define the same value for many different angles, the range (p. 113) of any inverse function (p. 52) must be carefully defined:

if $x = \sin y$, $y = \sin^{-1} x$ where $-\frac{1}{2}\pi \leqslant y \leqslant \frac{1}{2}\pi$,
if $x = \cos y$, $y = \cos^{-1} x$ where $0 \leqslant y \leqslant \pi$,
if $x = \tan y$, $y = \tan^{-1} x$ where $-\frac{1}{2}\pi < y < \frac{1}{2}\pi$,
if $x = \operatorname{cosec} y$, $y = \operatorname{cosec}^{-1} x$ where
$-\frac{1}{2}\pi \leqslant y < 0$ or $0 < y \leqslant \frac{1}{2}\pi$,
if $x = \sec y$, $y = \sec^{-1} x$ where
$0 \leqslant y < \frac{1}{2}\pi$ or $\frac{1}{2}\pi < y \leqslant \pi$,
if $x = \cot y$, $y = \cot^{-1} x$ where $0 < y < \pi$.

The notation arcsin x, arccos x, etc is also very common, and $\csc^{-1} x$ and arccsc x are found for $\operatorname{cosec}^{-1} x$. The inverse sine and tangent functions are particularly useful in integration.

inverse trigonometric relations defined as in the inverse trigonometric functions (↑) but without the ranges (p. 113) given there except that they are undefined when the corresponding trigonometric function (p. 132) is undefined. Some books write the relations as $\mathrm{Sin}^{-1} x$, $\mathrm{Cos}^{-1} x$ etc which is a useful convention (p. 9) to make the difference from the inverse trigonometric functions (↑) clear. *These relations should not be used in* integration (p. 152).

exponential constant the number usually called e defined by the infinite sum
$1 + 1/1! + 1/2! + 1/3! + \ldots + 1/n! + \ldots$
where $n!$ is n factorial. It is transcendental (p. 46) and its value is approximately 2.71828.

exponential function the function e^x or exp x defined by the infinite sum
$1 + x/1! + x^2/2! + x^3/3! + \ldots + x^n/n! + \ldots$.
It is also the solution of the differential equation (p. 159) $dy/dx = y$ for which $y = 1$ when $x = 0$. It is sometimes defined as the limit as n tends to infinity of $\{1 + (x/n)\}^n$. An important property is that if $y = e^x$, the derivative (p. 143) dy/dx is also e^x. If z is a complex number (p. 47), the function is defined in a similar way as $\Sigma\, z^n/n!$ for all integer values of n from 0 to infinity, but the notation exp z is more usual. Also known as **compound interest function**, **growth function** ($x > 0$) and **decay function** ($x < 0$).

exponential function

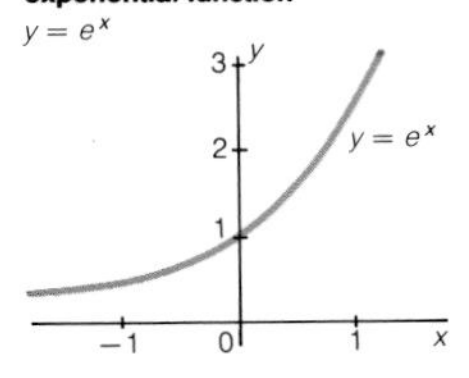

exponential curve

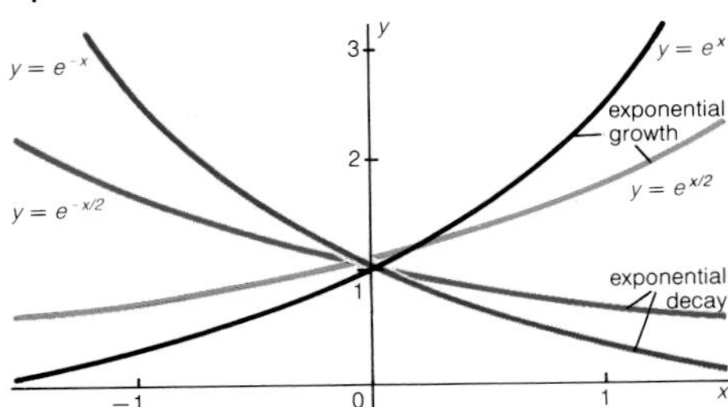

exponential curve usually a curve of the type $y = ae^{bx}$ where a and b are constants, but sometimes any curve whose equation contains a variable in the index (p. 35) or exponent.

exponential growth the growth of a population which is described by an exponential function (↑) and where the rate of change is proportional to the size of the population.

exponential decay similar to exponential growth (↑), but in which the size of the population becomes less. The exponential function (↑) now has a negative power.

logarithm
logarithmic functions $y = \log_e x$ and $y = \log_{10} x$

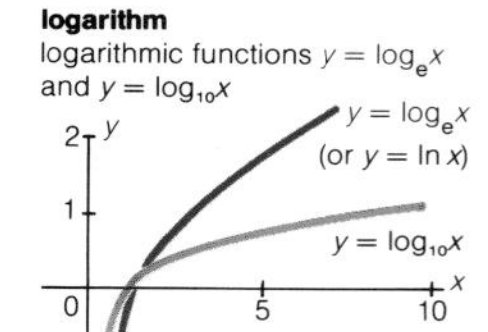

logarithm (*n*) if $x = a^y$, the inverse function (p. 52) $y = \log_a x$ is called the logarithm of x to the base a. *See also* base (p. 37). Usually a is either 10 (in calculations) or the exponential constant (↑) e (in analysis). **logarithmic** (*adj*), **log** (*abbr*).

common logarithms logarithms (↑) to base 10 usually used in calculation. Also known as **Briggsian logarithms**. Symbol: $\log_{10} x$.

natural logarithms logarithms (↑) whose base is the exponential constant (↑) e which form the inverse function (p. 52) to the exponential function (↑). Also known as **Naperian logarithms** or **hyperbolic logarithms**. Symbol: $\ln x$ or $\log_e x$.

binary logarithms logarithms (↑) with base 2. Symbol: $\log_2 x$.

characteristic[2] (*n*) the part of the logarithm (↑) of a number, usually in base 10, which is an integer. Thus $\log_{10} 500$ is approximately 2.6990 and has characteristic 2. *See also* p. 90, mantissa (↓).

mantissa (*n*) the part of a logarithm (↑) of a number, usually in base 10, which is less than 1 and comes after the decimal point (p. 32). Thus $\log_{10} 500$ is approximately 2.6990 and has mantissa 0.6990.

bar notation

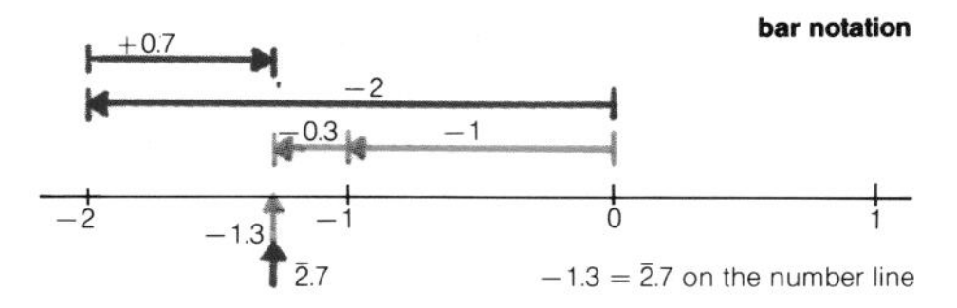

$-1.3 = \bar{2}.7$ on the number line

bar notation a notation where a negative number is written with negative integer less than the number and positive decimal part. Usually used in calculations with common logarithms (↑) of numbers less than 1, e.g. $\log_{10} 0.05$ is approximately -1.3010, usually written as $\bar{2}.6990$ which means $-2 + 0.6990$.

logarithmic tables tables giving logarithms (↑) of numbers, usually either to base 10 or base e. Tables of natural logarithms (↑) were first known as **Napier's analogies**.

modulus[3] (n) a multiplier (p. 23) used to change logarithms (p. 139) from one base to another. *See also* pp. 36, 49.

antilogarithm (n) the inverse function (p. 52) of logarithm (p. 139); if $y = \log_a x$, x is said to be the antilogarithm of y (x is also a^y). Usually used with calculations with common logarithms (p. 139).

complex logarithms if $z = \exp w$ where z and w are complex numbers (p. 47), $z \neq 0$ and exp is the exponential function (p. 138), then $w = \text{Ln } z$ is defined as $w = \ln |z| + i \text{Arg } z$ (for Arg *see* argument, p. 49, and principal value, p. 57) and Ln z is an inverse relation (p. 17). For an inverse function (p. 52) arg z rather than Arg z should be used, where $-\pi < \arg z \leqslant \pi$.

logarithmic scale

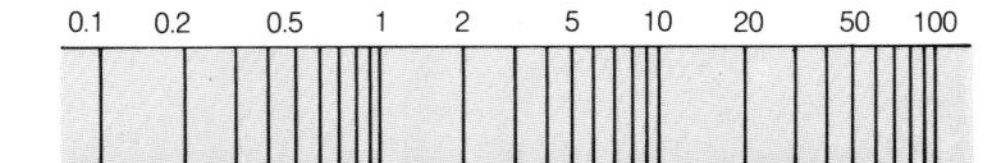

logarithmic scale a scale of numbers where lengths represent the logarithms (p. 139) of those numbers. Often used when points to be plotted have greater frequency for small numbers than for large numbers.

semi-logarithmic graph a cartesian graph (p. 215) in which one of the scales is a logarithmic scale (↑) and the other is a linear (p. 167) scale. Also known as **log-linear graph**.

log-log graph a cartesian graph (p. 215) in which both scales are logarithmic scales (↑).

logarithmic graph paper graph paper (p. 9) with one or both scales logarithmic (p. 139).

logarithmic growth growth of a population which is described by a logarithmic (p. 139) function.

Euler's constant the limit of $1 + (1/2) + (1/3) + \ldots + (1/n) - \ln n$ as n tends to infinity, which is approximately 0.577215. Symbol: γ.

prime number theorem if $\pi(n)$ is the number of prime (p. 32) numbers less than or equal to n, then the limit as n tends to infinity of $\{\pi(n). \ln n\} / n$ is 1.

semi-logarithmic graph
part of grid

20 18 16 14 12 10 8 6 4 2 0
1 2 5 10 20 50 100

log-log graph part of grid

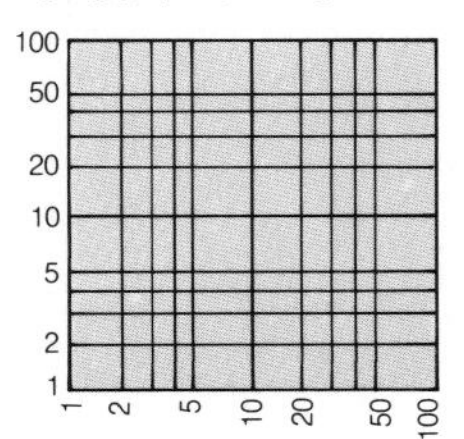

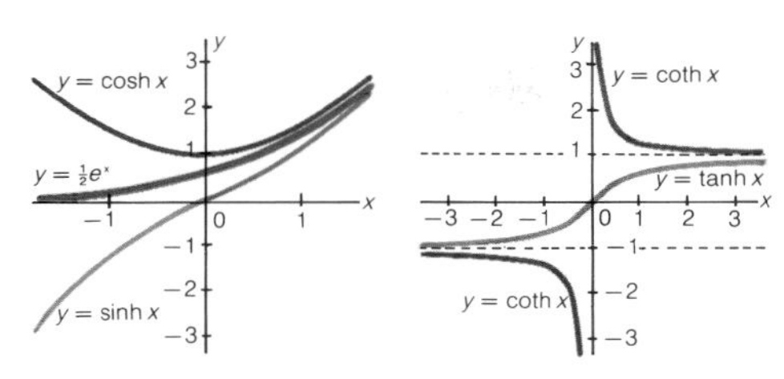

hyperbolic functions

left: graphs of $y = \sinh x$ and $y = \cosh x$ with $y = \frac{1}{2}e^x$
right: graphs of $y = \tanh x$ and $y = \coth x$

hyperbolic functions the functions called *hyperbolic sine*, *hyperbolic cosine* etc and usually written sinh, cosh, tanh, coth, sech and cosech which are similar in some ways to the corresponding trigonometric functions (p. 132). They are defined by:

$\sinh \theta = \frac{1}{2}(e^{\theta} - e^{-\theta})$, $\cosh \theta = \frac{1}{2}(e^{\theta} + e^{-\theta})$,
$\tanh \theta = \sinh \theta / \cosh \theta$, $\coth \theta = 1/\tanh \theta$,
$\operatorname{sech} \theta = 1/\cosh \theta$, $\operatorname{cosech} \theta = 1/\sinh \theta$,

for all values of θ except that $\coth \theta$ and $\operatorname{cosech} \theta$ are undefined when $\theta = 0$. The functions $\sinh \theta$, $\cosh \theta$, $\tanh \theta$ and $\operatorname{cosech} \theta$ are sometimes written $\operatorname{sh} \theta$, $\operatorname{ch} \theta$, $\operatorname{th} \theta$ and $\operatorname{csch} \theta$ respectively. They are useful in integration. *See also* power series (p. 149).

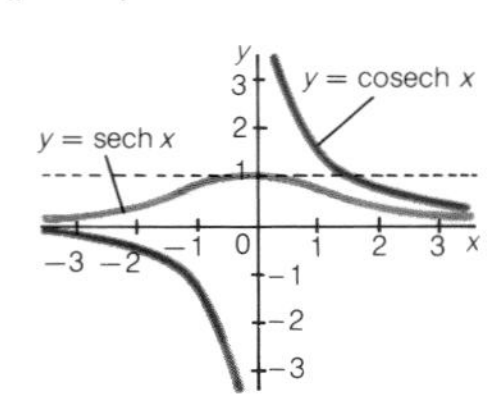

hyperbolic functions

graphs of $y = \operatorname{sech} x$ and $y = \operatorname{cosech} x$

Osborne's rule any identity (p. 57) which is true for trigonometric functions (p. 132) has a corresponding identity in hyperbolic functions (↑) in which the sign of any product of two sines (p. 131) (including tangent which is sine/cosine, p. 131, or cosecant, p. 132, which is 1/sine) is changed, e.g.
$\cos(A + B) = \cos A \cos B - \sin A \sin B$ gives
$\cosh(A + B) = \cosh A \cosh B + \sinh A \sinh B$,
and $1 + \tan^2\theta = \sec^2\theta$
gives $1 - \tanh^2\theta = \operatorname{sech}^2\theta$.

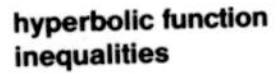

hyperbolic function inequalities

$\tanh x < x < \sinh x < \cosh x$ for $x > 0$

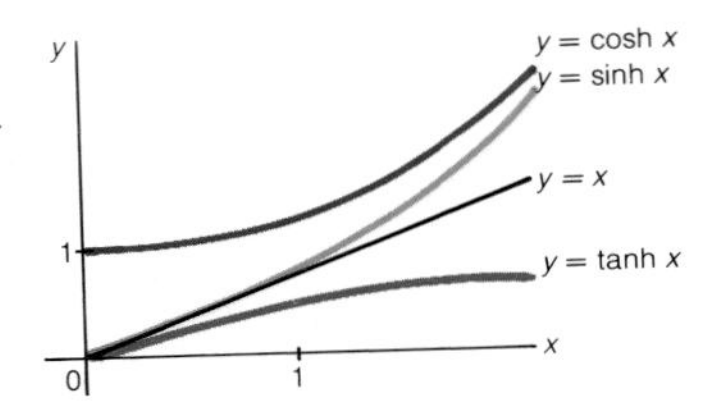

hyperbolic function inequalities for all $x > 0$,

$\tanh x < x < \sinh x < \cosh x$.

inverse hyperbolic functions if $x = \sinh y$, then y is defined to be $\sinh^{-1} x$ or arcsinh x. The remaining hyperbolic functions (p. 141) are similarly defined with the following restrictions (p. 291) on their ranges (p. 17):

$y = \cosh^{-1} x$: only defined for $y \geqslant 0$,
$y = \operatorname{cosech}^{-1} x$: not defined for $y = 0$,
$y = \operatorname{sech}^{-1} x$: only defined for $y \geqslant 0$,
$y = \coth^{-1} x$: not defined for $y = 0$. The other functions are defined everywhere.

inverse hyperbolic functions

left: $y = \sinh^{-1}x$ and $y = \cosh^{-1}x$ and the inverse hyperbolic relation $y = \operatorname{Cosh}^{-1}x$

centre: $y = \tanh^{-1}x$ and $y = \coth^{-1}x$

right: $y = \operatorname{sech}^{-1}x$ and $y = \operatorname{cosech}^{-1}x$ and the inverse hyperbolic relation $y = \operatorname{Sech}^{-1}x$

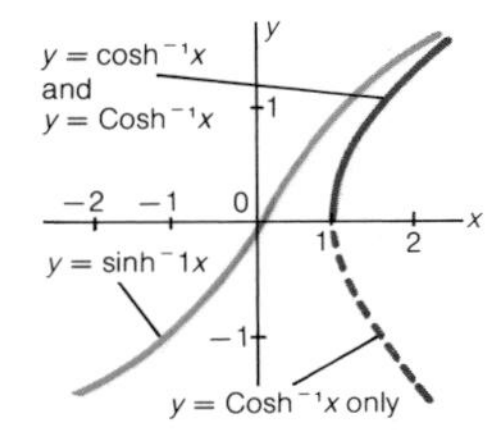

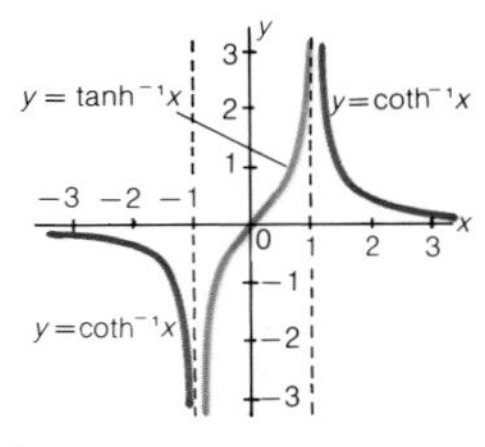

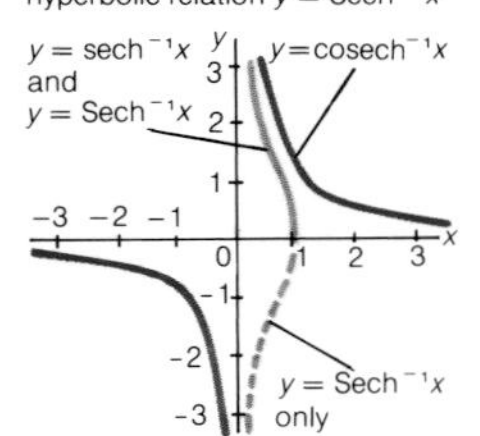

inverse hyperbolic relations these relations are the same as for the inverse hyperbolic functions (p. 141), except for the inverse hyperbolic cosine (p. 131) and secant (p. 132) where there are no limits to the value of y. If defined as relations without a range these are then sometimes written $\operatorname{Cosh}^{-1} x$ and $\operatorname{Sech}^{-1} x$ to make the difference from $\cosh^{-1} x$ and $\operatorname{sech}^{-1} x$ clear.

Gudermannian (*n*) the function $\tan^{-1}(\sinh \theta)$ written gd θ which can be used to calculate hyperbolic functions (p. 141) from trigonometric functions (p. 132). If $\phi = \text{gd}\ \theta$, $\tan \phi = \sinh \theta$, $\sec \phi = \cosh \theta$, and similarly for the remaining functions.

analysis (*n*) the part of mathematics which studies limiting processes, and includes infinitesimal calculus (↓); also used loosely for the study of an area of mathematics generally (as for example in vector analysis, p. 77, which is the study of vectors). **analytic** (*adj*), **analyse** (*v*).

calculus (*n*) a way of calculating particular problems, usually used to mean infinitesimal calculus (↓).

infinitesimal (*n*) an infinitely small quantity (used mainly in older books, and now usually replaced by the idea of limit).

infinitesimal calculus differential (p. 144) and integral calculus (p. 152), whose methods depend on the idea of a limit. Now usually called *calculus* (↑) or *analysis* (↑).

increment (*n*) a small but finite change, usually in a variable, sometimes used particularly for a positive change (*see also* decrement, ↓). A finite change in x may be written δx or Δx (regarded as a single symbol), or as h.

decrement[1] (*n*) a negative increment (↑). *See also* p. 278.

derivative

as the increment δx tends to zero, then $\{f(x+\delta x) - f(x)\}/\delta x$ tends to the derivative $f'(x)$ of $f(x)$

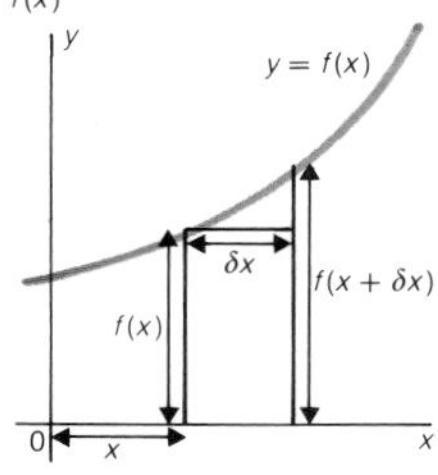

derivative (*n*) the function $f'(x)$ defined by

$$\lim_{\delta x \to 0} \frac{f(x + \delta x) - f(x)}{\delta x}.$$

This function gives the gradient (p. 219) of $y = f(x)$ at any point. Some books call $f'(x)$ the *derivative* and f' the *derived function*. For the derivative of a complex variable (p. 56) *see* analytic function (p. 150) Also known as **differential coefficient**, **derived function**, **gradient function**, **slope function**. *See also* differential (↓) and local scale factor (p. 235). **derive** (*v*) (also meaning calculate or obtain). Symbols: dy/dx, $D_x y$, $f'(x)$, y'.

derivation (*n*) the process of deriving (↑). *See also* differentiation (p. 144).

derivable (*adj*) able to be derived (↑).

differential (*n*) the differential of an independent variable (p. 59) x is written dx and is an arbitrary (p. 287) or freely chosen increment (↑) of x; the corresponding differential dy of the dependent variable (p. 59) y is defined by $dy = f'(x)\,dx$.

differentiate (*v*) to find a differential (p. 143), also used for derive or to find a derivative (p. 143), especially in older books.
differentiation (*n*) the process of differentiating (↑), also used unwisely to mean derivation (p. 143), especially in older books.
differentiable (*adj*) able to be differentiated (↑), often used for derivable (p. 143).
differential calculus the part of analysis which deals with the processes of derivation (p. 143) and differentiation (↑).
difference quotient the ratio $\{f(a) - f(b)\}/(a - b)$ for two values $f(a)$ and $f(b)$ of a function. In the limit as $b \rightarrow a$ this becomes the derivative (p. 143) of $f(x)$ at $x = a$.
order of a derivative the number of times it has been derived; second order derivatives of $y = f(x)$ can be written $f''(x)$, d^2y/dx^2 or D_x^2y; *n*th order derivatives can be written $f^{(n)}(x)$, d^ny/dx^n, $D_x^{(n)}y$ or $y^{(n)}$, or more shortly D^ny, and are often called just the *nth derivatives*. The order of a differential equation (p. 159) is the order of the highest derivative in it. *See also* pp. 20, 190.
Heaviside operator the operator (p. 14) D^n for an *n*th derivative (p. 143) such as is given by $D^ny = d^ny/dx^n$. It is useful in solving differential equations (p. 159).
right derivative a derivative (p. 143) defined with δx tending to zero only from positive values, used for points where no derivative exists except in this way. Similarly **left derivative**. A point of a function may have both left and right derivatives, but no derivative.
rate of change how one quantity changes with another. If $y = f(x)$, the rate of change of y with respect to x is the derivative (p. 143) $f'(x)$ of y with respect to x. Often used of a variable changing with respect to time.
instant (*n*) a point in time, often defined by the use of limits. **instantaneous** (*adj*).
delta notation the use of δ or Δ to show an increment (p. 143) in a variable. If x is the variable, δx or Δx are used for small but finite changes in x. *Note*: δx and Δx are each a single symbol.

difference quotient
the difference quotient for two points A and B on a curve gives the gradient of AB

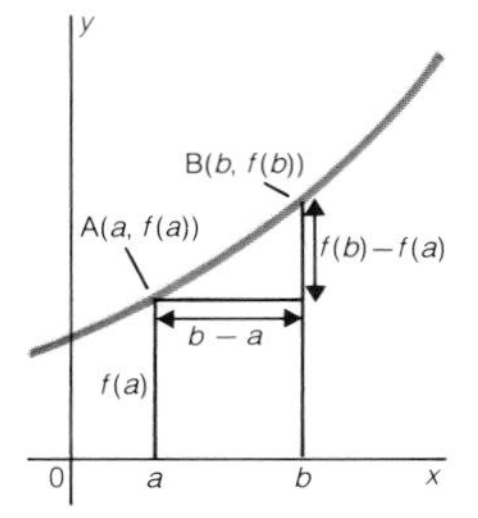

right derivative
similarly left derivative. The function $y = |x|$ has no derivative at the origin, but its right derivative is 1 and its left derivative is −1

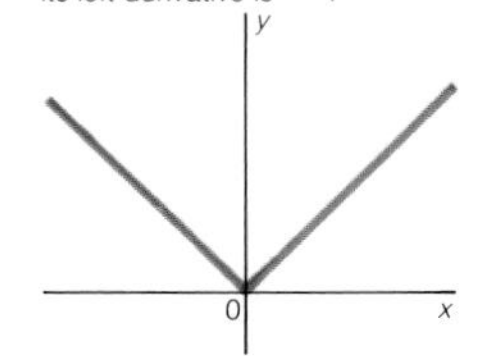

Leibnitz notation the notation dy/dx for derivatives (p. 143), and dx, dy for differentials (p. 143). Also known as **d-notation**.

Lagrange notation the notation y' for the derivative (p. 143) of y with respect to another variable.

dot notation the notation in which a dot or dots are placed over a variable to show derivatives (p. 143) with respect to time, e.g. $\dot{x} = \mathrm{d}x/\mathrm{d}t$, $\ddot{x} = \mathrm{d}^2x/\mathrm{d}t^2$.

fluxion (*n*) Newton's name for the derivatives (p. 143) of a function (the fluent) with respect to time, differential calculus (↑) being known as the *method of fluxions*.

stationary point

$y = x^3 - 3x$ has stationary points at (1, −2) and (−1, 2), both are turning points. (1, −2) is a local minimum and (−1, 2) is a local maximum. There is a point of inflection (which is not a stationary point) at (0, 0)

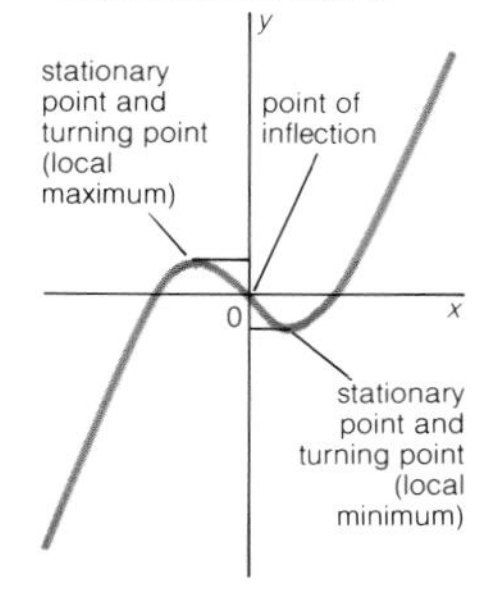

$y = x^3$ has a stationary point which is not a turning point, but which is a point of inflection at (0, 0)

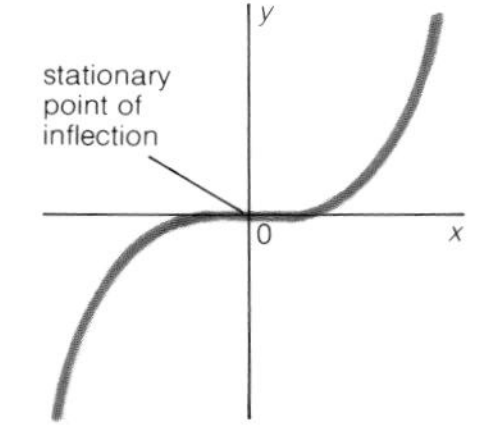

stationary point a point on a curve where the gradient (p. 219) is zero; if $f(x)$ is a function of x, and $f'(a) = 0$, then $f(x)$ has a stationary point at $x = a$. Also known as **critical point** which also has the wider meaning of any point at which some important change takes place.

stationary value the quantity $f(a)$ in the definition of stationary point (↑). Also known as **critical value**. *See also* p. 127.

turning point a stationary point (↑) at which the sign of $f'(x)$ changes. If the sign of $f'(x)$ does not change, it is a stationary point of inflection (p. 146) when both $f'(x)$ and $f''(x)$ are zero. Also known as **extremum** (*n*).

turning value the value of $f(x)$ at a turning point (↑).

maximum (*n*) of a function $f(x)$ at a point $x = a$ in an interval: a value $f(a)$ which is greater than or equal to the value of $f(x)$ at any other point in the interval. Very often used also for a local (↓) maximum. **maxima** (*pl*), **maximal** (*adj*).

minimum (*n*) of a function $f(x)$ at a point $x = a$ in an interval: a value $f(a)$ which is less than or equal to the value of $f(x)$ at any other point in the interval. Very often used also for a local (↓) minimum. **minima** (*pl*), **minimal** (*adj*).

local (*adj*) not absolute (p. 146), used of maxima (↑) and minima (↑) which remain so for arbitrarily (p. 287) small intervals on either side of the given point. At a local maximum or minimum the derivative (p. 143) is zero, but the converse (p. 250) is not necessarily true.

absolute (*adj*) (1) without limit or condition, e.g. used of maxima (p. 145) and minima (p. 145) which are not local (p. 145) but apply for the whole domain (p. 17) of a function; (2) not having any units of measurement and thus being independent of any system of units.

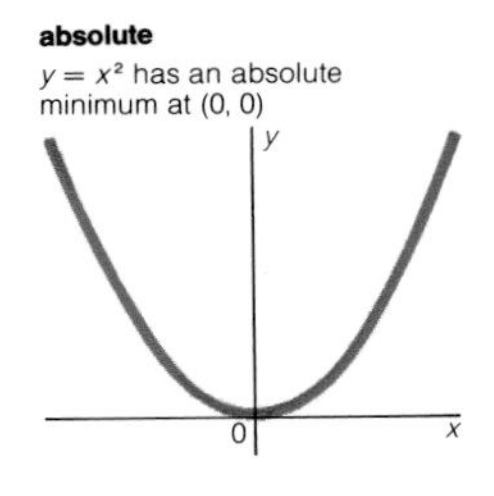

absolute

$y = x^2$ has an absolute minimum at (0, 0)

extreme value any local (p. 145) or absolute (↑) maximum (p. 145) or minimum (p. 145).

point of inflection of a function $f(x)$ of x at $x = a$: a point where $f'(x)$ is a local (p. 145) maximum (p. 145) or minimum (p. 145). *Note*: $f'(a)$ need not be zero as is sometimes supposed. The tangent at a point of inflection crosses the curve. Also known as **inflection** or **inflexion**.

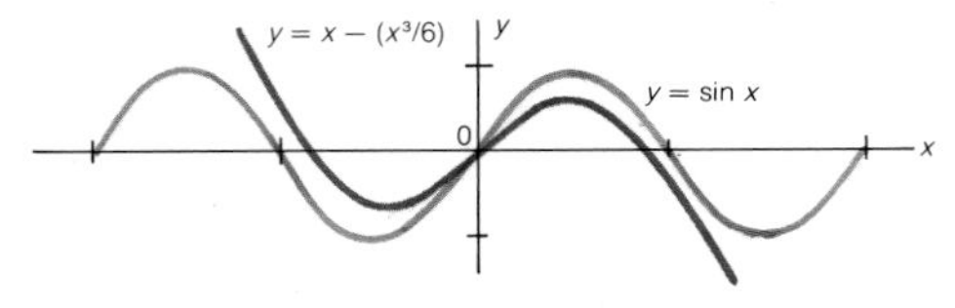

osculating polynomial

$y = x - (x^3/6)$ is an osculating polynomial of $y = \sin x$ at (0, 0)

osculating polynomial of order n at any point of a curve: a polynomial which has the values of its first n derivatives (p. 143) at that point the same as those of the curve's first n derivatives.

smooth function a function $f(x)$ is smooth for $a < x < b$ if for all x in the interval it is continuous, differentiable (p. 144) and the differences in $f(x)$ between one turning value (p. 145) and the next are not large compared with the differences in x at those values.

product rule of derivation (p. 143): if u and v are both derivable (p. 143) functions of x, then using the different notations:

$$\mathrm{d}(uv)/\mathrm{d}x = u(\mathrm{d}v/\mathrm{d}x) + v(\mathrm{d}u/\mathrm{d}x),$$
$$\mathrm{D}(uv) = u\,\mathrm{D}v + v\,\mathrm{D}u$$
$$h'(x) = f(x)g'(x) + g(x)f'(x)$$

where $u = f(x)$, $v = g(x)$ and $h(x) = f(x)g(x)$.

Leibnitz's theorem[2] a generalization of the product rule (↑) to the nth order derivative (p. 143), most easily written: $\mathrm{D}^n(uv) = u\mathrm{D}^n v + {}^nC_1\,\mathrm{D}u\,\mathrm{D}^{n-1}v + {}^nC_2\,\mathrm{D}^2u\,\mathrm{D}^{n-2}v + \ldots + {}^nC_r\,\mathrm{D}^r u\,\mathrm{D}^{n-r}v + \ldots + v\,\mathrm{D}^n u$.
See combination (p. 98) for ${}^nC_r$. *Do not mistake for* Leibnitz's theorem (p. 96).

quotient rule of derivation (p. 143): if u and v are both derivable (p. 143) functions of x, then:

$$\frac{\mathrm{d}(u/v)}{\mathrm{d}x} = \frac{v(\mathrm{d}u/\mathrm{d}x) - u(\mathrm{d}v/\mathrm{d}x)}{v^2}, \text{ or}$$

$$\mathrm{D}(u/v) = \{v\,\mathrm{D}u - u\,\mathrm{D}v\}/v^2.$$

chain rule if y is a function of x and both have derivatives (p. 143), then: $\frac{\mathrm{d}y}{\mathrm{d}x} = \frac{\mathrm{d}y}{\mathrm{d}u} \cdot \frac{\mathrm{d}u}{\mathrm{d}x}$, where y is a function of u and u is a function of x. The idea may be extended to a 'chain' of functions such as: $\frac{\mathrm{d}y}{\mathrm{d}x} = \frac{\mathrm{d}y}{\mathrm{d}u} \cdot \frac{\mathrm{d}u}{\mathrm{d}v} \cdot \frac{\mathrm{d}v}{\mathrm{d}x}$.
Also known as **function of a function rule**.

l'Hôpital's rule a rule for finding limits of quotients (p. 27) of functions which are derivable (p. 143): $\lim_{x \to a} \frac{g(x)}{f(x)} = \lim_{x \to a} \frac{g'(x)}{f'(x)}$, whenever the second limit exists, e.g.

$$\lim_{x \to 0} \frac{\sin x}{x} = \lim_{x \to 0} \frac{\cos x}{1} = 1.$$

Also known as **de l'Hôpital's rule**.

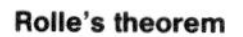
Rolle's theorem

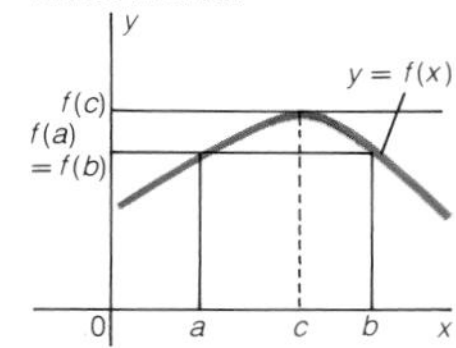

Rolle conditions if a function $f(x)$ is continuous for $a \leqslant x \leqslant b$ and derivable (p. 143) for $a < x < b$, it satisfies the Rolle conditions in $a \leqslant x \leqslant b$.

Rolle's theorem if a function $f(x)$ satisfies the Rolle conditions (↑) in $a \leqslant x \leqslant b$, and if $f(a) = f(b)$, then there is a number c for which $a < c < b$ and $f'(c) = 0$.

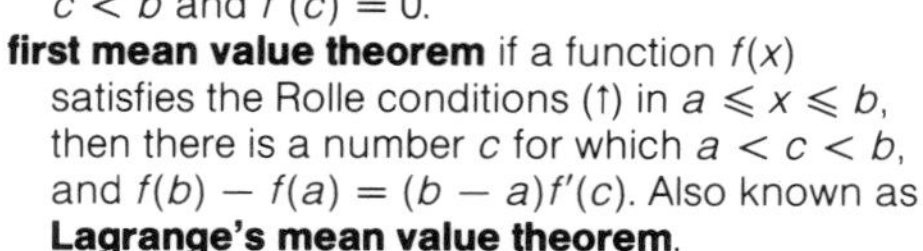

first mean value theorem if a function $f(x)$ satisfies the Rolle conditions (↑) in $a \leqslant x \leqslant b$, then there is a number c for which $a < c < b$, and $f(b) - f(a) = (b - a)f'(c)$. Also known as **Lagrange's mean value theorem**.

first mean value theorem

the tangent at $(c, f(c))$ is parallel to the line joining $(a, f(a))$ and $(b, f(b))$

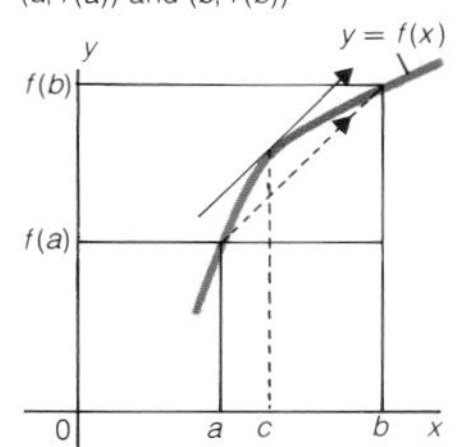

second mean value theorem if $f'(x)$ is continuous for $a \leqslant x \leqslant b$ and $f''(x)$ exists for $a < x < b$, then there is a number c for which $a < c < b$ and $f(b) = f(a) + (b - a)f'(a) + \frac{1}{2}(b - a)^2 f''(x)$.

intermediate mean value theorem if a function $f(x)$ is continuous for $a \leqslant x \leqslant b$ and if $f(a) \neq f(b)$, then for any number k which lies between $f(a)$ and $f(b)$ there is a number c between a and b for which $k = f(c)$.

Taylor's theorem if $f^{(n-1)}(x)$, the $(n-1)$th derivative (p. 143) of $f(x)$, is continuous for $a \leqslant x \leqslant b$ and $f^{(n)}(x)$ exists for $a < x < b$, then $f(x + a) = f(a) + f'(a)x + (1/2!)f''(a)x^2 + (1/3!)f^3(a)x^3 + \ldots + \{1/(n-1)!\}f^{(n-1)}(a)x^{n-1} + R_n$, where R_n is the remainder term (↓). This is a generalization of the mean value theorems above.

remainder term the last term R_n in the series (p. 91) for Taylor's theorem (↑) or any similar term in other series. Also known as **error term**.

Taylor's polynomial the polynomial given by Taylor's theorem (↑) apart from the remainder term (↑) R_n.

Taylor's series the polynomial and remainder term (↑) given by Taylor's theorem (↑).

Schloemilch's form the form of the remainder term (↑) of Taylor's theorem (↑):

$$R_n = \{1/(n-1)!p\}x^n(1-\theta)^{n-p}f^{(n)}(a + \theta x),$$

where $0 < \theta < 1$, $p \leqslant n$ and p is an integer.

Lagrange form the form of the remainder term (↑) of Taylor's theorem (↑):

$$R_n = (1/n!)x^n f^{(n)}(a + \theta x), \text{ where } 0 < \theta < 1.$$

This is obtained from Schloemilch's form (↑) by putting $p = n$.

Cauchy form the form of the remainder term (↑) of Taylor's theorem (↑):

$$R_n = \{(1-\phi)^{(n-1)}/(n-1)!\}x^n f^{(n)}(a + \varphi x)$$

where $0 < \varphi < 1$. This is obtained from Schloemilch's form (↑) by putting $p = 1$.

integral form the form of the remainder term (↑) of Taylor's theorem (↑):

$$R_n = \{1/(n-1)!\}\int_a^x (x-t)^{(n-1)}f^{(n)}(t)\,dt$$

where $a < t < x$.

Maclaurin's theorem the special case of Taylor's theorem (↑) when $a = 0$:

$$f(x) = f(0) + f'(0)x + (1/2!)f''(0)x^2 + \ldots + \{1/(n-1)!\}f^{(n-1)}(0) + R_n,$$

where R_n is a remainder term (↑).

Newton's method if $x = a$ is a first approximation to a solution of an equation $y = f(x)$, then in general a better solution is given by $a - \{f(a)/f'(a)\}$. It is proved from the first two terms of Taylor's series (↑). Often used also for Newton-Raphson method (↓).

Newton's method

since AP = $f(a)$ and tan AQ̂P = $f'(a)$, QP = $f(a)/f'(a)$ and in general Q is a better approximation to the point where $f(x) = 0$ than P is

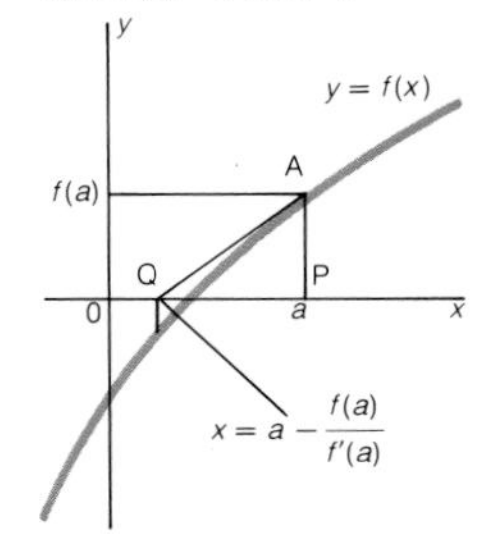

Newton-Raphson method an improved method similar to Newton's method (↑) but using further terms of Taylor's series (↑). Often used also for Newton's method (↑).

successive approximations any solution method which uses an approximate solution to obtain a better approximation, e.g. Horner's method (↓).

Horner's method a way to obtain a better approximation to the solution of an equation by adding a digit (p. 21) at a time to an approximate solution and repeating this as far as necessary.

power series a series (p. 91) whose terms form a power sequence (p. 93) such as those obtained from Taylor's theorem (↑), e.g.

$\sin x = x - x^3/3! + x^5/5! - x^7/7! + \ldots,$

$\cos x = 1 - x^2/2! + x^4/4! - x^6/6! + \ldots,$

$\tan^{-1} x = x - x^3/3 + x^5/5 - x^7/7 + \ldots,$

(*see* inverse trigonometric functions, p. 137),

$e^x = 1 + x + x^2/2! + x^3/3! + x^4/4! \ldots,$

(*see* exponential function, p. 138),

$\sinh x = x + x^3/3! + x^5/5! + x^7/7! + \ldots,$

$\cosh x = 1 + x^2/2! + x^4/4! + x^6/6! + \ldots$

(*see* hyperbolic functions, p. 141).

Power series may also consist of functions of a complex variable (p. 56). *See also* logarithmic series (p. 150), binomial theorem (p. 98).

Gregory's series the series (p. 91) for $\tan^{-1} x$ given under power series (↑).

Leibnitz's series the special case of Gregory's series (↑) when $x = 1$ which gives

$\pi/4 = 1 - (1/3) + (1/5) - (1/7) + \ldots.$

Euler's formula for π the formula

$\pi/4 = \tan^{-1}(1/2) + \tan^{-1}(1/3).$

Do not mistake for other Euler formulae.

Rutherford's formula for π the formula

$\pi/4 = 4\tan^{-1}(1/5) - \tan^{-1}(1/70) + \tan^{-1}(1/99).$

Machin's formula the formula

$\pi/4 = 4\tan^{-1}(1/5) - \tan^{-1}(1/239),$

which when Gregory's series (↑) is used for the two inverse tangents gives a formula for π (p. 192) which converges (p. 94) rapidly.

Wallis's formula the formula for π (p. 192):

$$\frac{\pi}{2} = \frac{2}{1}\cdot\frac{2}{3}\cdot\frac{4}{3}\cdot\frac{4}{5}\cdot\frac{6}{5}\cdot\frac{6}{7}\cdot\frac{8}{7}\cdot\frac{8}{9}\cdot\ldots.$$

logarithmic series the two power series (p. 149): $\log_e(1 + x) = x - x^2/2 + x^3/3 - x^4/4 + \ldots$, where $-1 < x \leqslant 1$ and $\log_e(1 - x) = -x - x^2/2 - x^3/3 - x^4/4 - \ldots$, where $-1 \leqslant x < 1$.

logarithmic inequality if $x \neq 0$ and $x > -1$, $x/(1 + x) < \log_e(1 + x) < x$.

logarithmic inequality
if $x > -1$ and $x \neq 0$,
$x/(1 + x) < \log_e(1 + x) < x$

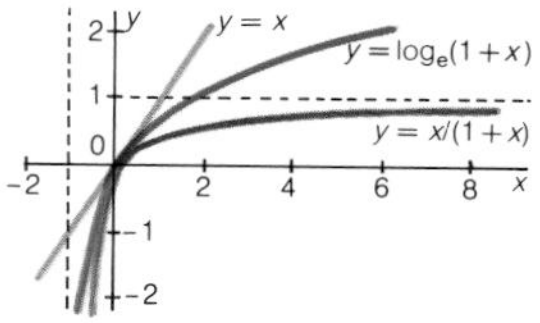

Fourier series the series (p. 91)

$$S(x) = a_0 + \sum_{n=1}^{\infty} (a_n \cos nx + b_n \sin nx),$$

which can be used to approximate to any continuous function for $-\pi \leqslant x \leqslant \pi$, or to any periodic (p. 276) function for all its values.

Fourier analysis a method of writing any periodic (p. 276) function or any continuous function over a given interval as an infinite sum of trigonometric functions (p. 132) by using Fourier series (↑). Also known as **harmonic analysis**.

Fourier transform the writing of a function as a sum of Fourier series (↑).

complex analysis the use of the methods of analysis with complex variables (p. 56).

complex function (1) a function of a complex variable (p. 56); (2) a complex-valued function (↓). These are better used to avoid misunderstanding.

complex-valued function a function of a real variable (p. 56) which has complex (p. 56) values.

Hilbert space the set of all functions of a complex variable (p. 56) which are integrable (p. 152).

analytic function a function of a complex variable (p. 56) which has a derivative (p. 143). (The definition of derivative is the same as for a real variable, p. 56, but z and δz are now complex and the real and imaginary parts, p. 48, of δz must tend to zero independently.)

singularity (*n*) the singularity of an analytic function (↑) is a point at which the function does not have a derivative (p. 143), but in every neighbourhood (p. 95) of the point there are points which have a derivative. Sometimes also used generally for singular point (p. 188).

entire function a function of a complex variable (p. 56) which is an analytic function (↑) for every point of its domain (p. 17). Also known as **integrable function**.

entire series if $f(z)$ is an entire function (↑) and if it can be written as $a_0 + a_1z + a_2z^2 + \ldots$, then this series is called an entire series.

calculus of variations the study of the complete behaviour of functions, e.g. the determination of the brachistochrone (p. 225).

fractal (*n*) a configuration, either a curve or a surface of higher dimension, which in one sense may be taken to have fractional dimension, e.g. any space-filling curve (↓) or surface.

space-filling curve any curve which fills a region or space in the sense that every point in the region or space lies on the curve. Such a curve is an example of a fractal (↑), e.g. the limiting case of the dragon curve (↓).

dragon curve one of a set of curves such that the *n*th curve has no derivative (p. 143) at $2^n - 1$ points, and obtained from the edge of a piece of paper which has been folded in half *n* times and then opened so that each fold makes a right angle. The limiting case as *n* tends to infinity is a space-filling curve (↑).

dragon curve first five cases

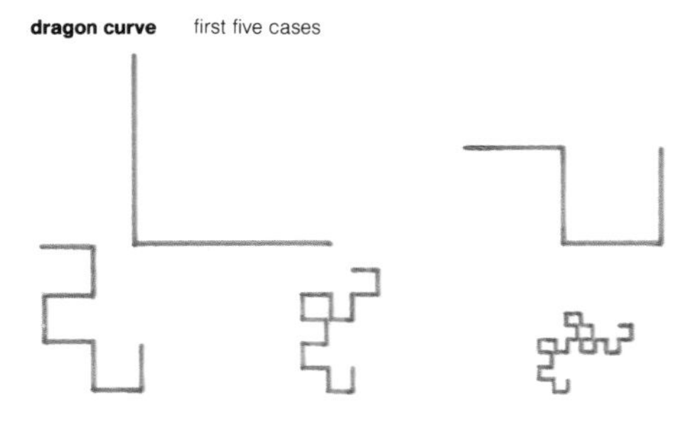

integral calculus the part of analysis which deals with the process of integration in its various forms.

integral (*n*) (1) the sum of an infinite (p. 11) series (p. 91) of infinitesimal (p. 143) quantities given as a limit; (2) the inverse function (p. 52) of derivation (p. 143) (the antiderivative, ↓). The relation between the two is the fundamental theorem of calculus (↓).

integration (*n*) the process of finding an integral. **integrable** (*adj*), **integrate** (*v*).

antiderivative (*n*) the antiderivative of $f(x)$ in an interval whose limits are a and b is $F(x)$ where $d(F(x))/dx = f(x)$. The function f must be continuous for $a \leqslant x \leqslant b$ and the function F must be derivable (p. 143) for $a < x < b$. Also known as **indefinite integral**. **antiderive** (*v*), **antidifferentiate** (*v*) (more common though less correct). Symbol: $F(x) = \int f(x)dx$.

antiderivation (*n*) the process of finding an antiderivative (↑); often less correctly known as **antidifferentiation**.

constant of integration if $F(x)$ is the antiderivative (↑) of $f(x)$, then so is $F(x) + c$ where c is the constant of integration. Also known as **arbitrary constant**, but this can also be used elsewhere.

definite integral the integral as the area (p. 191) under a curve. If $f(x)$ is defined for an interval $a \leqslant x \leqslant b$, and if the interval is divided into subintervals $\delta_1, \delta_2, \ldots \delta_n$, such that each δ_i is less than some value δ, and if x_i is a point in the subinterval δ_i, then

$$\lim_{\delta \to 0} \sum_{i=1}^{n} f(x_i)\delta_i \text{ is written } \int_a^b f(x)dx$$

and is the definite integral of $f(x)$ over the interval $a \leqslant x \leqslant b$.

integrand (*n*) a function which is to be integrated.

strip (*n*) a narrow band of width δx and height varying between $f(x)$ and $f(x + \delta x)$ into which an area (p. 191) is divided in the theory of the definite integral (p. 152). Also used for similar shapes elsewhere.

definite integral
the area under the curve. The number of strips in the interval is allowed to tend to infinity

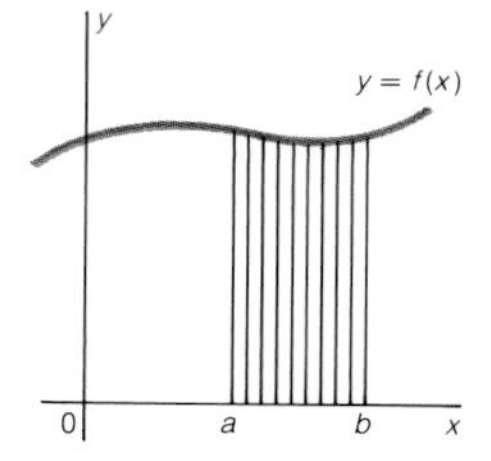

strip

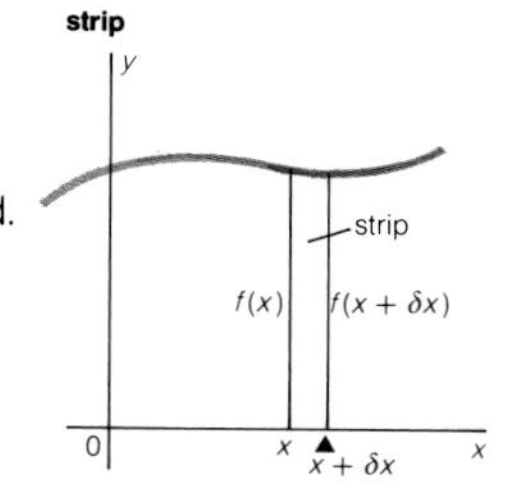

limits of integration the end values of the interval in definite integration (↑), the quantities a and b in $\int_a^b f(x)\mathrm{d}x$. They are known as the *lower* and *upper limits* respectively.

variable of integration the variable used in integration.

dummy variable a variable used in a formula to make writing it easier; the actual letter used does not matter. In definite integration (↑) the variable is used simply to give a general rule. Thus:
$\int_0^1 x^2\mathrm{d}x$ is the same as $\int_0^1 y^2\mathrm{d}y$.

planimeter (n) a machine which measures the area (p. 191) of any closed plane region, and thus performs integrations.

exhaustion (n) an early method of finding areas (p. 191) of closed curves by approximating to polygons and which corresponds to the idea of integration.

fundamental theorem of calculus the theorem relating definite (↑) and indefinite integrals (↑). If $f(x)$ can be integrated for $a \leqslant x \leqslant b$, and if $\mathrm{d}(F(x))/\mathrm{d}x = f(x)$, then
$\int_a^b f(x)\mathrm{d}x = F(b) - F(a)$.

Riemann integral a theoretical method of showing that definite integration (↑) is allowable in which the area (p. 191) under a curve is divided into narrow strips (↑) and the upper and lower bounds (p. 66) of their areas is considered as the number of strips tends to infinity. The sums of these bounds are called the *upper* and *lower Riemann sums* respectively.

Lebesgue integral a theoretical method of showing that definite integration (↑) is allowable which considers regions as measurable sets.

standard integrals integrals (↑) which allow various common functions to be integrated. Many of these integrals are listed on the endpapers.

integration by inspection the process of integrating a function by searching for another function with the correct derivative (p. 143).

substitution formula the form of the chain rule (p. 147) used in integration by substitution (↓), $\int f(x)\mathrm{d}x = \int g(t)(\mathrm{d}x/\mathrm{d}t)\mathrm{d}t$, where $g(t)$ is the function obtained from $f(x)$ by writing x in terms of another variable t.

integration by substitution the process of integrating a function by a change of variable using the substitution formula (↑), e.g. if $x = \tan\theta$, then since $\mathrm{d}x/\mathrm{d}\theta = \sec^2\theta$:

$$\int \mathrm{d}x/(1 + x^2) = \int \sec^2\theta\,\mathrm{d}\theta/(1 + \tan^2\theta) = \int \mathrm{d}\theta = \theta + c = \tan^{-1}x + c.$$

integration by parts the process of integration using the formula

$$\int u(\mathrm{d}v/\mathrm{d}x)\mathrm{d}x = uv - \int v(\mathrm{d}u/\mathrm{d}x)\mathrm{d}x,$$

which is obtained from the product rule (p. 146), e.g. if $u = \log x$ and $\mathrm{d}v/\mathrm{d}x = 1$, then:

$$\int \log x\,\mathrm{d}x = x\log x - \int \mathrm{d}x = x\log x - x + c.$$

reduction formula a formula which allows one step of a solution to be done and which is repeated again and again to complete the solution; many reduction formulae involve integrals, e.g. if $J_n = \int_0^{\pi/2}\cos^n\theta\,\mathrm{d}\theta$, then

$$J_n = \{(n-1)/n\}J_{n-2}.$$

A useful result of such formulae is

$$\int_0^{\pi/2}\sin^m x\cos^n x\,\mathrm{d}x = \frac{(m-1)(m-3)\ldots.(n-1)(n-3)\ldots.k}{(m+n)(m+n-2)(m+n+4)\ldots},$$

where m and n are positive integers. Each product ends at 1 or 2 and $k = \frac{1}{2}\pi$ if m and n are both even, otherwise $k = 1$.

infinite integrals (1) integrals where one or both of the limits of integration (p. 153) can tend to infinity; (2) integrals where the integrand (p. 152) may become infinite at some value. Both can be found in some cases if the limits give finite answers. Also known as **improper integrals**.

divergent integral an integral which, when one of its limits of integration (p. 153) tends to infinity, does not have a finite limit.

double integration integration in which the process is carried out twice; both integrations must, of course, be possible. Similarly in **repeated**, **multiple** or **iterated integration** the process is carried out more than once. Symbols: e.g. $\iint f(x,y)\mathrm{d}x\mathrm{d}y$.

surface integral if a three dimensional surface S has a function $f(x, y, z)$ defined at each point of S, then the limiting integral $\int f\,dS$ is, if it exists (p. 248), called a surface integral.

line integral if a two or three dimensional line C has a vector function $\mathbf{F}$ defined at each point on it, then the integral $\int_P^Q \mathbf{F} \cdot d\mathbf{r}$ which is the integral of the scalar product (p. 78) of vectors $\mathbf{F}$ and $d\mathbf{r}$ is the line integral of $\mathbf{F}$ along C from P to Q. Also known as **curvilinear integral**.

contour integral an integral in complex analysis (p. 150) around a plane closed curve (p. 190) which encloses a region. For an analytic function (p. 150) $\int_C f(z)dz = 0$. In real (p. 49) analysis contour integrals become line integrals (↑).

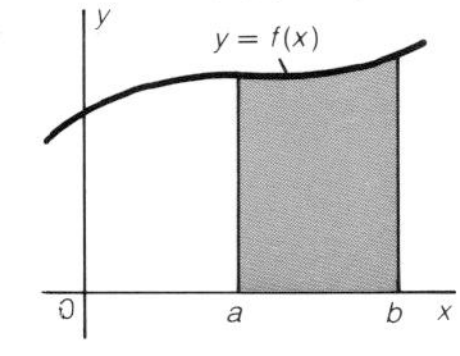

area under a curve
the area under $y = f(x)$ between $x = a$ and $x = b$

area under a curve the area (p. 191) bounded by the curve $y = f(x)$, the ordinates (p. 215) $x = a$ and $x = b$, where $a < b$, and the axis $y = 0$ has area $\int_a^b f(x)dx$ or $\int_a^b y dx$, provided $f(x)$ does not change sign for $a < x < b$. If $f(x) < 0$, this gives an area with negative sign. For its centre of mass (p. 261) *see* lamina (↓).

lamina (*n*) a thin, plane, stiff sheet. In applied mathematics (p. 8) it usually has constant mass per unit area. The centre of mass $(\bar{x}, \bar{y})$ of a lamina whose edges are $y = f(x)$, the ordinates $x = a$ and $x = b$ and the axis $y = 0$ is, under certain conditions, given by $A\bar{x} = \int_a^b xy\,dx$, $A\bar{y} = \int_a^b \frac{1}{2}y^2 dx$, where A is the area (p. 191) of the lamina.

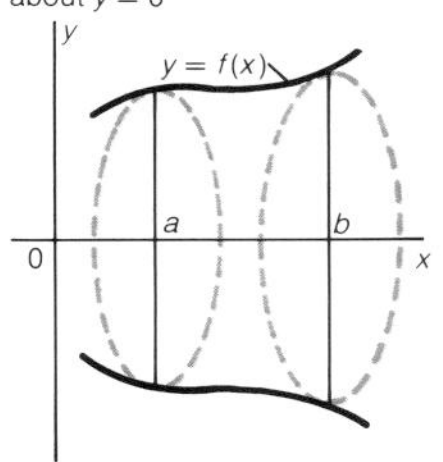

solid of revolution
formed by rotating $y = f(x)$ about $y = 0$

solid of revolution a solid which is obtained by rotating an area about a straight line on its edge or outside it. It is often taken to be the solid obtained when an area (p. 191), such as that defined in area under a curve (p. 187), is rotated about $y = 0$. This solid has volume $\int_a^b \pi y^2 dx$, and the surface area of the curved surface is $\int_a^b 2\pi y \{1 + (dy/dx)^2\} dx$. If it has uniform mass per unit volume, then the x-coordinate (p. 215) $\bar{x}$ of its centre of mass (p. 261) is given by $V\bar{x} = \int_a^b \pi x y^2 dx$ where V is the volume of the solid. All these formulae have certain conditions.

arc-length if δs is an increment (p. 143) of the length of an arc of a plane curve, then the total length is $\int_{s_1}^{s_2} ds = \int_{t_1}^{t_2} \sqrt{\{(dx/dt)^2 + (dx/dt)^2\}}dt$, where dx/dt and dy/dt are continuous, the parameter (p. 67) t is monotonic (p. 67) increasing from t_1 to t_2 and $ds/dt > 0$. It can also be written for a function $f(x)$ for which $a \leqslant x \leqslant b$ where $f(x)$ and $f'(x)$ are both continuous in the interval as:
$\int_a^b \sqrt{\{1 + [f'(x)]^2\}}dx$, or $\int_\alpha^\beta \sqrt{\{r^2 + (dr/d\theta)^2\}}d\theta$ in polar coordinates (p. 216), both between the given limits and under certain conditions.

Pappus-Guldin theorems (1) if a surface is obtained by rotating an arc of a plane curve about an axis in its plane, then its area (p. 191) is the length of the arc multiplied by the distance travelled by the centre of mass (p. 261) of the arc; (2) if a region is rotated about an axis in its plane which it does not cut, the volume of the solid obtained is the area of the region multiplied by the distance travelled by its centre of mass. *Do not mistake for* Pappus' theorem (p. 183). Also known as **Guldin's theorems**, **Guldinius' theorems**.

area of sector the area of a sector of a curve is the area (p. 191) bounded by the curve in polar coordinates (p. 216) $r = f(\theta)$ and the radius vectors (p. 216) $r = \alpha$ and $r = \beta$ given by $\int_\alpha^\beta \frac{1}{2}r^2 d\theta$ under certain conditions.

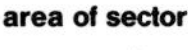

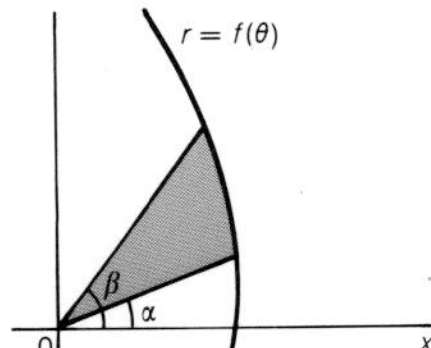

root mean square value the root mean square value of a function $f(x)$ between $x = a$ and $x = b$ is $\{[1/(b - a)]\int_a^b [f(x)]^2 dx\}^{\frac{1}{2}}$. *See also* root mean square deviation (p. 118).

elliptic functions functions which are obtained in trying to calculate the arc-length (↑) of an ellipse (p. 220) and which give rise to integrals which cannot be solved in terms of the usual functions. They are defined by the inverse functions (p. 52) of

$$\int_0^a \frac{dx}{\sqrt{\{(1 - x^2)(1 - e^2x^2)\}}} \text{ where } 0 < e < 1,$$

and their values may be found from tables.

elliptic integral an integral defining an elliptic function (↑).

gamma function the function $\Gamma(x) = \int_0^\alpha e^{-x}t^{x-1}dt$ for $x > 0$ and for which $\Gamma(x + 1) = x\Gamma(x)$. If n is a positive integer, $\Gamma(n) = (n - 1)!$ (*see* factorial, p. 97). For this reason it is sometimes called the **factorial function**.

beta function the function $B(s, t) = \{\Gamma(s) \cdot \Gamma(t)\}/\Gamma(s + t)$ (*see* gamma function, ↑) which is useful in integrating powers of trigonometric functions (p. 132).

zeta function the function $\zeta(x) = (1/1^x) + (1/2^x) + (1/3^x) + \ldots$, which converges (p. 94) for all $x > 1$, and may be a function of a real (p. 56) or complex variable (p. 56).

numerical methods methods of obtaining numerical answers to problems without first finding a general solution (in many cases such a solution cannot be found). Also called **numerical analysis**.

numerical integration integration by a numerical method (↑).

mid-ordinate rule

a typical strip whose area is approximately $(a_{i \times 1} - a_i)f(\frac{1}{2}(a_i + a_{i+1}))$

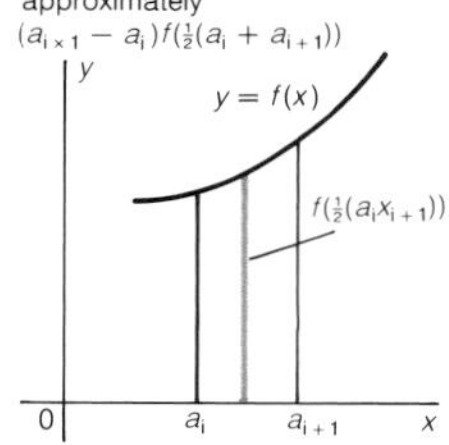

mid-ordinate rule a method of approximating to the area under a curve (p. 155) $y = f(x)$ by dividing it into strips (p. 152). If a strip is bounded by $x = a_i$ and $x = a_{i+1}$, the area of the strip is approximately the length of the mid-ordinate given by $f(\frac{1}{2}(a_i + a_{i+1}))$ multiplied by its width.

trapezoidal rule similar to the mid-ordinate rule (↑) except that the area (p. 191) of the strip is now taken to be the mean length of its two long sides which is given by $\frac{1}{2}\{f(a_i) + f(a_{i+1})\}$ multiplied by the width. If the area is divided into n strips of width h bounded by ordinates (p. 215) $y_0, y_1, y_2, \ldots, y_n$, the area is approximately $h\{\frac{1}{2}y_0 + y_1 + y_2 + \ldots + y_{n-1} + \frac{1}{2}y_n\}$. Also known as **trapezium rule**.

trapezoidal rule

a typical strip whose area is approximately $(a_{i+1} - a_i)\frac{1}{2}\{f(a_i) + f(a_{i+1})\}$

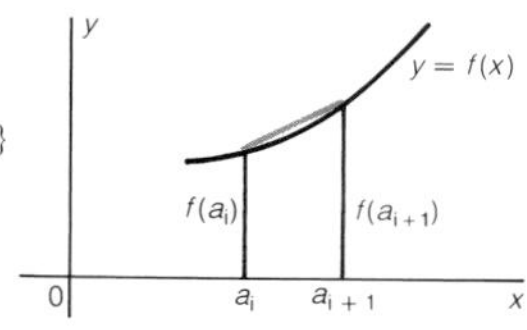

Simpson's rule a method of approximating to the area under a curve (p. 155) by dividing it into $2n$ strips (p. 152) and finding the areas (p. 191) under the parabolae (p. 220) which approximately define the curved ends of each pair of strips. If the ordinates (p. 215) are $y_0, y_1, y_2, \ldots, y_{2n}$, the area is approximately $(h/3)\{y_0 + 4(y_1 + y_3 + y_5 + \ldots + y_{2n-1}) + 2(y_2 + y_4 + y_6 + \ldots + y_{2n-2}) + y_{2n}\}$ where h is the width of each strip. Also known as **parabolic rule**.

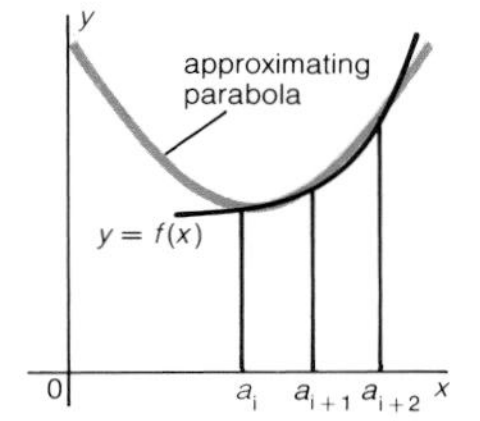

Simpson's rule
a typical pair of strips; their area is taken to be the area under the approximating parabola

Weddle's rule a rule similar to Simpson's rule (↑) but using sets of three strips (p. 152) instead of two. If the ordinates (p. 215) are $y_0, y_1, y_2, \ldots, y_{3n}$, the area (p. 191) is approximately $(3h/8)\{y_0 + 3(y_1 + y_2 + y_4 + y_5 + y_7 + \ldots + y_{3n-1}) + 2(y_3 + y_6 + \ldots + y_{3n-3}) + y_{3n}\}$, where h is the width of each strip.

quadrature an early method of finding the approximate areas of curved regions by using a limiting process of strips (p. 152) whose areas (p. 191) are known. It corresponds to integration.

difference operator a symbol showing the difference between two values of a function, usually $\Delta f(a) = f(a + h) - f(a)$. This is the *forward difference operator*, also found are the *backward difference operator* $\nabla f(a) = f(a) - f(a - h)$, and the *central difference operator* $\delta f(a) = f(a + \frac{1}{2}h) - f(a - \frac{1}{2}h)$. Higher order operators can be defined, e.g. $\Delta^2 f(a) = \Delta f(a + h) - \Delta f(a)$, etc. *Do not mistake for* del (p. 165).

finite difference one of the differences symbolized (p. 8) by a difference operator (↑) which is finite and unlike the increments (p. 143) in differential calculus (p. 144) will not need to tend to zero.

difference method a method using finite differences (↑). Also known as **method of differences**.

Gregory-Newton formula the formula

$$f(n) = f(0) + {}^nC_1\Delta f(0) + {}^nC_2\Delta^2 f(0) + \ldots + {}^nC_r\Delta^r f(0) + \ldots + \Delta^n f(0).$$

(*See* difference operator, ↑, and combination, p. 98, for meanings of symbols.) It is similar to the binomial theorem (p. 98) and used for finite differences (↑).

differential equation an equation in a set of variables which contains derivatives (p. 143) of some or all of those variables.

order of a differential equation the order of its highest derivative (p. 144).

degree of a differential equation the highest power of the derivative (p. 143) of the highest order in the equation, e.g. $(d^2y/dx^2)^2 + (dy/dx)^3 + y = 0$ has degree two.

boundary condition a condition given with a differential equation (↑) which allows an arbitrary constant (p. 152) or constants to be found.

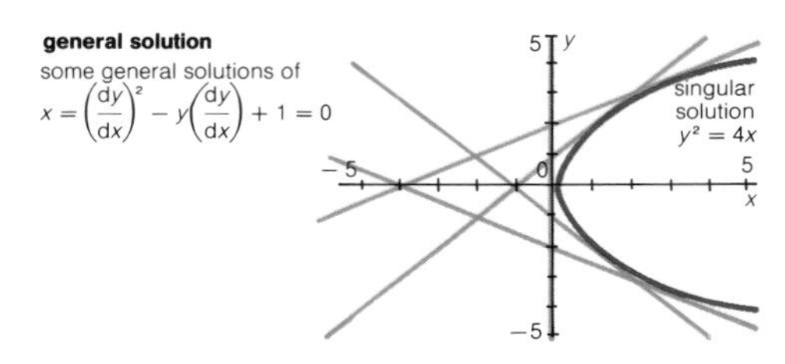

general solution a solution to a differential equation (↑) which contains as many arbitrary constants (p. 152) as the order of the equation and is as complete a solution as is possible. Also known as **complete primitive**. *See also* complementary function (p. 161).

initial condition any starting condition, especially a boundary condition (↑) of a differential equation (↑) at time zero.

initial value a value giving an initial condition (↑).

singular solution a solution of a non-linear first order differential equation (↑) which is not a special case of the general solution (↑), but is a solution, also similar solutions in other cases, e.g. $x(dy/dx)^2 - y(dy/dx) + 1 = 0$ has a general solution $x - Ay + A^2 = 0$ and a singular solution $y^2 = 4x$.

partial differential equation a differential equation (↑) containing partial derivatives (p. 163).

ordinary differential equation a differential equation (↑) which does not contain partial derivatives (p. 163).

separation of variables a method of solving some differential equations (p. 159). If the equation $y' = F(x, y)$ can be written down as $y' = f(x)g(y)$, where F is a function of x and y, f is a function of x and g is a function of y, then a solution may often be obtained from $\int\{1/g(y)\}\mathrm{d}y = \int f(x)\mathrm{d}x$. The variables are then known as **separable variables**.

exact equation a differential equation (p. 159) which is an immediate derivative (p. 143) of another equation.

integrating factor a function by which a differential equation (p. 159) is multiplied so that it becomes an exact equation (↑) and can be solved.

linear differential equation an equation of the type $\mathrm{d}y/\mathrm{d}x + P(x)y = Q(x)$ where P and Q are functions of x which can be solved by using the integrating factor (↑) $e^{\int P\mathrm{d}x}$.

Bernoulli's equation a differential equation (p. 159) of the type $\mathrm{d}y/\mathrm{d}x + P(x)y = Q(x)y^n$ where P and Q are functions of x, which may be solved by dividing by y^n and putting $z = y^{1-n}$, when it becomes a linear differential equation (↑).

Clairaut's equation a first order differential equation (p. 159) of the type $y = px + f(p)$ where $p = \mathrm{d}y/\mathrm{d}x$, whose general solution (p. 159) may be found by writing an arbitrary constant (*see* constant of integration, p. 152) A for p.

second order linear differential equation an important type of differential equation with order 2 (p. 159):

$$a(\mathrm{d}^2y/\mathrm{d}x^2) + b(\mathrm{d}y/\mathrm{d}x) + c = f(x)$$

where a, b and c are constants.

particular integral a particular solution of a differential equation (p. 159), usually of a second order linear differential equation (↑), which may often be found by trying solutions such as $y = pe^x + qe^{-x}$ and finding the values of the constants p and q which satisfy the resulting equation. Also known as **particular solution**.

complementary function the complementary function of the general second order linear differential equation (↑) may be found by putting $f(x) = 0$, obtaining $a(\mathrm{d}^2y/\mathrm{d}x^2) + b(\mathrm{d}y/\mathrm{d}x) + c = 0$. This can often be solved by substituting (p. 69) $y = Ae^{kx}$ and finding the possible values α and β of the constant k, giving the complementary function $y = Ae^{\alpha x} + Be^{\beta x}$. The solution of the general equation is then given by adding the complementary function to the particular integral (↑), and is usually known in this case as the complete primitive (p. 159). Can also be used with other orders of differential equations.

auxiliary equation the equation $ak^2 + bk + c = 0$ which is obtained when $y = Ae^{kx}$ is substituted in the second order linear differential equation (↑) $a(\mathrm{d}^2y/\mathrm{d}x^2) + b(\mathrm{d}y/\mathrm{d}x) + c = 0$. Can also be used with differential equations of other orders. Also known as **characteristic equation**. *See also* p. 90.

Euler's homogeneous differential equation an equation of the type

$$x^2(\mathrm{d}^2y/\mathrm{d}x^2) + ax(\mathrm{d}y/\mathrm{d}x) + by = f(x),$$

which may be solved by putting $x = e^t$.

first integral the first integral of a second order differential equation (p. 159) is a part solution which is a first order differential equation.

integral curves the set of curves obtained when the family (p. 190) of solutions of the differential equation (p. 159) $\mathrm{d}y/\mathrm{d}x = f(x, y)$ is drawn on a graph for different values of the arbitrary constant (p. 152). Also known as **contour map**.

integral curves
for $2xy\frac{\mathrm{d}y}{\mathrm{d}x} = y^2 - x^2$ which are $x^2 - 2ax + y^2 = 0$ for different values of a

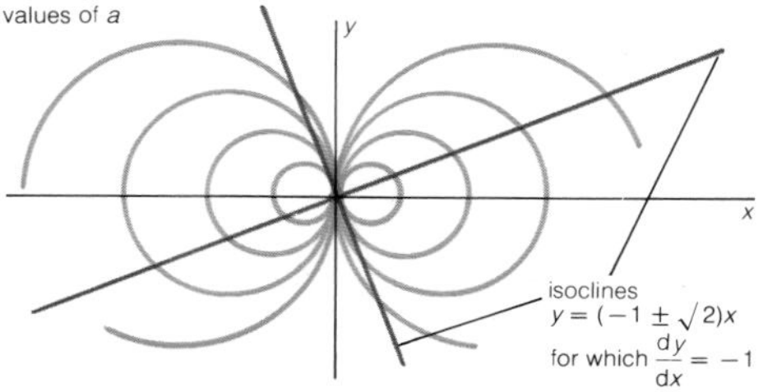

isoclines $y = (-1 \pm \sqrt{2})x$ for which $\frac{\mathrm{d}y}{\mathrm{d}x} = -1$

isocline (*n*) a curve for which dy/dx is constant when $dy/dx = f(x, y)$. It is the locus (p. 187) of points of equal gradient (p. 219) of a set of integral curves (p. 161). **isoclinal** (*adj*).

differential analyser a machine used for obtaining numerical solutions to differential equations (p. 159). Also known as **integrator**.

Legendre polynomials the set of polynomials which are solutions of

$$(1 - x^2)(d^2y/dx^2) - 2x(dy/dx) + n(n + 1)y = 0$$

where n is a non-negative integer. *See also* Rodrigue's formula (↓).

Rodrigue's formula a general formula:

$$P_n(x) = \{1/(2^n n!)\}\{d^n(x^2 - 1)^n/dx^n\}$$

for the Legendre polynomials (↑).

Bessel functions the functions $J_n(x)$ which are the solutions of the second order differential equations (p. 159) $x^2y'' + xy' + (x^2 - n^2)y = 0$, where n is a natural number (p. 21). Lists of numerical solutions may be obtained.

Laplace transform if $f(t)$ is an integrable (p. 152) function, its Laplace transform is $\int_0^\infty e^{-pt}f(t)\,dt$, where $p > n$, and n is a positive integer depending on $f(t)$. It may be used in solving differential equations (p. 159), and lists of transforms (p. 55) of functions may be obtained.

predictor-corrector method a method of obtaining a numerical solution by obtaining a first approximate value and then using a formula to obtain a better approximation. It is used in the numerical solution of differential equations (p. 159), and also in statistics where a sample statistic (p. 110) has its bias (p. 105) corrected to estimate (p. 109) a population parameter (p. 110).

Runge-Kutta method a method of obtaining numerical solutions to differential equations (p. 159). The Runge-Kutta method is an iterative (p. 44) method which avoids the need to find higher order derivatives (p. 143) and is useful in computing.

partial derivation if $y = f(x_1, x_2, \ldots)$, then partial derivation is the operation (p. 14) in which y is derived (p. 143) with respect to one of the variables $x_1, x_2, \ldots$, the remaining variables being held constant, provided that the derivative exists. Often known less correctly as **partial differentiation**. *See also* derivative (p. 143). **partially derive**, **partial derivative** are better than **partially differentiate**, **partial differential**. Symbol: the partial derivative of y with respect to x is written $\partial y/\partial x$.

total derivative if $z = f(x, y)$ and x and y are both functions of t, then provided the various functions are derivable (p. 143), the total derivative of z with respect to t is $\mathrm{d}z/\mathrm{d}t = (\partial z/\partial x)(\mathrm{d}x/\mathrm{d}t) + (\partial z/\partial y)(\mathrm{d}y/\mathrm{d}t)$. The definition can easily be extended to further variables. Also known as **total differential coefficient**.

total partial derivative if $z = f(x, y)$ and x and y are both functions of u, v then provided the various functions are derivable (p. 143), the total partial derivative of z with respect to u is $\partial z/\partial u = (\partial z/\partial x)(\partial x/\partial u) + (\partial z/\partial y)(\partial y/\partial u)$; similarly for $\partial z/\partial v$, and further variables. Also known as **total partial differential coefficient**.

saddle point a point on a surface shaped like a saddle or seat on a horse, which is a maximum point along one line and a minimum along another. A point on $z = f(x, y)$ for which $\partial z/\partial x = 0$ and $\partial z/\partial y = 0$ is a saddle point if for every circle drawn round the point, no matter how small, there are some values of z which are greater than and some values of z which are less than the value at the saddle point.

saddle point
P on a surface

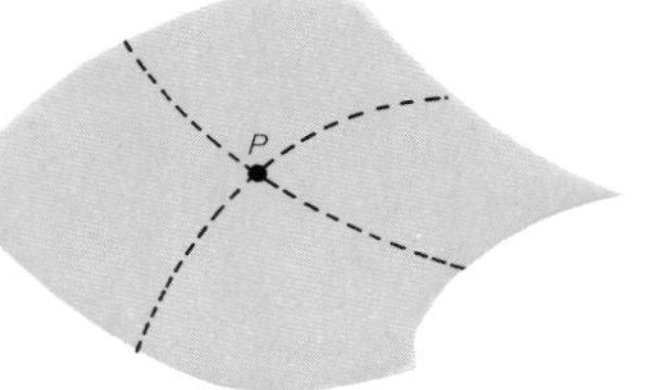

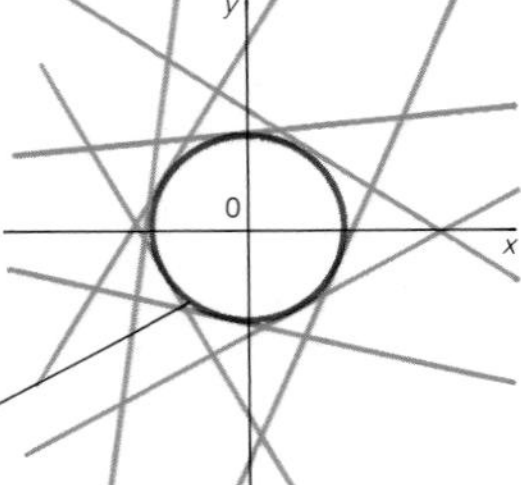

c-discriminant

for $x \cos c + y \sin c + 1 = 0$ (for several values of c) is found by eliminating c with the equation $-x \sin c + y \cos c = 0$ which is its partial derivative with respect to c giving $x^2 + y^2 = 1$

circle $x^2 + y^2 = 1$ which is the envelope of the lines $x \cos c + y \sin c + 1 = 0$

c-discriminant (*n*) the *c*-discriminant of $f(x, y, c) = 0$ is found by eliminating (p. 69) c from $f = 0$ and $\partial f/\partial c = 0$. The result is the envelope (p. 190) of the family (p. 190) of curves $f(x, y, c) = 0$.

chain rule of partial derivation if $f(x, y, z)$ is transformed (p. 55) to $g(u, v, w)$ by $x = x(u, v, w)$, $y = y(u, v, w)$ and $z = z(u, v, w)$, then with certain conditions

$$\frac{\partial g}{\partial u} = \frac{\partial f}{\partial x}\cdot\frac{\partial x}{\partial u} + \frac{\partial f}{\partial y}\cdot\frac{\partial y}{\partial u} + \frac{\partial f}{\partial z}\cdot\frac{\partial z}{\partial u},$$

and similarly for $(\partial g/\partial v)$ and $(\partial g/\partial w)$.

Euler's theorem on homogeneous functions if φ is a homogeneous (p. 61) function of degree n in x, y and z, then $x(\partial\varphi/\partial x) + y(\partial\varphi/\partial y) + z(\partial\varphi/\partial z) = n\varphi$. Similar equations apply for other numbers of variables.

Laplace's equation the partial differential equation (p. 159)

$$(\partial^2 f/\partial x_1^2) + (\partial^2 f/\partial x_2^2) + \ldots + (\partial^2 f/\partial x_n^2) = 0$$

where $f(x_1, x_2, \ldots, x_n)$ is a continuous function in n variables (n is often three).

Lagrange multiplier if k is an extreme value (p. 146) of $f(x, y, \ldots)$, and if there is also a constraint (p. 57) $\varphi(x, y, \ldots) = 0$, then

$$\partial f/\partial x + \lambda(\partial\varphi/\partial x) = 0,\ \partial f/\partial y + \lambda(\partial\varphi/\partial y) = 0, \ldots$$

at the extreme value. The constant λ is the Lagrange multiplier.

directional derivative if a line has direction ratios (*see* direction cosines, p. 80) α, β and γ respectively with the three axes, the directional derivative of $f(x, y, z)$ along it is

$$(\partial f/\partial x)\cos\alpha + (\partial f/\partial y)\cos\beta + (\partial f/\partial z)\cos\gamma.$$

del (*n*) the operator (p. 14) ∇ defined by:

$$\nabla = \mathbf{i}\frac{\partial}{\partial x} + \mathbf{j}\frac{\partial}{\partial y} + \mathbf{k}\frac{\partial}{\partial z},$$

where **i**, **j**, **k** are the unit vectors (p. 79) in three dimensions. *Do not mistake for* the left difference operator (p. 158). Also known as **nabla**, **vector operator**, **differential operator**, **Hamilton-Tait operator**.

grad (*n*) the greatest rate of change of a scalar (p. 77) *S* at a point in a scalar field (p. 274) given by

$$\nabla S = \mathbf{i}\frac{\partial S}{\partial x} + \mathbf{j}\frac{\partial S}{\partial y} + \mathbf{k}\frac{\partial S}{\partial z},$$

where ∇ is the vector operator del (↑), and **i**, **j** and **k** are the unit vectors (p. 79) in three dimensions. This vector is normal (p. 188) to the equipotential (p. 274) surface at any point. Also known as **gradient of a field**, **potential gradient**.

div (*n*) if **V** is a vector field (p. 274), then div **V** is

$$\nabla . \mathbf{V} = \frac{\partial V_x}{\partial x} + \frac{\partial V_y}{\partial y} + \frac{\partial V_z}{\partial z},$$ where ∇ is the vector operator del (↑), the dot (**.**) shows the scalar product (p. 78) and V_x, V_y, V_z are the components of **V** in the directions of the axes *Ox, Oy, Oz*, respectively. Also known as **divergence of a field**.

curl (*n*) if **V** is a vector field (p. 274) as defined under div (↑), then curl **V** is the vector

$$\nabla \times \mathbf{V} = \mathbf{i}\{(\partial V_z/\partial y) - (\partial V_y/\partial z)\} + \mathbf{j}\{(\partial V_x/\partial z) - (\partial V_z/\partial x)\} + \mathbf{k}\{(\partial V_y/\partial x) - (\partial Vx/\partial y)\}$$

where ∇ is the vector operator del (↑), **i**, **j** and **k** are the unit vectors (p. 79) in three dimensions and × shows the vector product (p. 82). Also known as **rot**, **rotation of a field**.

Laplacian operator the operator (p. 14) ∇^2 (read as del, ↑, squared or nabla squared) which is del used twice. In the usual three variables, Laplace's equation (↑) may be written:

$$\nabla^2 f = \frac{\partial^2 f}{\partial x^2} + \frac{\partial^2 f}{\partial y^2} + \frac{\partial^2 f}{\partial z^2} = 0.$$

Also known as **del squared**, **nabla squared**, **Laplace's operator**.

circulation (*n*) the quantity $\int \mathbf{v} \cdot d\mathbf{r}$, the scalar product (p. 78) of the tangential (p. 187) velocity (p. 258) and the perimeter (p. 191) $d\mathbf{r}$ for each element of a fluid (p. 280) flowing round a closed contour (p. 208).

Green's theorem a theorem relating the triple integral of a function over a solid with the double integral over its surface. It can also be applied in two dimensions to the double integral over a surface and the line integral (p. 155) around its edge.

Stoke's theorem a theorem obtained from Green's theorem (↑) which shows how to calculate the circulation(↑).

Kronecker delta the symbol δ_{ij} which has the value zero except when $i = j$ when it has the value unity, e.g. the elements of the unit matrix (p. 83) of order n can be written as δ_{ij} where i and j are the positive integers from 1 to n.

substitution operator the Kronecker delta (↑) symbol δ_{ij} used to substitute (p. 69) one index (p. 35) for another, since, e.g.

$$\Sigma\delta_{ij}a_j = \Sigma_{j=1}^{n}\delta_{ij}a_j = a_i.$$

step function a function whose graph has horizontal (p. 174) steps with vertical (p. 174) jumps between, and is thus discontinuous (p. 64), e.g. the greatest integer function (↓).

Heaviside step function the step function (↑) $H(t)$ which takes the value 1 for $t > 0$ and the value 0 for $t < 0$. The value for $t = 0$ is undefined.

Dirac delta function the function $\delta(u - v)$ which is infinite when $u = v$, but is otherwise zero. It is related to the derivative (p. 143) of the Heaviside step function (↑).

greatest integer function the step function (↑) $[\![x]\!]$ or $[x]$ which for any real x is the greatest integer less than x. The characteristic (p. 139) of a logarithm (p. 139) is an example of the use of this function.

signum *x* the step function (↑) $[x]$ which is -1 when $x < 0$, 0 when $x = 0$ and 1 when $x > 0$. **sgn *x*** (*abbr*).

Dirichlet's function the function $\psi(x)$ which has the value 1 if x is rational (p. 26) and 0 if x is irrational (p. 46).

Heaviside step function
undefined at $x = 0$

greatest integer function

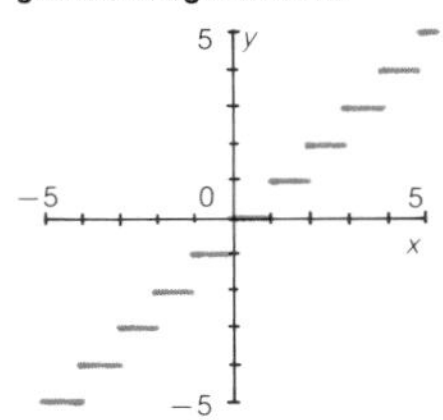

signum *x*
y = signum x

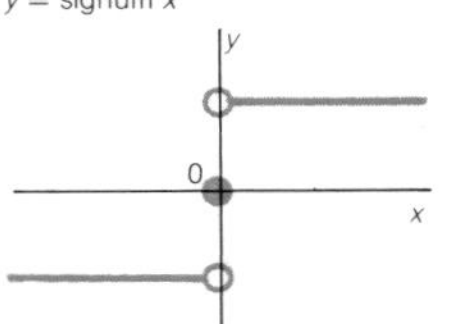

geometry (*n*) the part of the study of mathematics which looks at points, lines, surfaces and solids and their properties. **geometric** (*adj*).

point (*n*) an intuitive (p. 243) concept, a position or place in space having no size; an ordered number pair, triplet (p. 19) or set of *n* coordinates where *n* is the dimension of the space. *See also* node (p. 239).

line (*n*) an intuitive (p. 243) concept which can be thought of as a dense (p. 65) set of points having no thickness, so that each point has exactly two points next to it. Often loosely used for straight line (↓) and for line segment (↓). **linear** (*adj*).

straight line a set of points satisfying an equation such as $lx + my + n = 0$; a line drawn by a ruler (p. 185) and imagined to continue for ever in both directions. Often called simply *line* (↑). *See also* curve (p. 187).

rectilinear (*adj*) relating to a straight line (↑). Also known as **straight**.

ray

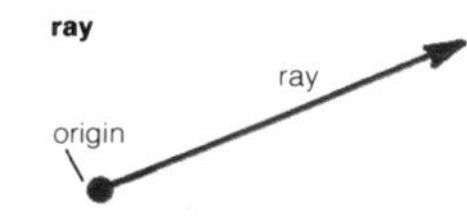

ray (*n*) a half-line; a straight line (↑) which starts at a point and goes to infinity in one direction only. Also known as **part-line**.

origin of a ray that point from which the ray (↑) starts. *See also* vertex (p. 174), p. 215.

line segment a part of a straight line (↑) between two fixed points on the line, often called simply a *line*. *See also* interval (p. 65).

produce (*v*) to make longer, extend; used when a line segment (↑) is to be drawn longer to form a longer line segment, ray (↑) or straight line (↑).

pencil of conics

three members of the pencil of conics through four points

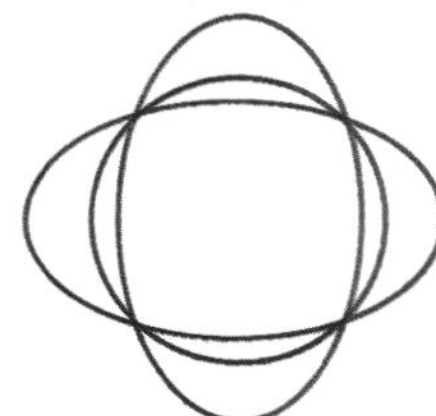

range[3] (*n*) a set of points, all of which lie on a given straight line (↑) or curve. *See also* pp. 17, 113, 268.

base[2] (*n*) the straight line (↑) or curve on which a range (↑) lies. *See also* pp. 37, 179.

pencil (*n*) a set of lines in a plane all of which pass through a given point; a set of curves in a plane passing through given points, e.g. the set of conics (p. 220) passing through four points; a set of planes all having a common line. *See also* sheaf (p. 176).

surface (*n*) an intuitive (p. 243) concept, a set of points in three dimensions whose coordinates (x, y, z) satisfy an equation of the form $f(x, y, z) = 0$; the set of points on all or part of the boundary of a solid.

solid (*n*) an intuitive (p. 243) concept, a set of points in space whose boundaries are defined by parts of surfaces, e.g. cube, cylinder.

plane (*n*) a surface on which straight lines (p. 167) may be drawn between any two points belonging to it, and which is defined by any three points which are not on a straight line. In three dimensions its equation is $lx + my + nz = c$. **planar** (*adj*).

plane curve a curve which lies in a given plane.

region[1] (*n*) a subset (p. 12) of the points in a plane or in space, any two of which can be joined by a line (not necessarily straight) which is completely inside the region. *See also* p.239.

configuration (*n*) any set of points, lines, planes or other objects which is being studied in a given geometrical problem.

figure (*n*) (1) a configuration, a picture of what is being studied, a diagram; (2) a closed shape in a plane (such as a polygon) or in space (such as tetrahedron, p. 201). *See also* digit (p. 21).

sketch (*n*) a rough figure (↑); one meant to show only a general idea of what is being studied.

plan (*n*) a carefully drawn figure (↑), especially one seen from above, looking down.

elevation (*n*) a carefully drawn figure (↑), showing a view from the front, sides or back.

directed (*adj*) used of lines which have a direction usually shown by an arrow. A directed line segment (p. 167) represents a vector in geometrical form.

oriented (*adj*) used of plane and other surfaces where the two sides are to be distinguished (p. 288). The positive direction of the plane of (x, y) coordinates is oriented upwards from the paper so that a right-hand screw turned in that direction gives anticlockwise (p. 174) turning from Ox to Oy as positive. A spherical shell (p. 204) with inside and outside is an example of an oriented surface. **orientable** (*adj*).

plan
scale drawing of a simple house; the red lines show the construction

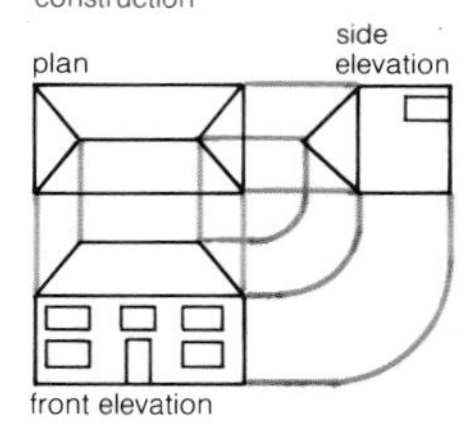

directed
directed line segment

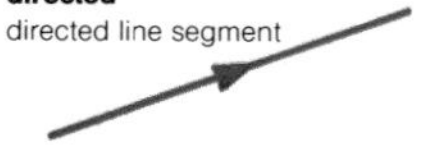

oriented
oriented plane showing positive direction at right angles to the plane, so that a right-hand screw turns from Ox to Oy

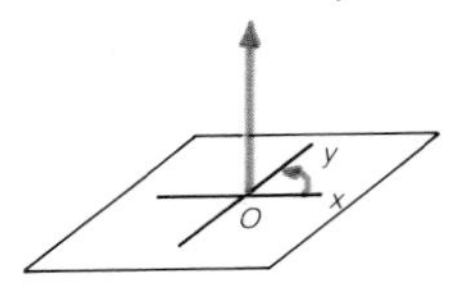

non-orientable (*adj*) used of surfaces such as the Möbius strip (p. 242) and the Klein bottle (p. 242) where the inside and outside cannot be distinguished (p. 288).

distance (*n*) an intuitive (p. 243) concept obtained from the idea of measuring how far apart two points are. The distance between two elements P and Q of a set is a function $D(P, Q)$ of the two elements for which:

I $D(P, Q)$ is positive;
II $D(P, Q) = D(Q, P)$;
III $D(P, Q) = 0$ if and only if $P = Q$;
IV $D(P, R) \leqslant D(P, Q) + D(Q, R)$.

The last condition is known as the triangle inequality (p. 177). In Euclidean geometry (p. 171) the shortest distance is intuitively a straight line (p. 167). *See also* metric (↓).

metric
in Euclidean geometry

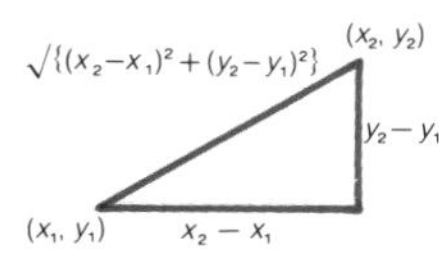

metric (*n*) the defining form of the distance (↑) function in a geometry. In Euclidean geometry (p. 171) the metric is given in two dimensions by $\sqrt{\{(x_2 - x_1)^2 + (y_2 - y_1)^2\}}$ where (x_1, y_1) and (x_2, y_2) are the cartesian (p. 214) coordinates of the two points, and a similar formula applies in three dimensions. This metric is a form of Pythagoras' theorem (p. 180). In some other geometries other metrics are used, e.g. in Manhattan geometry (p. 172). **metrical** (*adj*).

pseudometric (*n*) similar to a metric (↑) except that $D(P, Q) > 0$ need not be true when P and Q are different. *See also* distance (↑).

metric space a set of elements together with a metric (↑).

compact (*adj*) used of metric spaces (↑) having a cluster point (p. 65).

equidistant (*adj*) lying at the same distance (↑).

length (*n*) of a line segment (p. 167): the distance between its two end points; also used roughly for the longest measurement of a plane figure or a solid. For a curved line, the length can be defined as the least upper bound (p. 66) of all possible line segments (p. 167) joining sets of points lying on the curve, provided that such a quantity exists. *See also* dimension (p. 170).

rectifiable (*adj*) used of a curve for which the length is defined. **rectify** (*v*), **rectifiability** (*n*).

measure theory[2] the part of mathematics which deals with the ideas of length, area and volume. The Lebesgue integral (p. 153) is an example of the theory which deals with measurable sets in topology (p. 238).

breadth (*n*) the measurement of a plane figure which is perpendicular to its length. Also known as **width** (*n*). *See also* dimension (↓).

height (*n*) the distance of the top of an object from the bottom; often *vertical height* is used to show the difference from slant height (p. 210).

depth (*n*) the distance of the bottom of an object from the top, or of the bottom of a liquid from the surface.

dimension (*n*) (1) a basic quantity from which other quantities are obtained, e.g. force has the dimensions of mass (p. 259) multiplied by length and divided by the square of time (MLT^{-2}); (2) the number of measures needed to give the place of any point in a given space, the number of coordinates needed to define a point in it. A line has one dimension, a plane has two (sometimes called *length* and *breadth*) and the space in which we live has three (sometimes called *length, breadth* and *height*). A point itself has zero dimension; (3) the measure of the size of an object, the value of its length, breadth and height. *See also* p. 77.
dimensional (*adj*).

***n*-space** (*n*) a space of *n* dimensions.

multidimensional (*adj*) having many (usually more than three) dimensions.

hyperspace (*n*) an imaginary space having more than three dimensions.

hyperplane (*n*) a general point $(x, y, z, \ldots, t)$ in *n*-space (↑) given by the equation $a_1x + a_2y + a_3z + \ldots + a_nt = b$. If $n = 3$ this gives an ordinary plane.

bounded region a region in *n*-space (↑) which can be contained in a finite hypersphere (p. 203) of that space, e.g. part of a plane which can be contained by a finite circle. Also known as **closed region**.

unbounded region a region which is not bounded.

breadth

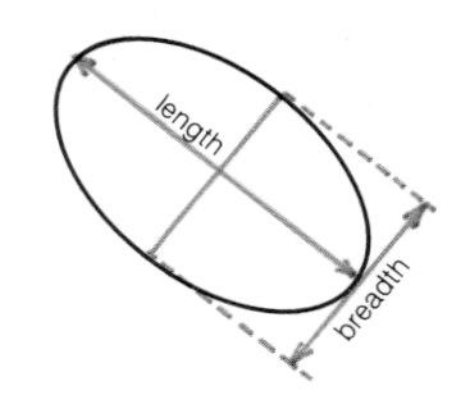

bounded region
a bounded plane region in a finite circle

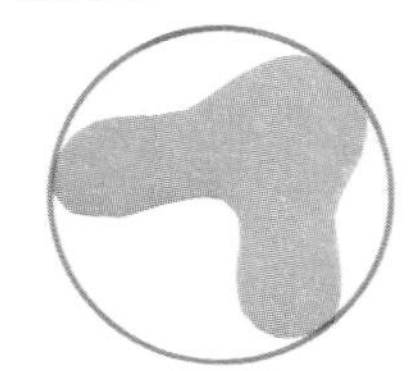

unbounded region
an unbounded region whose boundary curve is a parabola

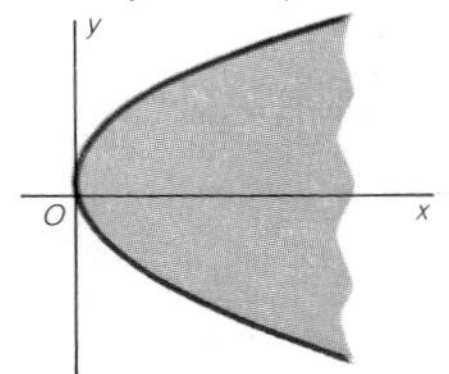

simplex

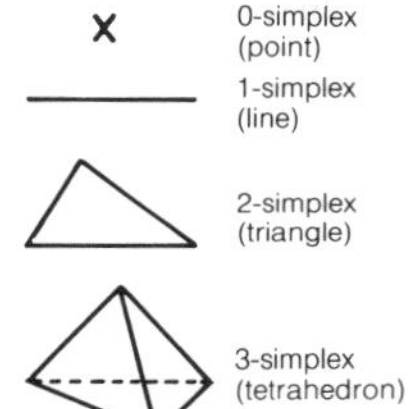

simplex (*n*) a generalization in *n*-space (↑) of the concept that a triangle or 2-simplex is the simplest bounded region (↑) in a plane, and a tetrahedron (p. 201) or 3-simplex is the simplest bounded region in 3-space. A 0-simplex is a point, and a 1-simplex is a line segment (p. 167).

join (*v*) the straight line (p. 167) passing through two points is said to join them.

collinear (*adj*) lying in the same straight line (p. 167). **collinearity** (*n*).

meet (*v*) two configurations passing through the same point are said to meet at that point.

concurrent (*adj*) passing through the same point, often used of three or more curves or lines. **concurrence** (*n*).

coplanar (*adj*) lying in the same plane.

coincident (*adj*) lying at the same place, having the same value. **coincide** (*v*), **coincidence** (*n*).

adjacent and opposite
naming of sides in a right triangle with respect to acute angle θ

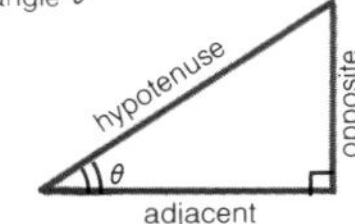

adjacent (*adj*) lying next to, in a right triangle (p. 179) used of the side between a given acute angle (p. 173) and the right angle.

opposite (*adj*) lying opposite to, or at the other end of a diameter to; in a triangle (p. 179) used of the side not lying next to a given angle (p. 173).

bisect (*v*) to cut in two equal parts (used of line segments, p. 167, angles, etc.). Similarly **trisect**, for three parts. **bisector** (*n*), **bisection** (*n*).

mid-point (*n*) that point which bisects (↑) a straight line (p. 167).

incident (*adj*) lying upon, so that a point may be incident upon a line, and a point or line may be incident upon a plane. **incidence** (*n*).

Euclidean geometry the geometry built upon the postulates (p. 251) of Euclid, where distance (p. 169) is defined by Pythagoras' theorem (p. 180), and where Playfair's axiom (↓) is taken to be true. It is usually taken to be the geometry of everyday life. Also known as **parabolic geometry**. Hence **non-Euclidean geometry**.

Playfair's axiom
through point P there is one and only one line parallel to *l*

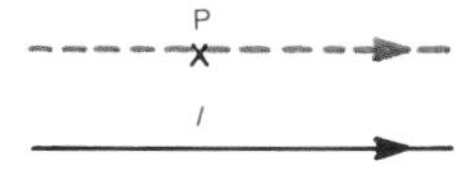

Playfair's axiom the axiom (p. 250) or postulate (p. 251) of Euclidean geometry (↑) that through any point not on a given line, one and only one line parallel to the given line may be drawn.

hyperbolic geometry a geometry in which, through any point not on a given line, an infinite number of lines parallel to the given line may be drawn.

elliptic geometry a geometry in which, through any point not on a given line, no line parallel to the given line may be drawn.

plane geometry a geometry (usually Euclidean, p. 171) which deals only with the points in a plane. Also known as **two dimensional geometry**.

synthetic geometry a geometry which uses the methods of proof described by Euclid as opposed to the algebraic (p. 56) methods of cartesian geometry (p. 214).

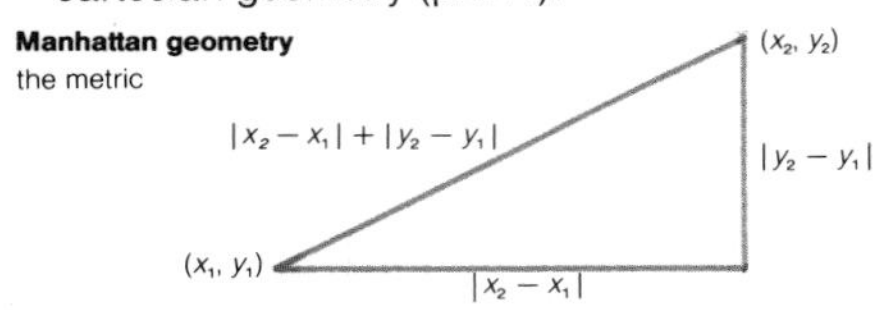

Manhattan geometry
the metric

Manhattan geometry a plane geometry where distances are taken by measuring only along two perpendicular axes (from Manhattan where most of the streets meet at right angles). If the axes are the coordinate axes and the two points are (x_1, y_1) and (x_2, y_2), then the metric (p. 169) is $|x_2 - x_1| + |y_2 - y_1|$. *See also* modulus (p. 49).

metrical geometry a geometry which uses the concept of distance (p. 169), e.g. Euclidean geometry (p. 171), Manhattan geometry (↑).

intrinsic geometry a geometry using only its own space (often a surface) in its definitions and proofs and not any points outside that space.

lattice (*n*) a set of points spaced by some rule in a regular pattern, e.g. of triangles or squares.

gnomon (*n*) an L-shape, a shape added to a square to give a larger square; the corresponding shape for other figures. Also used of the numbers of points added to such figures drawn in a lattice (↑) to produce the next larger figure of the same type in the lattice. *See also* gnomonic (p. 208).

lattice
equilateral triangles

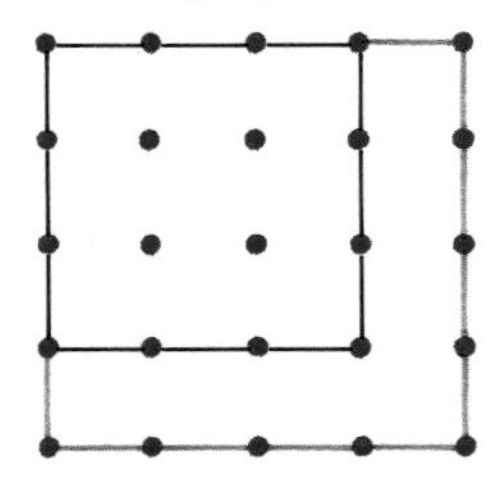

gnomon
an L-shape, the number 9, which when added to 16 gives the next larger square of 25

angle
the positive angle from ray *l* to ray *m*

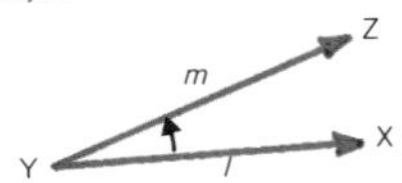

angle (*n*) a measure of change of direction or turning about a point in a plane; or about a line (or axis) in space; the amount of turning between two rays (p. 167) with a common origin. *See also* p. 81, directed angle (p. 174). **angular** (*adj*). Symbols: $\angle XYZ$ or $X\hat{Y}Z$ for the angle between rays *YX* and *YZ*.

degree[2] (*n*) a measure of angle which is one 360th of a complete turn. The word is also used for other units such as those of temperature (p. 284). *See also* pp. 60, 73. Symbol: °.

minute (*n*) (1) a minute of arc is a unit of angle, one sixtieth of a degree. Symbol: ′; (2) a unit of time equal to 60 seconds (↓).

second (*n*) (1) a second of arc is a unit of angle, one sixtieth of a minute (↑). Symbol: ″; (2) the fundamental unit (p. 284) of time in the SI system (*see* Appendix 1, p. 293). s (*abbr*).

gradian (*n*) a measure of angle equal to one hundredth of a right angle or 0.9 of a degree. Also known as **grade**. *See also* p. 107.

circular measure the measurement of angles, especially in radians (p. 133).

equiangular
an equiangular hexagon

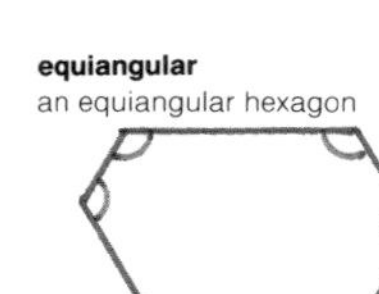

equiangular (*adj*) having equal angles, often used of polygons.

acute angle an angle of less than 90 degrees.

right angle an angle of 90 degrees, a quarter turn.

obtuse angle an angle of between 90 and 180 degrees (an angle of 180° is a straight line, p. 167).

reflex angle an angle of between 180 and 360 degrees.

revolution (*n*) a complete turn, an angle of 360 degrees.

oblique angle an angle which is not a right angle, usually one between 0 and 180 degrees.

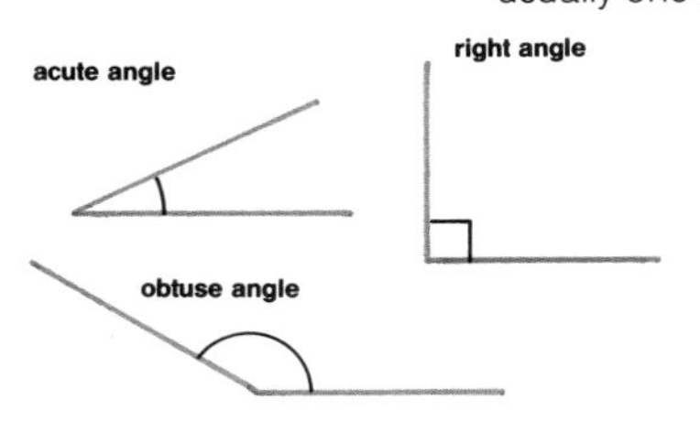

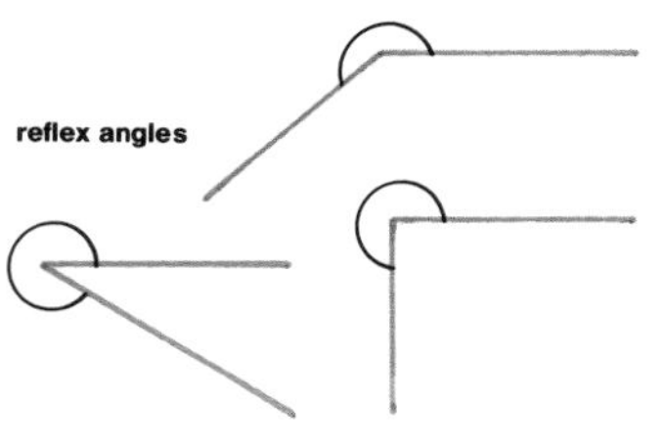

vertex (*n*) (1) a point where two or more lines meet, used especially of the common origin of the pair of rays (p. 167) defining an angle, of pencils (p. 167) of lines and of corners of polygons (*see also* origin of a ray, p. 167); (2) a point where three or more planes meet, used especially of the corners of polyhedra (p. 200); (3) the point on a parabola (p. 220) where the axis cuts the curve; one of the points on the ellipse (p. 220) where the major axis (p. 221) cuts the curve or on the hyperbola (p. 221) where the transverse axis (p. 222) cuts the curve. **vertices** (*pl*), **vertical** (*adj*). *See also* vertical (↓).

arm (*n*) the arm of an angle is one of the two rays (p. 167) with origin at the vertex which define the angle.

clockwise (*adj, adv*) the direction of turning in the plane as of the hands of a clock. *See also* anticlockwise (↓). Symbol: ↻

anticlockwise (*adj, adv*) the opposite direction to clockwise (↑), taken as the positive direction of turning in mathematics. Also known as **counter-clockwise**. Symbol: ↺

sense (*n*) one of the two directions along a line, or of ways of turning of an angle. The positive sense of turning is taken to be anticlockwise (↑). *Do not mistake for* bearing (p. 205).

directed angle an angle together with a positive or negative symbol showing its sense (↑).

horizon (*n*) the circle formed by the edge of the earth's surface seen from a point above the earth (*but note the meaning of* horizontal, ↓).

horizontal (*adj*) in the plane which touches the earth's surface at the given point; at a fixed level; across the paper in a drawing. Also known as **level**.

vertical (*adj*) at right angles to the horizontal; up and down the paper in a drawing. *See also* vertex (↑).

perpendicular (*adj*) at right angles (of two lines, a line and a plane, two planes, etc). Not used in mathematics to mean vertical. Symbol: ⊥ for 'is perpendicular to'.

clockwise and anticlockwise
directions of turning in a plane

perpendicular
two perpendicular lines

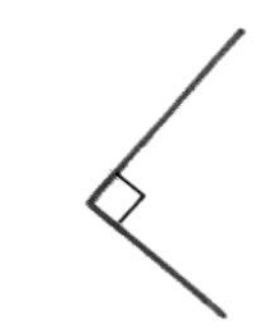

a perpendicular line to a plane

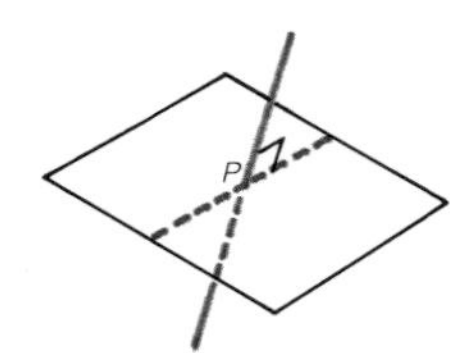

a perpendicular *p* raised from a point *P* on a line

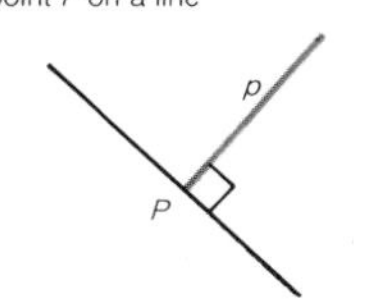

a perpendicular *p* dropped from a point *P* to meet a line *l* at its foot *N*

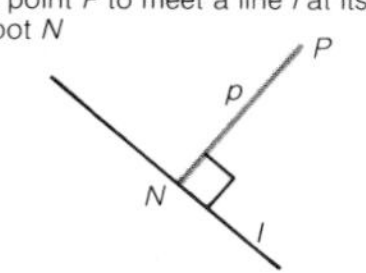

orthogonal
curves with perpendicular tangents at their intersection

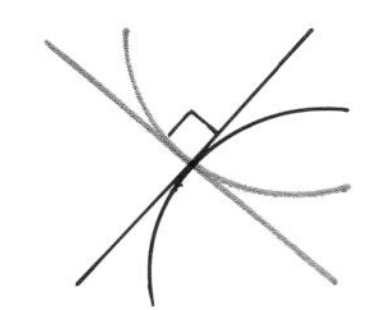

subtend
line segment AB subtends angle APB at point P

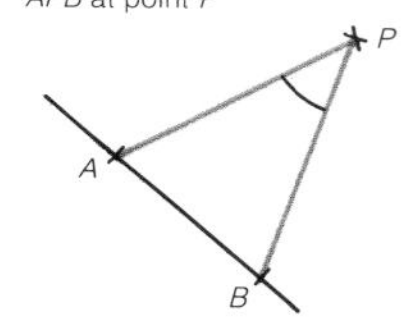

angular distance
for two stars S_1 and S_2 this is the angle $S_1 ES_2$

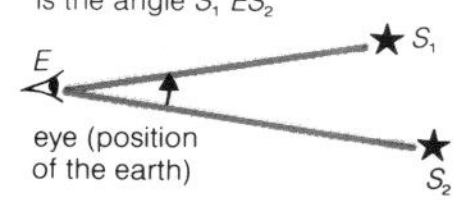

angle of elevation
angle of depression
for a person at point E

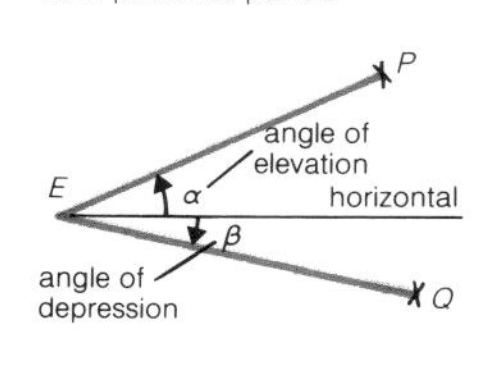

complementary angles
$\theta + \varphi = 90°$

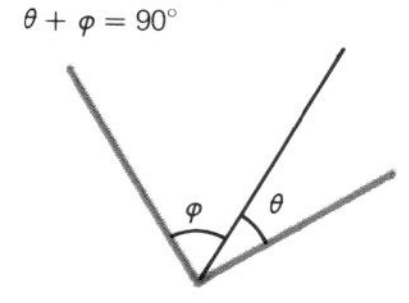

raise[2] (*v*) (of a perpendicular) a straight line drawn perpendicular to a given line or curve through a given point on that line or curve is said to be raised from that point on the line or curve. *See also* p. 36.

drop (*v*) (of a perpendicular) a straight line drawn perpendicular to a given line or curve through a given point not on that line or curve is said to be dropped to the line or curve from that point. *See also* normal (p. 188).

foot (*n*) (of a perpendicular) that point to which a perpendicular is dropped (↑), or to which a normal (p. 188) to a curve is drawn. **feet** (*pl*).

orthogonal (*adj*) at right angles, used of lines and also of curves, where the tangents at the point of intersection form the arms (↑) of the angle.

subtend (*v*) if AB is a line segment (p. 167) and P is a point not on it, then AB is said to subtend an angle APB at the point P.

angular distance the angle between two points subtended (↑) at the person viewing the two points by the line segment (p. 167) joining them.

angular diameter the angle subtended (↑) by an object at a given point (used in astronomy, p. 287).

angle of elevation the angle measured in a vertical plane above the horizontal. In astronomy (p. 287) known as the **altitude**.

angle of depression the angle measured in a vertical plane below the horizontal.

complementary angles two angles whose sum is 90°.

supplementary angles two angles whose sum is 180°.

conjugate angles two angles whose sum is 360°.

adjacent angles two angles with a common vertex and a common arm (↑).

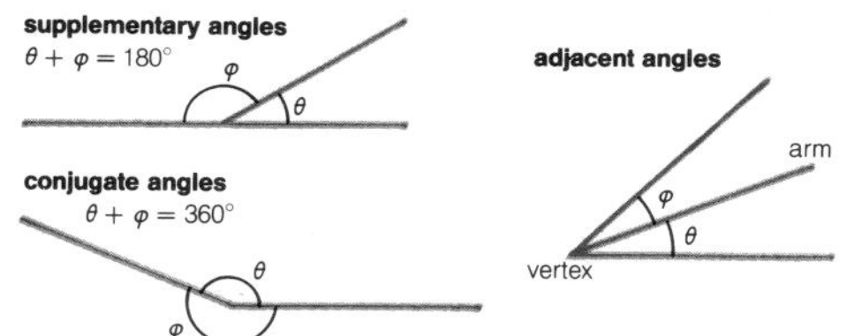

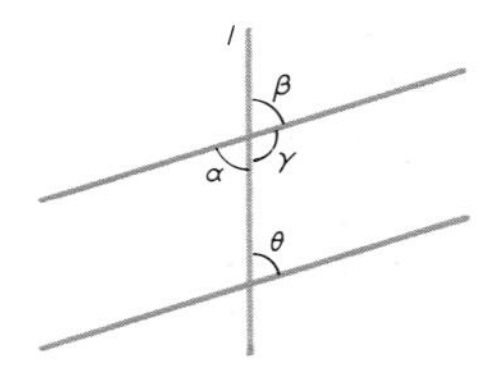

transversal
a pair of parallel lines and a transversal *l* with various types of angle pairs

$\alpha = \beta$: vertically opposite angles

$\alpha = \theta$: alternate angles

$\beta = \theta$: corresponding angles

$\gamma + \theta = 180^\circ$: interior angles

parallel (*adj*) used of lines or planes in Euclidean geometry (p. 171) which either have no point in common or coincide (p. 171) completely. Symbols: arrows are drawn on parallel lines, || for 'is parallel to'.

antiparallel (*adj*) parallel but having the opposite sense (p. 174) or direction.

transversal (*n*) a line which intersects a set of other lines or planes, often parallel lines or planes.

vertically opposite angles when two lines intersect, the pairs of angles which are opposite one another and which are equal.

alternate angles when a transversal (↑) crosses two parallel lines, the pairs of angles on either side of the transversal which are equal. Also known as **Z-angles**.

corresponding angles when a transversal (↑) crosses two parallel lines, the pairs of angles on the same side of the transversal which are equal. Also known as **F-angles**. *See also* corresponding (p. 17).

interior angles when a transversal (↑) crosses two parallel lines, the pairs of angles on the same side of the transversal and between the parallel lines which are supplementary (p. 175). Also known as **allied angles**, **conjoined angles**. *See also* interior angle (↓).

sheaf (*n*) variously a set of lines or planes in three dimensions which are all parallel or all pass through a given point, or a set of planes containing a given line.

star (*n*) of planes: a set of planes which all pass through a given point.

paraplanar (*adj*) used of a set of lines in three dimensions which are all parallel to a given plane.

polygon

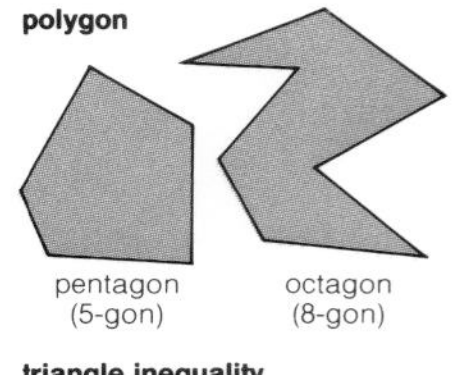

triangle inequality
$AB + BC > AC$

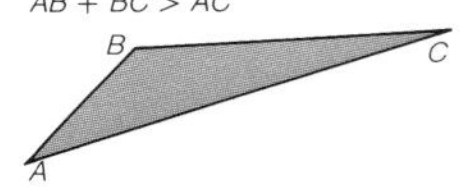

diagonal
a hexagon has 9 diagonals

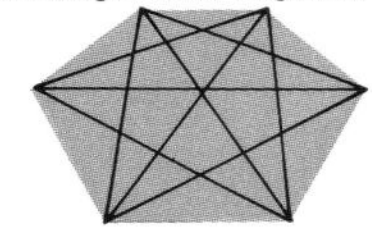

equilateral
an equilateral pentagon

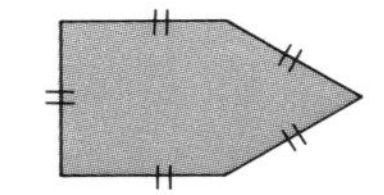

regular polygon
a regular heptagon

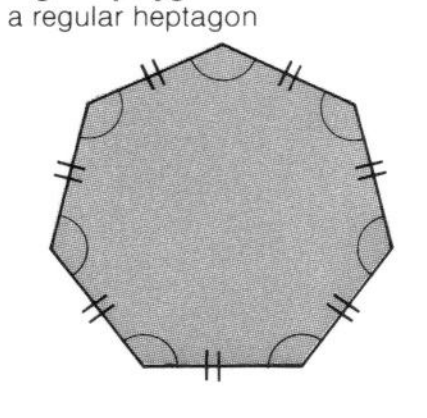

interior and exterior angles

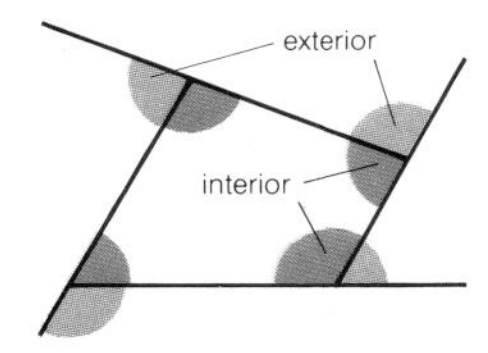

polygon (*n*) a configuration of straight lines (p. 167) meeting in vertices, usually in a plane. Usually taken to be the line segments (p. 167) between adjacent (p. 171) vertices and forming a closed figure so that no two sides cross (*but see also* star polygon, p. 178). **polygonal** (*adj*).

side (*n*) any straight line or line segment (p. 167) joining two adjacent (p. 171) vertices of a polygon. Also known as **edge**. *See also* p. 201.

triangle (*n*) a polygon having three sides. Also known as **trilateral**. **triangular** (*adj*). Symbol: $\triangle$ (also used for area of a triangle, p. 135).

triangle inequality[2] the sum of the lengths of any two sides of a triangle is greater than the length of the third side. *See also* p. 79, distance (p. 169).

quadrilateral (*n*) a polygon having four sides. Also known as **quadrangle**.

***n*-gon** (*n*) a polygon having *n* sides and *n* vertices, thus a triangle is a 3-gon.

pentagon (*n*) a polygon with 5 sides. This name is formed using the Greek prefix (*see* Appendix 2, p. 294). Similarly **hexagon**, **heptagon**, **octagon**, **enneagon**, **decagon**, **hendecagon**, **dodecagon**. Also common but less correct are **septagon**, **nonagon**, **endecagon** for 7-gon, 9-gon, 11-gon (*see n*-gon, ↑) respectively. **trigon**, **tetragon** for triangle, quadrilateral respectively are uncommon. **pentagonal** (*adj*).

diagonal (*n*) a line joining two non-adjacent (p. 171) vertices of a polygon; not horizontal or vertical. **diagonal** (*adj*).

equilateral (*adj*) having equal sides, especially used of triangles and other polygons.

regular polygon a polygon which is both equilateral (↑) and equiangular (p. 173).

interior angle an angle inside a polygon between two adjacent (p. 171) sides. The sum of all interior angles of an *n*-gon (↑) is $2n - 4$ right angles. *See also* interior angles (↑).

exterior angle an angle between one side of a convex polygon and the adjacent (p. 171) side produced (p. 167). The sum of all the exterior angles of a convex (p. 188) *n*-gon is four right angles.

convex polygon a polygon in which no interior angle (p. 177) is greater than two right angles.

re-entrant polygon a polygon which has at least one interior angle (p. 177) which is reflex (p. 173) (i.e. greater than two right angles). Also known as **concave polygon**.

re-entrant polygon

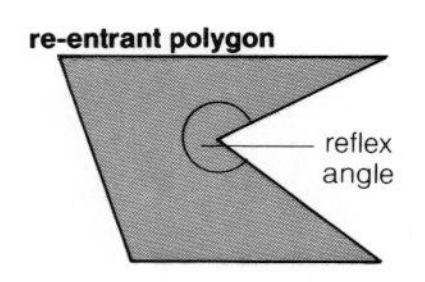

star polygon if the n vertices of a polygon are joined each to the rth vertex around the polygon in the same direction ($1 < r < n - 1$), a star polygon is formed which is denoted by $\{n/r\}$, provided that n and r are relatively prime (p. 32). Star polygons are often assumed (p. 251) to be regular polygons.

star polygon
regular star heptagons $\{\frac{7}{2}\}$ and $\{\frac{7}{3}\}$

***n*-gram** (*n*) a star polygon (↑) with n sides.

pentagram (*n*) the star polygon (↑) $\{5/2\}$. Also known as **pentacle**.

pentagram

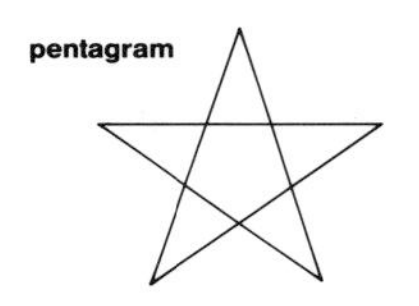

polyomino (*n*) a polygon whose shape is that of a number of squares joined edge to edge. Similarly **domino**, **triomino**, **tetromino**, **pentomino**, **hexomino**, etc (2, 3, 4, 5, 6, ... squares) are formed using the Greek prefixes (*see* Appendix 2, p. 294).

polyomino

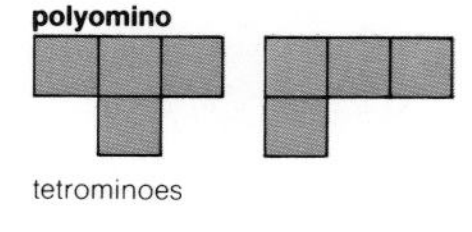

tetrominoes

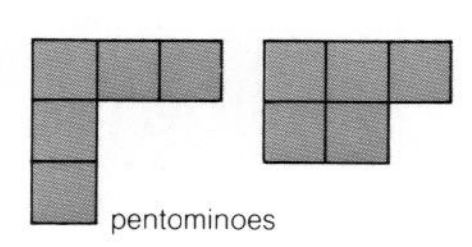

pentominoes

dissect (*v*) to cut into pieces. Used of solids as well as polygons and other plane figures, e.g. the tangram (↓). **dissection** (*n*).

tangram

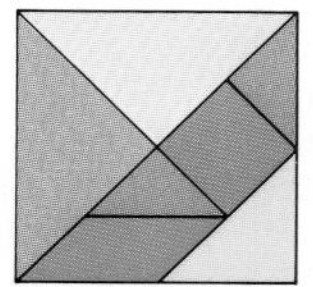

the dissection of a square to form a tangram

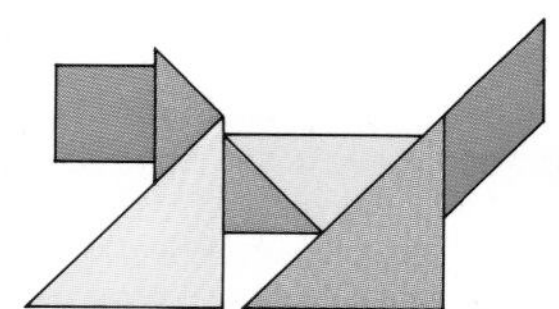

a tangram of a dog formed from the seven pieces of the square

tangram (*n*) a dissection (↑) of a square into seven parts. The dissected parts are used to make pictures, an old Chinese game.

isosceles (*adj*) having two sides equal in length. Usually used of triangles (which then also have two equal angles) and also of trapezia (p. 181) with the non-parallel sides equal.

isosceles

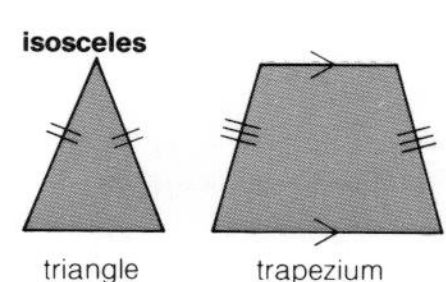

triangle trapezium

scalene (*adj*) used of triangles having three sides of which no two are equal.

scalene
a triangle with no two sides equal

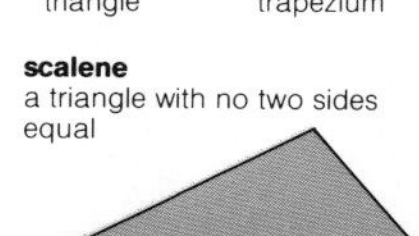

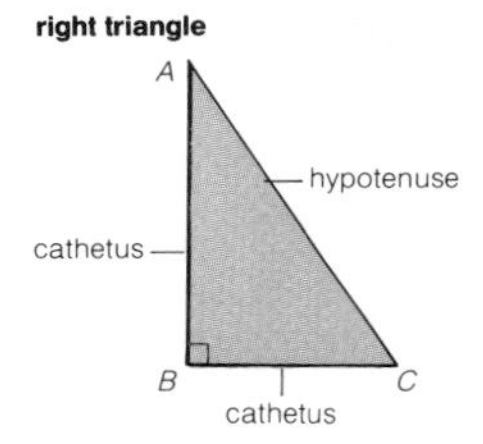

right triangle a triangle having one angle a right angle. Also known as **right angled triangle**.

hypotenuse (*n*) that side of a right triangle (↑) which is opposite the right angle, and is the longest side.

cathetus (*n*) a line perpendicular to another line; one of the two sides of a right triangle (↑) which is not the hypotenuse (↑). (These are often related to one of the acute angles, p. 173, of the triangle as the adjacent and opposite sides, p. 171, of that angle.) **catheti** (*pl*).

base[3] (*n*) (1) that part of a configuration at the bottom (any part of a configuration which has been turned round can be regarded as the base for the purposes of a proof); (2) anything from which a theory is started or from which something is proved. *See also* pp. 37, 167.

base (*v*) to set as a starting point.

vertical angle that angle of a triangle opposite a given base.

apex (*n*) that vertex of a polyhedron (p. 200) furthest from a given base. **apexes** (*pl*), **apices** (*pl*).

congruent (*adj*) having the same shape and size. Used as one of the main concepts in Euclidean geometry (p. 171). *See also* isometry (p. 233). **congruence** (*n*), **congruency** (*n*). Symbol: ≡ for 'is congruent to'.

congruent

SSS
AB = *XY*, *BC* = *YZ*, *CA* = *ZX*

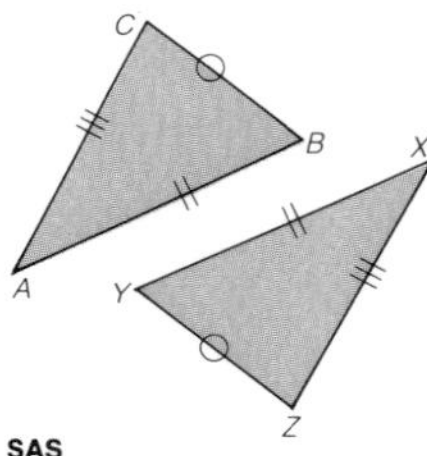

SAS
AB = *XY*,
angle *CAB* = angle *ZXY*,
BC = *YZ*

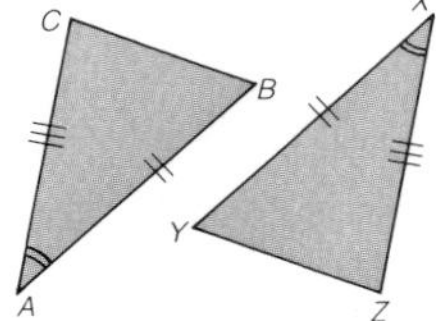

AAS
angle *CAB* = angle *ZXY*,
angle *ABC* = angle *XYZ*,
AB = *XY*

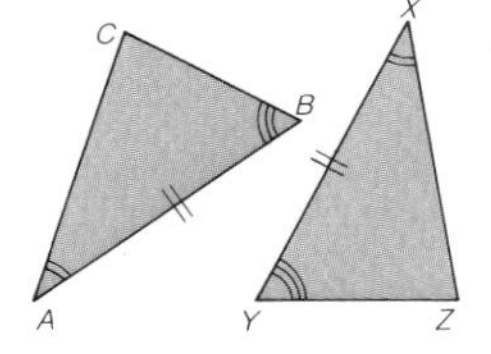

SSS (*abbr*) used of a proof of congruence (↑) of two triangles in which all three sides (SSS) of one triangle are shown to be equal to the respective three sides of the other triangle.

SAS (*abbr*) used of a proof of congruence (↑) of two triangles in which two sides of one are shown to be equal to the respective two sides of the other and the respective angles between those two sides (SAS) are also shown to be equal.

AAS (*abbr*) used of a proof of congruence (↑) of two triangles in which two angles of one triangle are shown to be equal to the respective two angles of the other (and hence the third angles must also be equal), and a corresponding pair of sides (AAS) is also shown to be equal.

RHS congruency
angle ABC = angle $XYZ = 90°$
$AC = XZ$
$AB = XY$

RHS (*abbr*) used of a proof of congruence (p. 179) of two right triangles (p. 179) in which the two sides opposite the right angles are shown to be equal and another pair of sides is also shown to be equal.

direct and indirect congruence

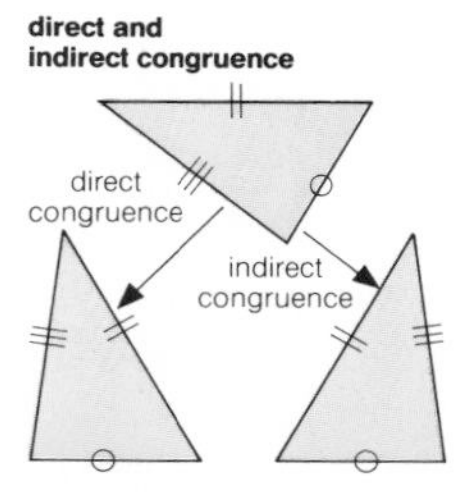

direct congruence congruence (p. 179) in a plane in which one configuration can be moved to the place of the other by translation (p. 233) or rotation without any reflection (p. 234), i.e. the configuration need not be taken out of the plane to make the movement.

indirect congruence congruence (p. 179) in a plane in which one configuration can be moved to the place of the other only if a reflection (p. 234) is used (possibly together with a direct isometry, p. 233), that is the configuration must be taken out of the plane to make the movement.

superposition (*n*) the placing of one configuration on top of another to show congruence (p. 179), used as a method of proof in Euclidean geometry (p. 171). **superpose** (*v*).

Pythagoras' theorem
if A is a right angle, the two red squares have the same total area as the blue one

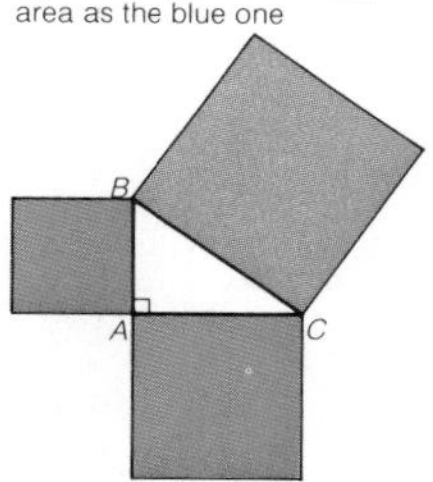

Pythagoras' theorem the area (p. 191) of the square on the hypotenuse (p. 179) of a right triangle (p. 179) is equal to the sum of the areas of the squares on the other two sides. This important theorem gives the metric (p. 169) for Euclidean geometry (p. 171) and also gives the trigonometric (p. 131) relation $\cos^2\theta + \sin^2\theta = 1$ between the sine (p. 131) and the cosine (p. 131) for angles θ such that $0° < \theta < 90°$.

Pythagorean triples
some of the simpler sets

3	4	5	12	16	20
5	12	13	12	35	37
6	8	10	14	48	50
7	24	25	16	30	34
8	15	17	20	21	29
9	40	41	20	48	52
10	24	26	24	32	40
11	60	61	27	36	45

Pythagorean triples sets of three positive integers such that the sum of the squares of the two smallest equals the square of the largest. The integers can then, by Pythagoras' theorem (↑), form the sides of a right triangle (p. 179).

Perigal's dissection
the red squares are congruent, as are the pairs of quadrilaterals

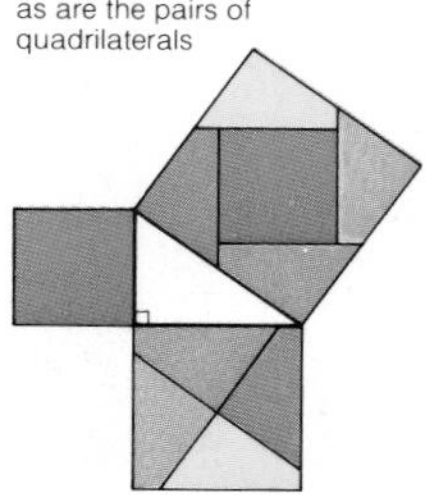

Perigal's dissection a dissection (p. 178) of the squares on the three sides of a right triangle (p. 179) which may be used to show that Pythagoras' theorem (↑) appears to be true.

extensions to Pythagoras' theorem
1 in triangle ABC where angle A is acute
$BC^2 = AB^2 + AC^2 - 2AC.AM$

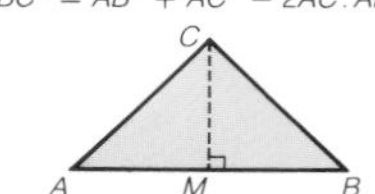

2 in triangle XYZ where angle X is obtuse
$YZ^2 = XY^2 + XZ^2 + 2XZ.XN$

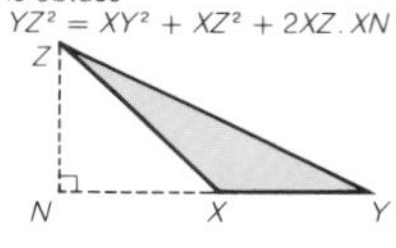

complete quadrilateral

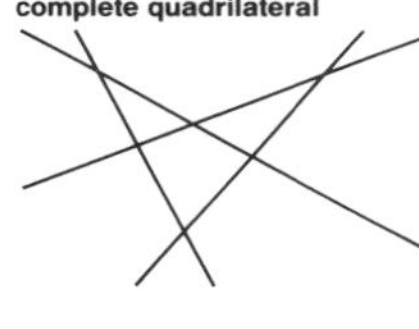

complete quadrangle

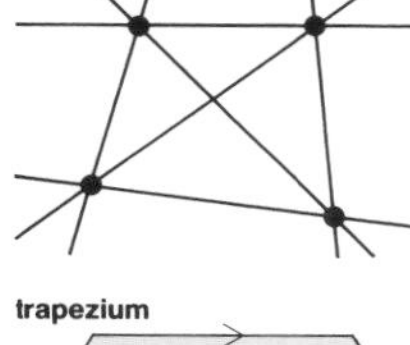

trapezium

parallelogram

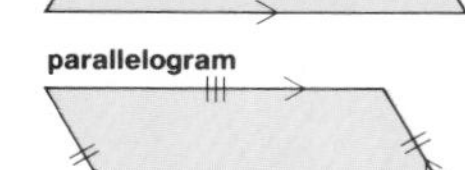

extensions to Pythagoras' theorem a generalization of Pythagoras' theorem (↑) to all triangles, the geometrical form of the cosine formula (p. 135).

complete quadrilateral the quadrilateral (p. 177) regarded as four infinite lines meeting in six separate points. Also known as **four-line**.

complete quadrangle the quadrilateral (p. 177), or better the quadrangle (p. 177), regarded as four points joined by six separate infinite lines. Also known as **four-point**.

trapezium (*n*) a quadrilateral (p. 177) in which one pair of opposite sides is parallel. Also known as **trapezoid**. **trapezia** (*pl*), **trapezoidal** (*adj*).

parallelogram (*n*) a quadrilateral (p. 177) in which both pairs of opposite sides are parallel (and thus also equal).

rhombus (*n*) a parallelogram (↑) in which all four sides are equal in length. The words *diamond, lozenge* are also used in everyday life. **rhombuses** (*pl*), **rhombi** (*pl*), **rhombic** (*adj*), **rhomboidal** (*adj*).

rectangle (*n*) a parallelogram (↑) in which all four angles are right angles. **rectangular** (*adj*).

oblong (*n*) an everyday word for rectangle, not usually used in mathematics. Sometimes used to mean a rectangle which is not a square.

square[2] (*n*) a rectangle in which all four sides are equal in length; a rhombus (↑) in which all four angles are right angles. *See also* p. 34. **square** (*adj*) is also used with units of area.

kite (*n*) a convex (p. 188) quadrilateral (p. 177) in which two pairs of adjacent (p. 171) sides are equal.

arrowhead a re-entrant (p. 178) quadrilateral (p. 177) in which two pairs of adjacent sides are equal and the interior angle between the shorter pair is reflex (p. 173).

rhombus

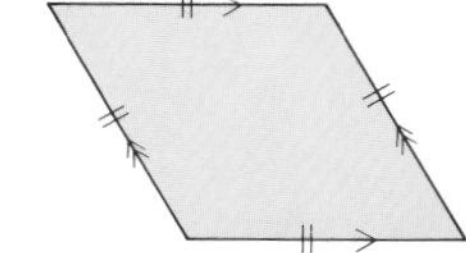

rectangle **square**

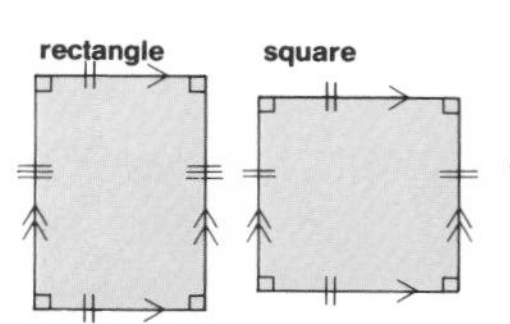

kite

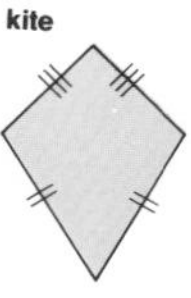

arrowhead

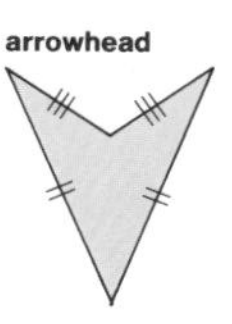

intercept (*n*) part of a line (or plane) cut off by lines (or planes) crossing it, especially the distance from the origin to a point on the axis in cartesian (p. 214) coordinates.

intercept theorem if one transversal (p. 176) divides a set of parallel lines in a plane into equal line segments (p. 167), then all transversals will do so.

internal division an interval on a line has internal division by a point if that point lies on the interval.

external division an interval on a line has external division by a point if that point lies on the line but not on the interval. The ratio of the lengths made by the division is then often taken to be negative, e.g. as in Menelaus' theorem (↓).

golden ratio the ratio $(1 + \sqrt{5}):2 = 1.618\ldots:1$. Lines divided in this way are supposed to give the sides of rectangles pleasing to the eye. If a line *AB* has internal division (↑) at *G* in this ratio then $AG^2 = GB.AB$. Also known as **golden section**.

golden rectangle a rectangle whose sides are in the golden ratio (↑).

Modulor (*n*) a set of numbers obtained from the height of a man using the golden ratio (↑) and used by le Corbusier (the architect or planner of buildings).

Ceva's theorem if three concurrent (p. 171) straight lines (p. 167) are drawn through the vertices *A*, *B* and *C* of a triangle *ABC* to meet the opposite sides at *X*, *Y* and *Z* respectively, then

$$\frac{BX}{XC} \cdot \frac{CY}{YA} \cdot \frac{AZ}{ZB} = +1.$$

Ceva's theorem
$\frac{BX}{XC} \cdot \frac{CY}{YA} \cdot \frac{AZ}{ZB} = +1$

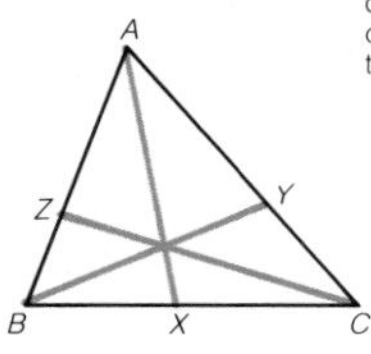

the case when the lines are concurrent outside the triangle

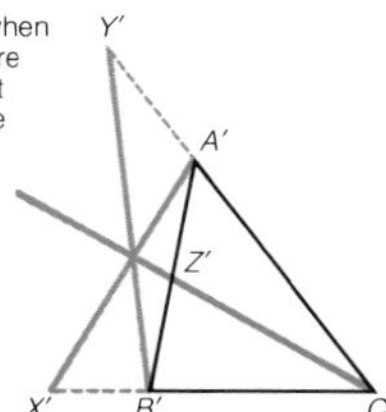

intercept
the intercepts on the *x* and *y* axes

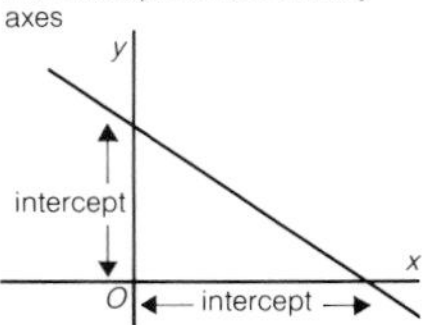

intercept theorem
if *AB* = *BC*, then *XY* = *YZ*

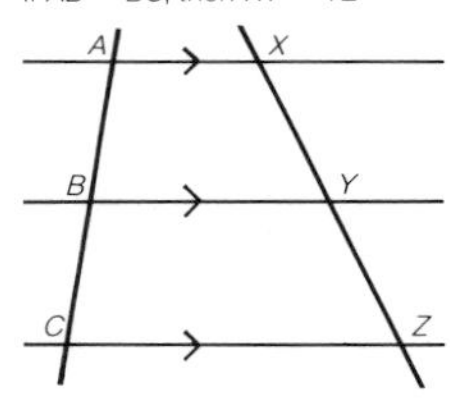

internal division
AB is internally divided at *I* in the golden ratio $(1 + \sqrt{5}):2$

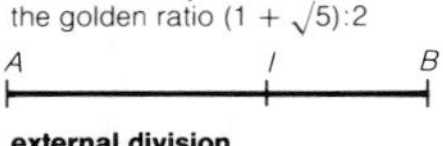

external division
XY is externally divided at *E* in the golden ratio so that $XE:EY = (1 + \sqrt{5}):2$ (ignoring signs)

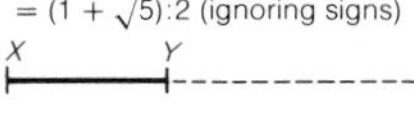

golden rectangle

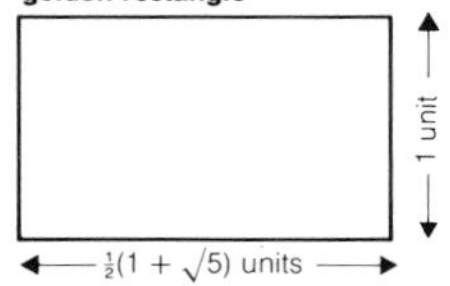

Menelaus' theorem
$\frac{BX}{XC} \cdot \frac{CY}{YA} \cdot \frac{AZ}{ZB} = -1$

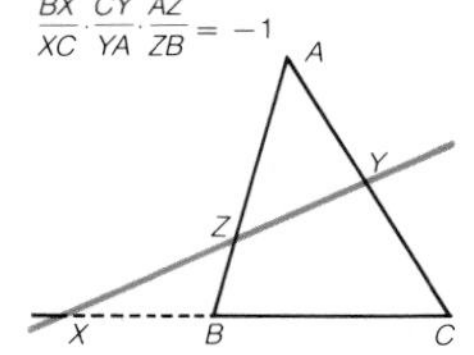

perspective
a rectangular block drawn in artistic perspective

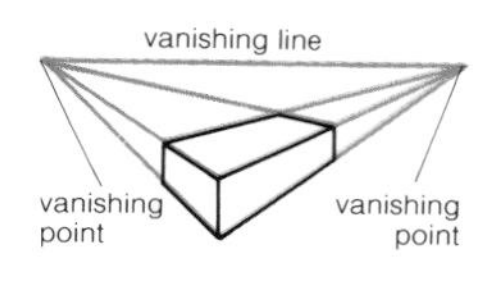

perspectivity
quadrilaterals *ABCD, A′B′C′D′* are in perspective with centre of perspectivity *O*

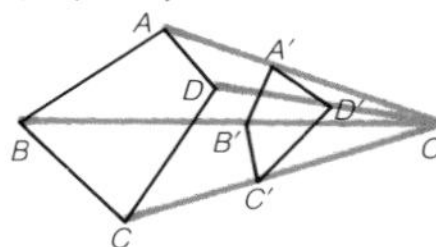

projectivity
the ranges *A,B,C,D* and *A′,B′,C′,D′* form a projectivity

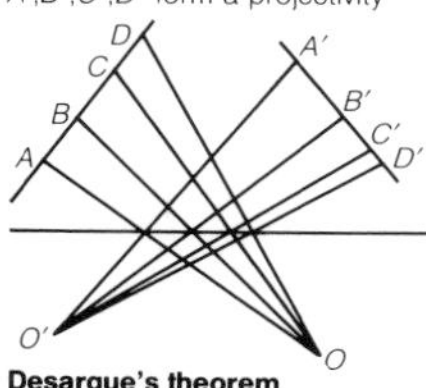

Desargue's theorem
L,M,N lie on the axis of perspective

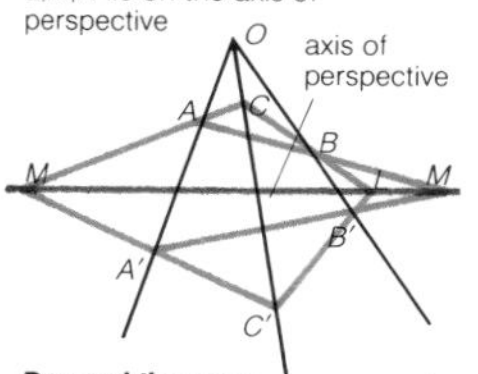

Pappus' theorem
L,M,N lie on Pappus' line

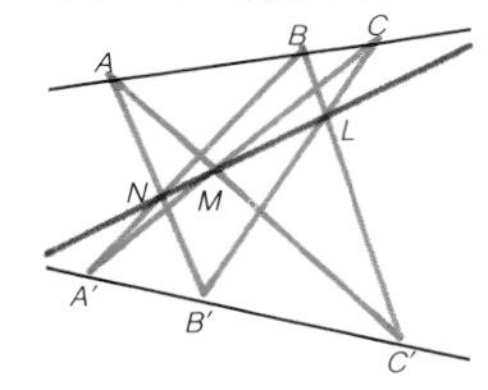

Menelaus' theorem if a transversal (p. 176) meets the sides *BC, CA* and *AB* (produced when necessary) of a triangle *ABC* at *X, Y* and *Z* respectively, then

$$\frac{BX}{XC}\cdot\frac{CY}{YA}\cdot\frac{AZ}{ZB} = -1,$$ (for the meaning of the negative sign *see* external division, ↑).

perspective (*n*) (1) a convention (p. 9) in art where lines which are really parallel are drawn as if to meet in a point inside or outside a picture; (2) two configurations are in perspective if the lines joining corresponding points all meet at a single point.

perspectivity (*n*) a perspective (↑) relation between two sets of collinear (p. 171) points.

vanishing point a point where perspective (↑) lines in art meet.

vanishing line the locus (p. 187) of the vanishing points (↑).

centre of perspectivity the point at which the joins of corresponding points in a perspective (↑) relation meet, e.g. the point O defined under Desargue's theorem (↓).

projectivity (*n*) a relation between two sets of points which can be obtained from one or more perspectivities (↑).

Desargue's theorem if *ABC* and *A′B′C′* are two triangles in two or three dimensions such that *AA′, BB′* and *CC′* meet at O; and if *BC, B′C′* meet at *L*; *CA, C′A′* meet at *M* and *AB, A′B′* meet at *N*, then *L, M* and *N* are collinear (p. 171).

axis of perspective the line *LMN* in Desargue's theorem (↑).

Pappus' theorem if *A, B, C* and *A′, B′, C′* are two sets of collinear (p. 171) points and *BC′, B′C* meet at *L*, *CA′, C′A* meet at *M* and *AB′, A′B* meet at *N*, then *L, M* and *N* are collinear (p. 171). *Do not mistake for* Pappus' theorem (p. 220) or Pappus-Guldin theorems (p. 156).

Pappus' line the line *LMN* in Pappus' theorem (↑).

cross-axis theorem a generalization of Pappus' theorem (↑): if $P_1, P_2, \ldots$ and $Q_1, Q_2, \ldots$ are two homographic (p. 53) ranges (p. 167) of points on two lines, then the intersection of P_1Q_2 and P_2Q_1 lies on a fixed line.

cross-axis (*n*) the fixed line in the cross-axis theorem (p. 183).

cross-centre (*n*) if two homographic (p. 53) pencils (p. 167) $p_1, p_2, \ldots$ and $q_1, q_2, \ldots$ have different vertices, then the join of the intersection of p_1 and q_2 to the intersection of p_2 and q_1 passes through a fixed point called the cross-centre.

cross-ratio (*n*) the ratio $(AB.CD)/(AD.CB)$ of a range (p. 167) of four points *A, B, C* and *D*. The cross-ratio of a pencil (p. 167) of four lines is the cross-ratio of the four points in which any line not passing through the vertex of the pencil meets the lines of the pencil. Also known as **anharmonic ratio**. Symbols: {AB, CD}, (ABCD).

equicross (*adj*) having the same cross-ratio (↑).

related[2] (*adj*) used of ranges (p. 167) or pencils (p. 167) which are homographic (p. 53) and where the cross-ratios (↑) of corresponding sets of elements are equal. *See also* p. 16.

cobasal (*adj*) having the same base, used of related ranges (p. 167) of points on the same line.

covertical (*n*) having the same vertex, used of related pencils (p. 167) of lines with the same vertex.

harmonic[1] (*adj*) used of ranges (p. 167) of four points or pencils (p. 167) of four lines which have a cross-ratio (↑) equal to -1. (For negative sign *see* external division, p. 182.) *See also* p. 277.

harmonic tetrad the four points of a harmonic (↑) range (p. 167).

harmonic conjugates *A* and *C* are said to be harmonic conjugates with respect to *B* and *D* if *A, B, C* and *D* form a harmonic (↑) range (p. 167); similarly for lines forming a harmonic pencil (p. 167).

involutory hexad the six points in which a line meets the sides of a complete quadrangle (p. 181). The pairs of meets (p. 171) with the opposite sides form mates (p. 55) in an involution (p. 55).

cross-ratio
for points *A, B, C, D* on a straight line

$$\text{if } (AC,BD) = \frac{AB.CD}{AD.CB} = k,$$

$$\text{then } (AB,CD) = \frac{AC.BD}{AD.BC} = 1 - k$$

$$(AC,DB) = \frac{AD.CB}{AB.CD} = \frac{1}{k}$$

$$(AD,BC) = \frac{AB.DC}{AC.DB} = \frac{k}{k-1}$$

$$(AB,DC) = \frac{AD.BC}{AC.BD} = \frac{1}{1-k}$$

$$(AD,CB) = \frac{AC.DB}{AB.DC} = \frac{k-1}{k}$$

cobasal
A,B,C,D and *A′,B′,C′,D′* are cobasal ranges

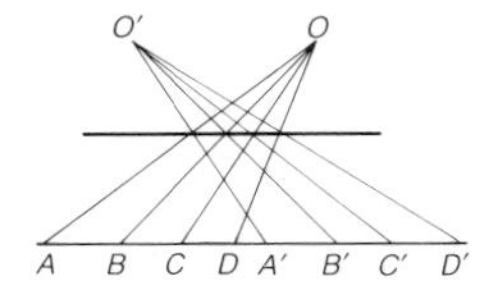

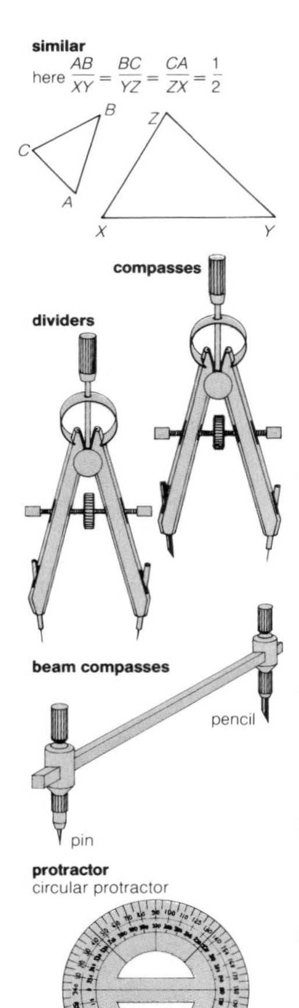

similar (*adj*) used of configurations having the same shape and with corresponding lengths in the same ratio (but areas are in the square of that ratio and volumes are in its cube). Similar triangles *ABC* and *XYZ* are equiangular (p. 173) and have $AB/XY = BC/YZ = CA/ZX$. **similarity** (*n*).

instrument (*n*) a thing which helps in drawing configurations, often in geometry, but also in mechanics and elsewhere.

straight edge a guide for drawing straight lines. In Euclidean geometry (p. 171) it is supposed that it is possible to make a straight edge.

ruler (*n*) a straight edge with a set of marks on it for measuring lengths. Also known as **rule**. *See also* p. 291.

compasses (*n.pl.*) (often called *pair of compasses*) an instrument for drawing arcs of circles, usually having two arms which are jointed and can be placed at different angles to each other, one arm ending in a pin and the other holding a pencil. *Do not mistake for* compass which is an instrument for finding the direction of north.

dividers (*n.pl.*) (often called *pair of dividers*) an instrument similar to a pair of compasses (↑), but with pins at the ends of both arms and used for taking the lengths of line segments (p. 167) from one part of a diagram to another.

beam compasses compasses (↑) in which a horizontal beam with sliding pieces replaces the jointed arms, used for drawing circular arcs of large radius.

scale (*n*) (1) a set of numbered marks used in measuring; (2) the ratio between a length on a map or drawing and that on the real thing. *See also* scale factor (p. 235).

protractor (*n*) an instrument with a scale of angles marked on it, usually semicircular (p. 192) or circular in shape and used to draw or measure angles.

set square an instrument in the shape of a right triangle (p. 179) used for drawing parallel lines and for other purposes. The common types have angles 90°, 45° and 45° or 90°, 30° and 60°.

French curve an instrument used to draw curves, perhaps best avoided in mathematical work. Also known as **artist's curve**.

Flexicurve (*n*) a bendable bar which will keep the shape of any curve into which it is bent and used for drawing curves, perhaps best avoided in mathematical work.

pantograph (*n*) an instrument made of six jointed beams used for copying drawings to a larger or smaller scale.

French curve

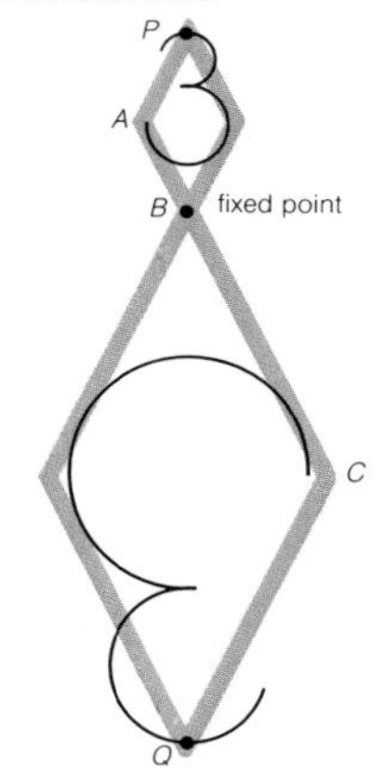

pantograph
as *P* moves, *Q* draws a similar figure (upside down) enlarged in the ratio *AB*:*BC*

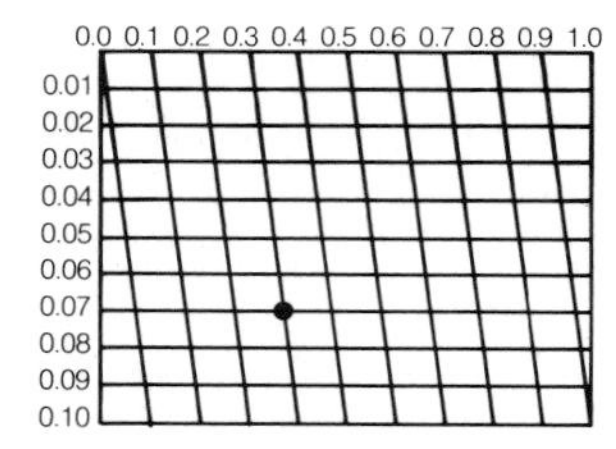

diagonal scale
a length of 0.37 units is measured from the left along the horizontal line 0.07 to where it meets the line 0.3

diagonal scale a scale with two sets of parallel lines which can be used with dividers (p. 185) to measure lengths more accurately (p. 43) than a ruler (p. 185).

construct (*v*) (1) to add lines and/or points to a given figure (p. 168) to allow a proof to be made more easily in geometry; (2) to draw accurately (p. 43), often using the Platonic restriction (↓). *See also* concept (p. 288).

construction (*n*) (1) lines or points added to a figure so that a proof can be made more easily; (2) a method of drawing various concepts in geometry, e.g. bisecting (p. 171) angles or line segments (p. 167), and drawing perpendiculars.

constructible (*adj*) that which can be constructed (↑), usually that which can be constructed using only the Platonic restriction (↓).

Platonic restriction a rule sometimes used in Euclidean geometry (p. 171) which only allows a straight edge (p. 185) (not a ruler, p. 185) and a pair of compasses (p. 185) to be used for constructions (↑). The idea probably began as the use of a piece of stretched string and two sticks to draw diagrams in sand.

curve (*n*) an intuitive (p. 243) concept; a continuous set of points (often in a plane) which is defined by a law or rule, either by an equation or by geometrical or physical properties; a line which is not straight.

locus (*n*) a set of points on a curve satisfying particular conditions. Also known as **trace**. **loci** (*pl*).

class[3] (*n*) the class of a locus (↑) which satisfies an algebraic expression (p. 60) is the degree of the polynomial which defines the locus. *See also* pp. 13, 110.

arc[1] (*n*) an interval or continuous set of points on a curve (or a space-curve, p. 212). It may be open (p. 65) or closed. *See also* p. 239.

path (*n*) the curve or track followed by a moving point.

chord (*n*) a straight line (p. 167) passing through two ends of an arc, usually taken as the line segment (p. 167) defined by the two ends of the arc. *See also* secant (↓).

secant[2] (*n*) a straight line (p. 167) passing through two ends of an arc. *See also* p. 132, chord (↑).

tangent[2] (*n*) when the two ends of an arc are allowed to move towards each other, the limiting position of the secant (↑) through the ends of the arc as these ends become coincident (p. 171) is known as the tangent. Except in the case of a few special points, the tangent meets the curve at its point of contact (↓) but does not cross it. *See also* p. 131. **tangential** (*adj*).

common tangent a straight line (p. 167) which is a tangent to two or more curves, not necessarily at the same point.

touch (*v*) a tangent and a curve, or two curves with a common tangent (↑) at a given point are said to touch. The word is also used similarly for space-curves (p. 212) and for surfaces and planes. Also known as **contact**.

point of contact the point at which a tangent and a curve or two curves touch, also used similarly for space-curves (p. 212) and for surfaces and planes. Also known as **contact**.

secant
often the word chord is also used for the whole line

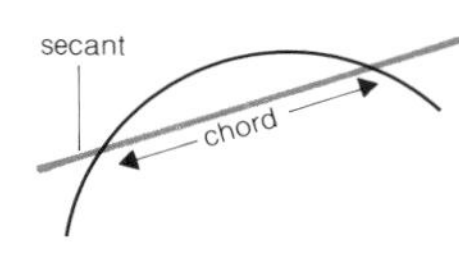

tangent
the limit of the set of secants $PQ_1, PQ_2, PQ_3, PQ_4, \ldots$ as Q_i tends to P is the tangent at P

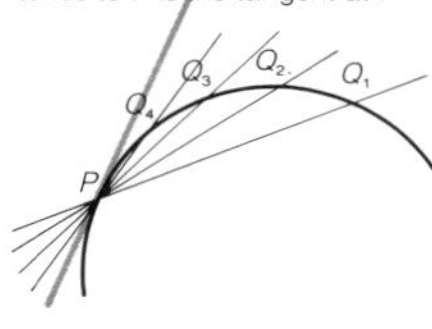

common tangent

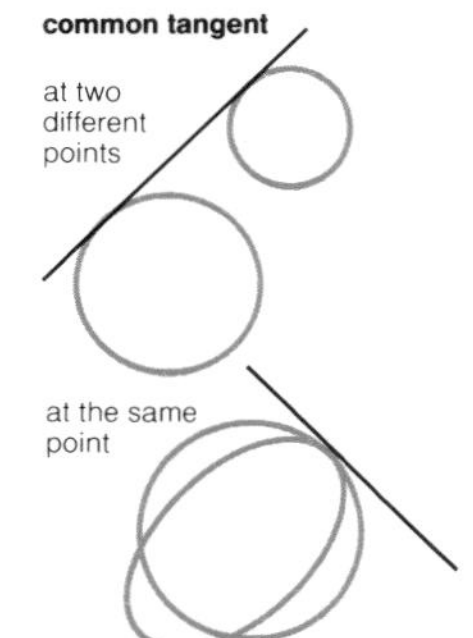

internal contact used of curves (or surfaces) which touch and both lie on the same side of the common tangent (p. 187) (or tangent plane) at the point of contact (p. 187). Similarly in **external contact** the curves lie on opposite sides of the tangent (or tangent plane).

normal (n) the straight line perpendicular to the tangent to a plane curve (or to the tangent plane to a surface) and passing through the point of contact (p. 187); also used to mean standard (p. 292) or usual. *See also* p. 120, principal normal (p. 212). **normal** (*adj*).

subtangent (n) if the tangent to $y = f(x)$ at $x = a$ meets the axis $y = 0$ at T, then the subtangent at P $(a, f(a))$ is the length TK where K is the point $(a, 0)$.

internal contact

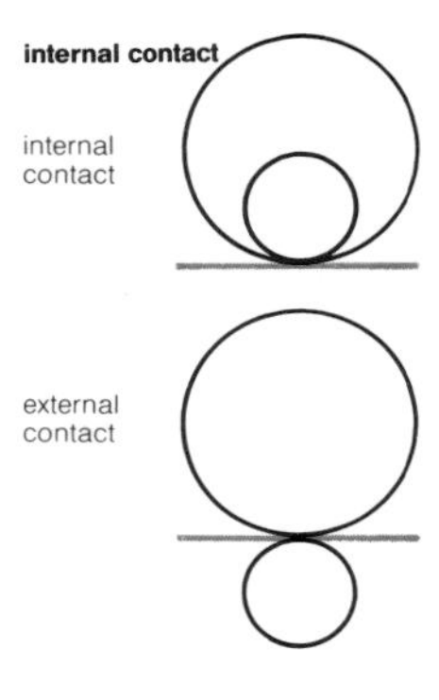

subtangent
the subtangent of $y = f(x)$ at P is the length TK and the subnormal is KN

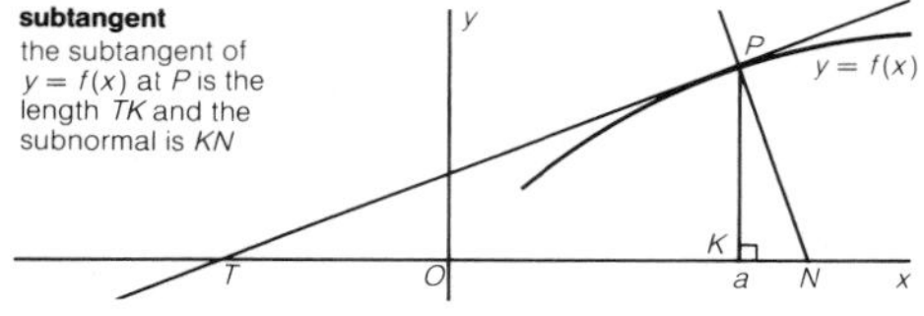

normal

tangent
normal
curve

subnormal (n) if the normal (↑) to $y = f(x)$ at $x = a$ meets the axis $y = 0$ at N, then the subnormal at P $(a, f(a))$ is the length NK where K is the point $(a, 0)$.

asymptote (n) a straight line (p. 167) which is the limit of the tangents to a curve as the point of contact (p. 187) tends to infinity along the curve. The asymptote and the curve come closer and closer as points on the curve tend to infinity. **asymptotic** (*adj*).

concave (*adj*) curved inwards, hollow. Used of both curves and surfaces. **concavity** (n).

convex (*adj*) curved outwards, the opposite of concave (↑). Used of both curves and surfaces. **convexity** (n).

singular point a point on a curve which has special properties, a point such as a node (↓), acnode (↓) etc.

regular point a point on a curve which has no special properties, a point which is not a singular point (↑).

asymptote
a hyperbola showing two asymptotes

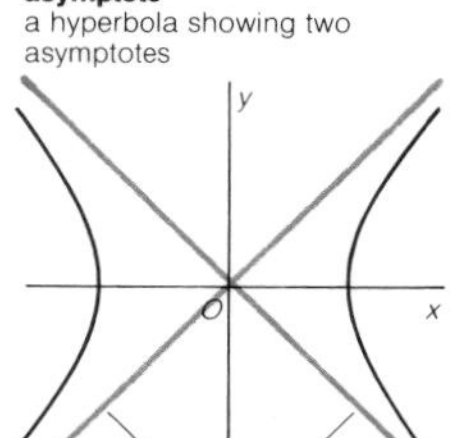

cusp (*n*) when a point moving along a curve arrives at a fixed point and leaves it in the direction from which it came, a cusp is formed, e.g. in a cycloid (p. 225). **cusped** (*adj*).

branch (*n*) a part of a curve, often a part ending at or passing through a singular point (↑), or a part separate from the rest of the curve. *See also* arc (p. 239).

node[1] (*n*) a point on a curve where two or more branches (↑) meet. *See also* pp. 239, 279.

double point[2] a node (↑) at which just two branches of a curve meet. Similarly **multiple point** for two or more branches. *See also* p. 53.

loop[2] (*n*) a branch (↑) of a curve which has both ends at the same node (↑). Also known as **folium**. *See also* p. 69, closed curve (p. 190).

crunode (*n*) a double point (↑) which has two distinct (p. 60) tangents on the two branches (↑) of the curve.

tacnode (*n*) a node (↑) or cusp (↑) where two branches (↑) of a curve have the same tangent. Also written **tachnode**.

acnode (*n*) a point on a curve separate from all other points on the curve. Also known as **isolated point**.

stop point a point on a curve at which there is a discontinuity (p. 64) (as in a step function, p. 166). Also known as **terminus point**.

osculating circle at a given point of a curve, the limiting circle of all circles cutting the curve at three points as all three points tend to the given point. It is the circle which 'fits' the curve best at that point. Also known as **circle of curvature**.

centre of curvature of a given point of a curve: the centre of the osculating circle (↑) at that point. Similarly **radius of curvature** is the radius of the osculating circle at that point.

curvature[1] (*n*) if the radius of curvature (↑) of a curve at a given point is r, then the curvature is its reciprocal (p. 27) $1/r$. In the intrinsic equation (p. 217) form of the curve it is the derivative (p. 143) $d\psi/ds$ of ψ with respect to s. In parametric (p. 67) form with parameter t it is given by $(\ddot{y}\dot{x} - \ddot{x}\dot{y})/(x^2 + y^2)^{(3/2)}$ where the dots show derivation with respect to t. *See also* p. 213.

cusp
nephroid with two cusps joined by two branches

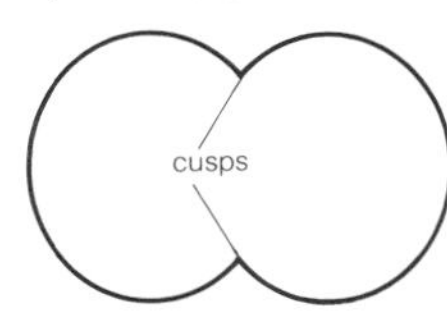

node
a lemniscate with two loops and a node which is both double point and crunode

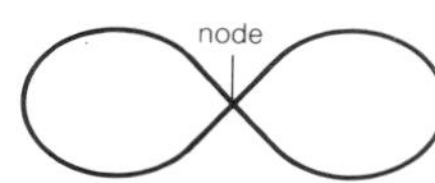

tacnode
$y^4 = x^6$ which has a tacnode at (0, 0); the tangent is $y = 0$

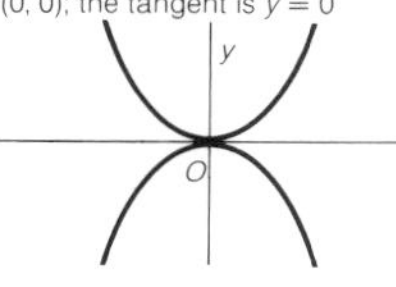

acnode
$y = (x + 1)\sqrt{x}$ has an acnode at (−1, 0)

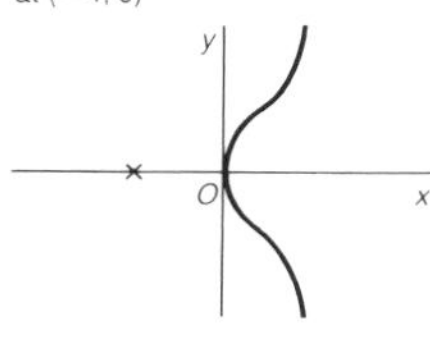

osculating circle
the parabola $y^2 = 4ax$ has osculating circle $x^2 + y^2 - 4ax = 0$ at the vertex. The radius of curvature is $2a$ and the centre of curvature is $(2a, 0)$

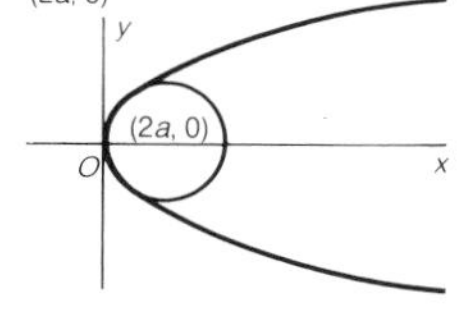

mean curvature the quantity $\delta\psi/\delta s$ obtained from the intrinsic equation (p. 217) of a curve for a finite length of the curve, where δs is the length of an arc of the curve and $\delta\psi$ is the angle between the two tangents at either end of the arc. Its limiting value as δs tends to 0 is the curvature (p. 189).

evolute (*n*) the evolute of a given curve is the locus (p. 187) of the centres of curvature (p. 189) of that curve; it is also the envelope (↓) of the normals (p. 188) to that curve.

involute (*n*) the involute of a given curve is one of the family (↓) of curves which have the given curve as evolute (↑). It is also the locus (p. 187) of a point on a thread stretched along the curve as it is undone from the curve.

family (*n*) a set of configurations, often lines or curves which have a common property. Also known as **system**. *See also* p. 292.

envelope (*n*) a curve or surface which touches every member of a family of lines, curves, planes or surfaces. Also known as **tangent curve**, **tangent surface**.

order² (*n*) the order of an envelope (↑) which satisfies an algebraic expression (p. 60) is the degree of the polynomial (p. 60) defining it. *See also* pp. 20, 72, 83.

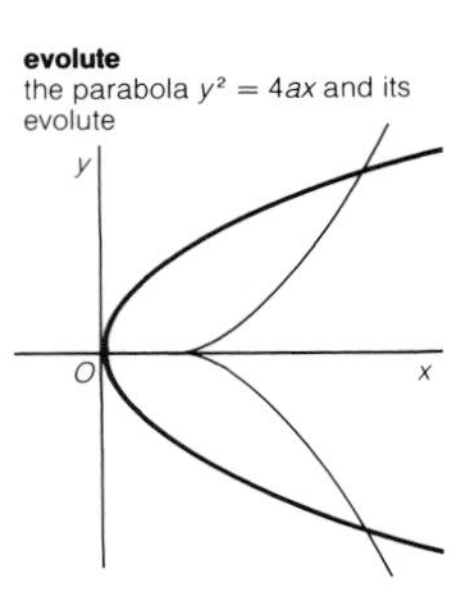

evolute
the parabola $y^2 = 4ax$ and its evolute

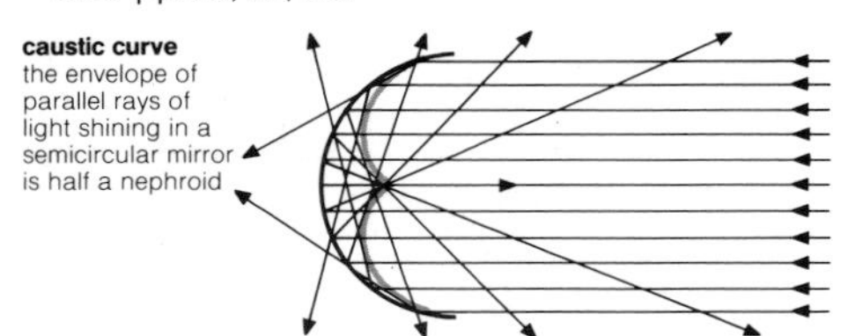

caustic curve
the envelope of parallel rays of light shining in a semicircular mirror is half a nephroid

caustic (*n*) an envelope (↑) formed by light shining or in a similar way, e.g. the nephroid (p. 227) on the surface of a cup of liquid with the sun shining on it. **caustic** (*adj*).

closed curve a continuous curve of finite length with no stop (or end) points (p. 189). If it is a plane curve it separates the plane into regions of which only one is not finite. *See also* loop (p. 189).

normal region
any line parallel to the axes meets its boundary in not more than two points

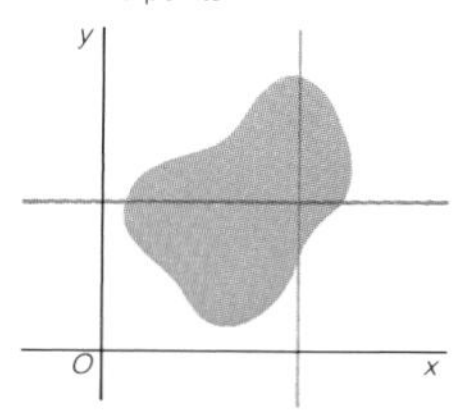

Reuleaux polygon
any chord through the centre of symmetry has a constant length

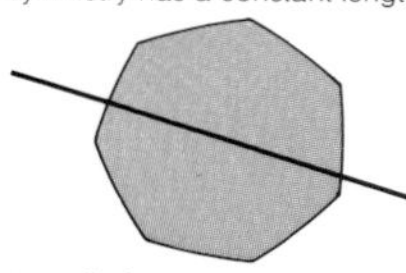

tessellation
part of a regular tessellation of rectangles. It is regular because a translation through a unit of the pattern leaves it unaltered

concentric circle **eccentric** circles

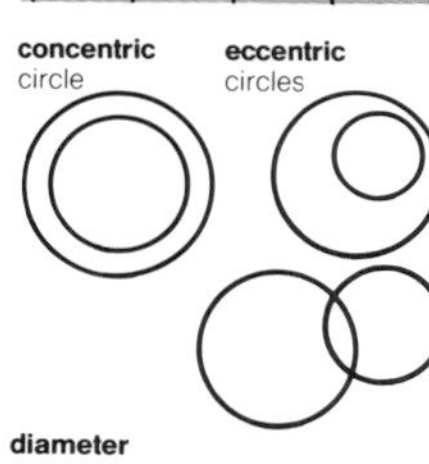

diameter
the diameter of parabola

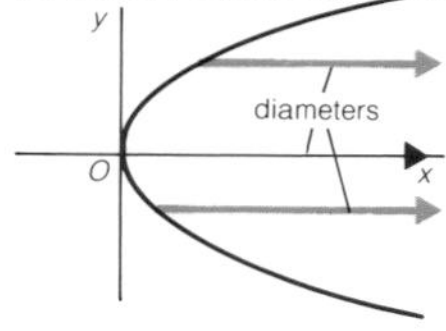

simple closed curve a closed curve (↑) without singular points (p. 188), e.g. circle, ellipse (p. 220).

normal region a region whose boundary is a simple closed curve (↑) in the plane of rectangular cartesian (p. 214) coordinates (x, y) which has not more than two values of x for any value of y and not more than two values of y for any value of x.

Reuleaux polygon a normal region (↑) of constant breadth shaped like a polygon but with curved sides; the shape is used in some countries for pieces of money.

tessellation (*n*) a covering of any surface with simple closed curves (↑), often polygons. Also known as **tiling**.

interior point a point inside a polygon, closed curve (↑) or similar figure. Similarly **exterior point**.

perimeter (*n*) the total length of a simple closed curve (↑), polygon or other similar figure.

area (*n*) the measure of the amount of space inside a finite region (either a closed curve, ↑, or a polygon or similar figure), obtained by approximating to a sum of areas of small rectangles whose areas are defined to be length × breadth (as in integration).

centre (*n*) the point which is equidistant (p. 169) from any point on a circle or sphere; the point about which a curve or polygon has rotational symmetry (p. 232); the point of a solid where an axis of symmetry (p. 232) meets a plane of symmetry (p. 232) perpendicular to it. **central** (*adj*).

concentric (*adj*) having the same centre.

eccentric (*adj*) not having the same centre.

diameter (*n*) (1) a straight line (p. 167) passing through the centre of a curve or solid, especially a circle, sphere, central conic (p. 221) or central quadric (p. 211); the length of that line; (2) (of a parabola, p. 220) a line parallel to the axis.

diametral plane a plane passing through the centre of a solid.

circle (*n*) a set of points in a plane which are equidistant (p. 169) from a fixed point (the centre), sometimes also used with the meaning of disc (↓). Three points are needed to define any given circle. **circular** (*adj*).

round² (*adj*) circular, cylindrical (p. 209) or spherical (p. 204), or approximately so. The word has no exact meaning in mathematics. *See also* p. 44.

disc (*n*) the region consisting of all the points inside a circle. A closed disc includes the boundary points, an open (p. 65) disc does not.

annulus (*n*) the region between two concentric (p. 191) circles in a plane.

circumference (*n*) the perimeter (p. 191) of a circle, and sometimes of other closed curves (p. 190).

pi (*n*) the ratio of the circumference (↑) of a circle to the length of its diameter. It is a transcendental number (p. 46) which is approximately 3.14159265.... *See also* radius (↓). Symbol: π.

radius (*n*) the distance from the centre of a circle to any point on its circumference (↑); the distance from the centre of a sphere to any point on its surface. It is half the length of the diameter. The circumference of a circle is $2\pi r$ and its area (p. 191) is πr^2 where r is the radius. The surface area of a sphere is $4\pi r^2$ and its volume is $4\pi r^3/3$ where r is its radius. Any radius is perpendicular to the tangent or tangents through the point of contact (p. 189). *See also* radius vector (p. 216). **radii** (*pl*), **radial** (*adj*).

semicircle (*n*) half a circle, with a diameter as part of the boundary, also often used for half a disc (↑). The angle subtended (p. 175) by the ends of the diameter at any other point on the boundary is a right angle.

minor arc an arc of a circle whose length is less than half the circumference (↑). The length of an arc of a circle is $r\theta$ where r is the radius of the circle and θ is the angle in radians (p. 133) which it subtends (p. 175) at the centre. Similarly **major arc** is more than half the circumference.

circle

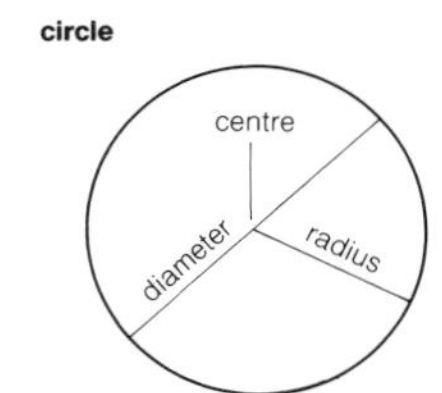

annulus

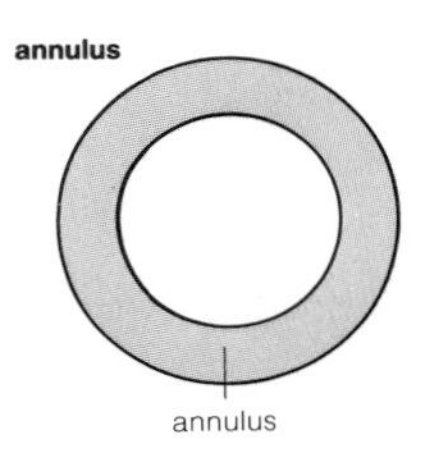

semicircle

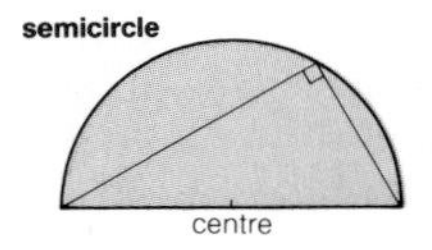

minor arc

major arc

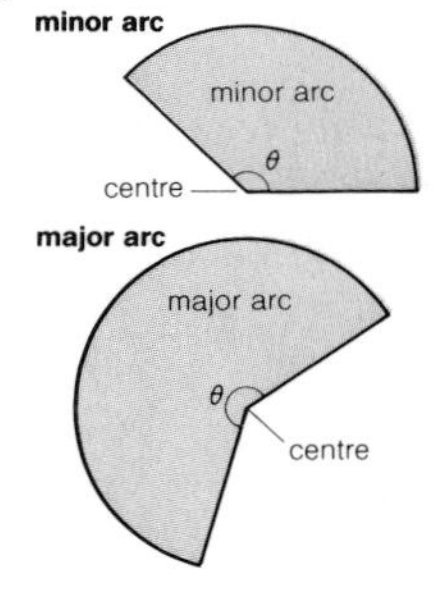

quadrant

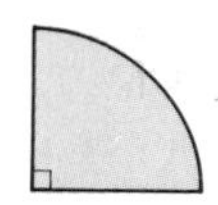

the four quadrants defined by the coordinate axes

y

second quadrant	first quadrant
third quadrant	fourth quadrant

O x

octant

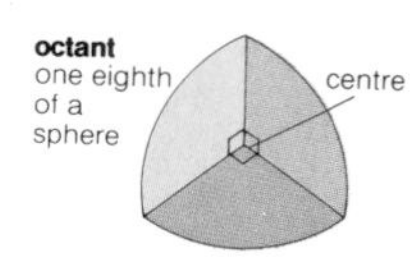

sector

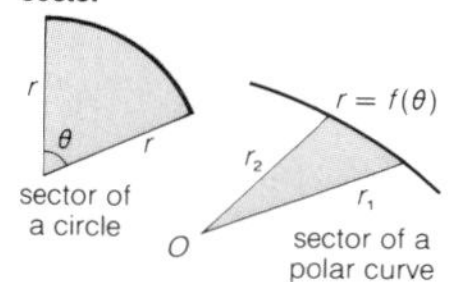

segment

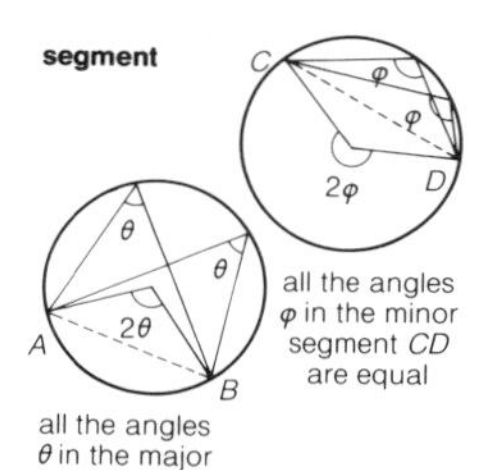

cyclic quadrilateral

$\alpha + \gamma = 180^\circ \quad \beta + \delta = 180^\circ$

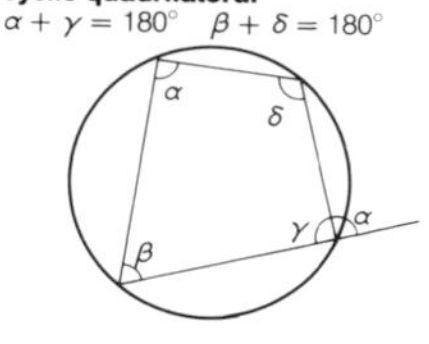

quadrant (*n*) (1) a quarter of a circle or disc (↑), an angle of 90°; (2) in plane rectangular cartesian coordinates (p. 214) a quarter of the plane bounded by the axes. In the first quadrant, *x* and *y* are both positive, in the second only *y* is positive, in the third both *x* and *y* are negative and in the fourth only *x* is positive.

sextant (*n*) a sixth of a circle or disc (↑), an angle of 60°.

octant (*n*) (1) an eighth of a circle or disc (↑), an angle of 45°; (2) an eighth of a sphere bounded by three diametral planes (p. 191) all at right angles to each other.

sector (*n*) (1) part of a disc (↑) bounded by an arc and two radii and having area $\frac{1}{2}\pi r^2\theta$ where *r* is the radius of the disc, and θ is the angle in radians (p. 133) between the radii; (2) the area (p. 191) which is bounded by an arc of a polar (p. 223) curve and two radius vectors (p. 216).

segment (*n*) the region bounded by an arc, particularly a circular arc, and a chord (p. 187); part of a solid cut off by a plane. Sometimes used to mean a part of a line, curve or surface. (Major, minor segments of a circle are defined as for major, minor arcs, ↑). The angles subtended (p. 175) by the ends of a segment of a circle at any other point on the segment are all equal and are half the angle subtended at the centre of the circle (for a minor segment the reflex angle, p. 173, at the centre must be taken).

cyclic (*adj*) (1) (of a polygon) having its vertices lying on a circle; (2) able to be arranged around a circle so that the starting point does not matter, as in a cyclic group (p. 74) or cyclic permutation (p. 97).

cyclic quadrilateral a quadrilateral whose vertices lie on a circle. It has opposite angles supplementary (p. 175), and each exterior angle (p. 177) is equal to the opposite interior angle (p. 177).

concyclic (*adj*) (of points) lying on the same circle.

direct common tangent a common tangent (p. 187) of two circles which does not pass between their centres. Similarly **indirect common tangent** which does pass between their centres, also known as **transverse common tangent**.

alternate segment theorem the angle between a tangent to a circle and a chord (p. 187) through the point of contact (p. 187) equals the angle subtended (p. 175) by the chord in the segment (p. 193) of the circle on the opposite side of the chord.

direct and indirect common tangents

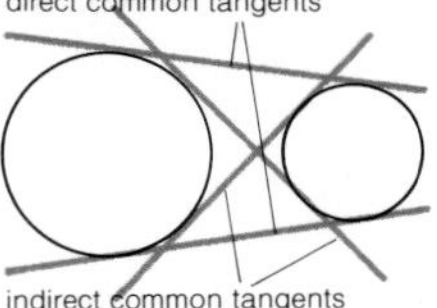

alternate segment theorem

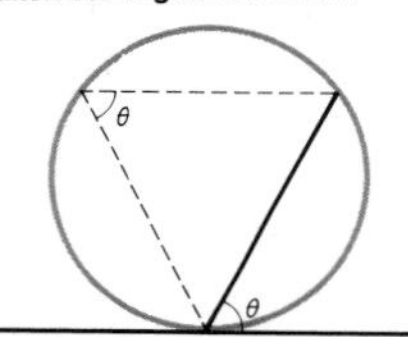

intersecting chord theorems

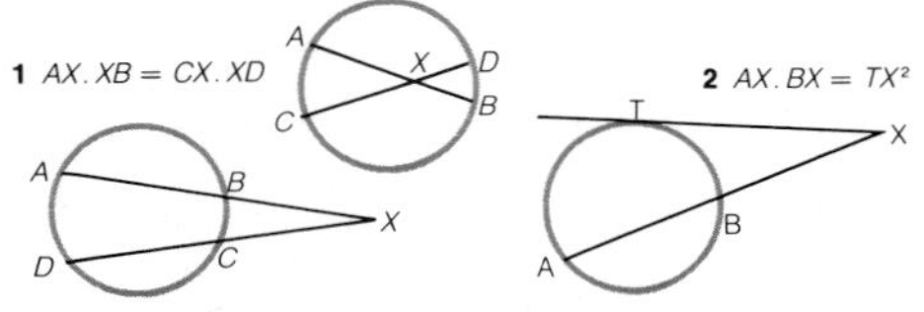

intersecting chord theorems (1) if two chords (p. 187) *AB* and *CD* of a circle meet at a point *X* inside or outside the circle, $AX.XB = CX.XD$; (2) if a chord *AB* meets the tangent at *T* in a point *X*, then $AX.BX = TX^2$.

Appolonius' circle if *A* and *B* are two fixed points, then the locus (p. 187) of the point *P* moving in a plane containing *A* and *B*, so that *PA/PB* is constant, is a circle. *Do not mistake for* Appolonius' theorem (p. 196).

Appolonius' circle
in this case $PA/PB = 2$ for every point *P* on the circle

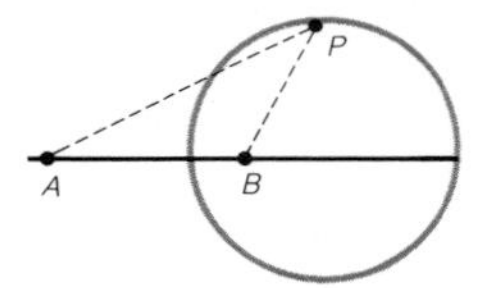

Ptolemy's theorem if *ABCD* is a cyclic quadrilateral (p. 193), $AB.CD + BC.DA = AC.BD$.

Ptolemy's theorem
$AB.CD + BC.DA = AC.BD$

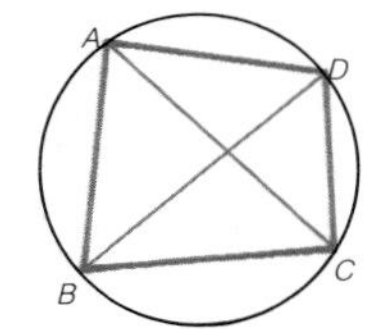

squaring the circle constructing (p. 186) a square equal in area (p. 191) to a given circle. It is now known that this is impossible with the Platonic restrictions (p. 186).

inverse points two points *P* and *Q* are inverse with respect to a point *O* if $OP.OQ = k^2$ where *k* is a constant and *O*, *P* and *Q* are collinear (p. 171).

centre of inversion the point *O* in the definition of inverse points (↑).

radius of inversion the length *k* in the definition of inverse points (↑).

inverse points
if $OP.OQ = k^2$, then *P* and *Q* are inverse points with respect to the circle centre *O* radius *k*

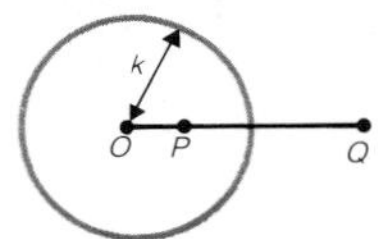

Steiner's circles

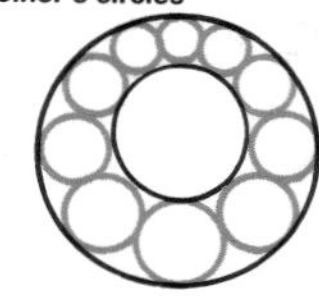

Peaucellier's cell
$AB = BC = CD = DA$ and $OB = OD$. All are jointed and only O is a fixed point. If C moves on a circle through O, then A moves on a straight line

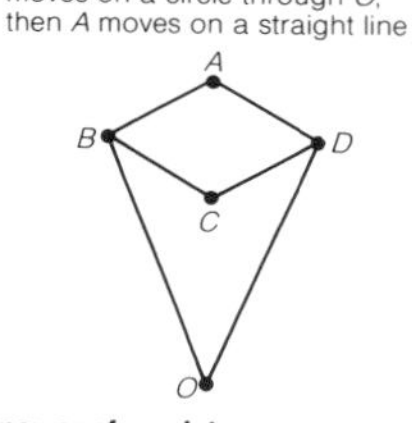

power of a point
the power of P with respect to the circle is $OP^2 - r^2 = PT^2 = PA.PB$

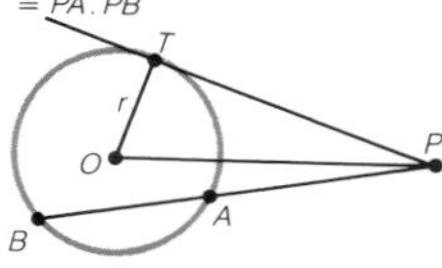

coaxal circles
two sets of coaxal circles: the red set has $y = 0$ as radical axis and two limiting points. These are the common points of the black set which has $x = 0$ as radical axis. Every red circle is orthogonal to every black one

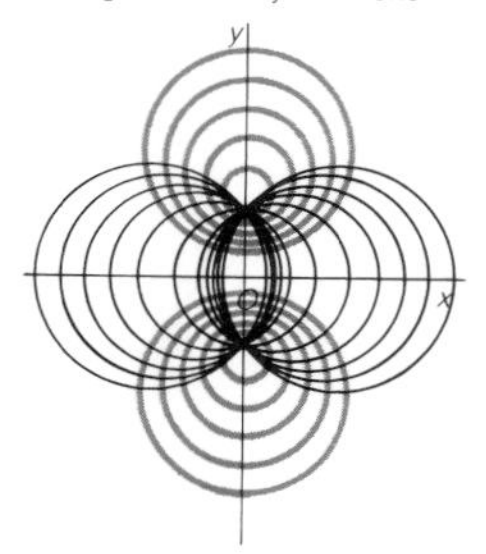

circle of inversion the circle centre O and radius k in the definition of inverse points (↑).

inverse curves curves for which corresponding points are inverse points (↑) with respect to some given point.

inversion (*n*) the process of finding a configuration whose points are the inverse points (↑) with respect to the points in a given configuration. Inversion preserves angle and is thus a conformal (p. 235) transformation (p. 54). **invert** (*v*).

Steiner's circles if a circle C_1 lies wholly inside a circle C_2, and if one set of circles can be drawn between them so that each touches the next and also touches C_1 and C_2, then an infinite number of such sets can be drawn. This can easily be proved by inverting (↑) the circles C_1 and C_2 into a pair of concentric (p. 191) circles. Also known as **Steiner's porism**.

Peaucellier's cell a set of jointed bars which allows movement in a straight line to be changed into movement in a circle, obtained from the idea of inversion (↑). Also known as **Peaucellier's linkage**.

power of a point if a circle or sphere has centre O and radius r, the power of a point P with respect to the circle or sphere is $OP^2 - r^2$. If a chord (p. 187) through P cuts the circle or sphere at A and B this is equal to $PA.PB$ (*see* intersecting chord theorems, ↑). The power is negative if the point is inside the circle.

radical axis of two circles: the locus (p. 187) of a point whose powers with respect to the two circles are equal. **radical axes** (*pl*).

coaxal circles infinite sets or systems of circles, any two of which have the same radical axis (↑). Any set of coaxal circles either has two common points, or none of the circles meet at all. Also known as **coaxal systems**.

point circle a circle of zero radius which is a special case, e.g. the limiting points of a coaxal system (↓).

limiting points of a coaxal system a set of coaxal circles (↑) which does not have two common points contains two point circles (↑) called its limiting points.

radical centre of three circles whose centres do not lie on a straight line: the point at which their three radical axes (p. 195) meet.

radical centre
the point where the three radical axes meet

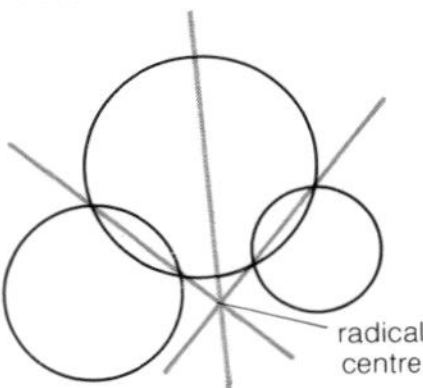

radical plane the locus of points having equal powers (p. 195) with respect to two spheres; if the spheres intersect, it is their plane of intersection.

orthogonal circles pairs of circles which intersect at right angles so that the centres of each lie on the tangents to the other through the point of intersection. If A and B are the limiting points of a coaxal system (p. 195), then every circle of the second coaxal system whose common points are A and B is orthogonal (p. 175) to every circle of the first set.

orthogonal circles
the tangents to one at the points of intersection pass through the centres of the other

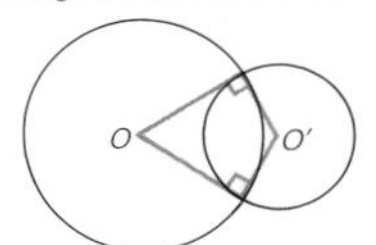

median² (*n*) a line joining the vertex of a triangle to the mid-point of the opposite side, or the vertex of a tetrahedron (p. 201) to the centroid (↓) of the opposite face. *See also* p. 114.

centroid (*n*) the point at which the three medians (↑) of a triangle or the four medians of a tetrahedron (p. 201) intersect. It gives the centre of mass (p. 261) of a uniform triangular lamina (p. 155) or of a uniform tetrahedron. The word is often used with the meaning of centre of mass for other bodies. *See also* centre of mass (p. 261).

median
AA', BB', CC' are the medians and meet at the centroid G so that $AG:GA' = 2:1$ etc

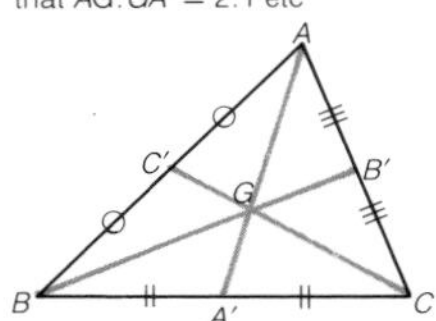

Appolonius' theorem if AA' is a median (↑) of a triangle ABC, then $AB^2 = AC^2 + 2AA'^2 + 2BA'^2$. *Do not mistake for* Appolonius' circle (p. 194).

Stewart's theorem if B, D and C are three collinear (p. 171) points and A is any other point not on the line, then

$$AB^2 . DC + AD^2 . CB + AC^2 . BD + DC . CB . BD = 0$$

(*note*: lengths along BC take account of the signs). If D is the mid-point of BC this becomes Appolonius' theorem (↑).

Stewart's theorem
$AB^2 . DC + AD^2 . CB + AC^2 . BD + DC . CB . BD = 0$

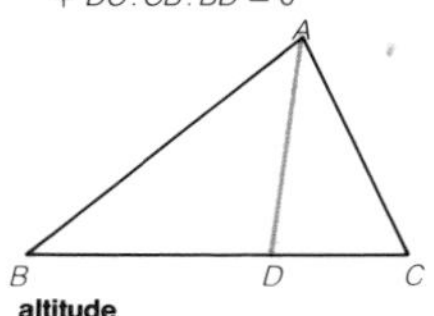

altitude (*n*) a line through the vertex of a triangle perpendicular to the opposite side. *See also* angle of elevation (p. 175).

orthocentre (*n*) the point at which the three altitudes (↑) of a triangle meet.

pedal triangle the triangle formed by the feet (p. 175) of the altitudes (↑) of a triangle.

altitude
AD, BE, CF are the altitudes

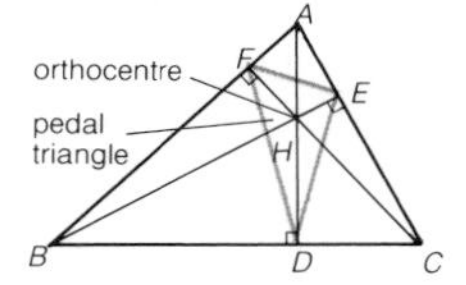

circumcircle
O is the circumcentre, *OA* = *OB* = *OC* is the circumradius, *OA′*, *OB′*, *OC′* are the mediators of the sides of triangle *ABC*

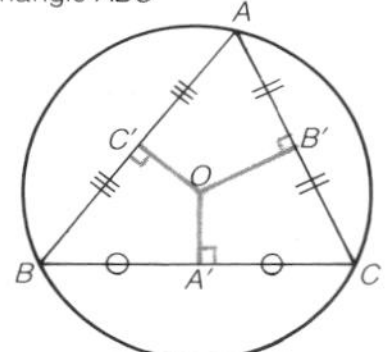

circumscribe (*v*) to draw a closed curve (p. 190) (especially a circle) through all the vertices around the outside of a polygon, or to take a closed surface (p. 209) (especially a sphere) through all the vertices around the outside of a polyhedron (p. 200).

circumcircle (*n*) a circumscribed (↑) circle, often of a triangle. Similarly **circumcentre**, its centre, and **circumradius**, its radius.

mediator (*n*) a line through the mid-point of a given line and perpendicular to it. The three mediators of a triangle meet at the circumcentre (↑). Also known as **perpendicular bisector**, **right bisector**.

Simson's line

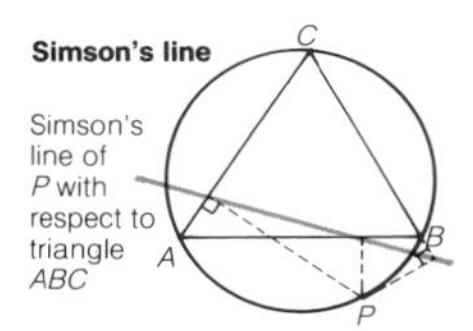

Simson's line the straight line (p. 167) through the feet (p. 175) of the perpendiculars from any point on the circumcircle (↑) of a triangle to its three sides. Also known as **pedal line**.

Euler points
if *H* is the orthocentre of triangle *ABC*, the Euler points *P*, *Q* and *R* are the mid-points of *AH*, *BH* and *CH*

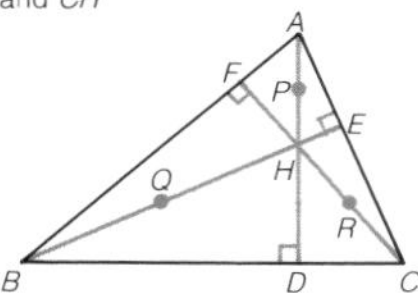

Euler points of a triangle, the three points which are the mid-points of the lines joining the vertices to the orthocentre (↑).

nine-point circle in triangle *ABC*
D, *E*, *F* are the feet of the altitudes,
A′, *B′*, *C′* are the mid-points of the sides,
P, *Q*, *R* are the Euler points (the mid-points of *AH*, *BH*, *CH*)
N is the nine-point centre
H is the orthocentre
G is the centroid
O is the circumcentre

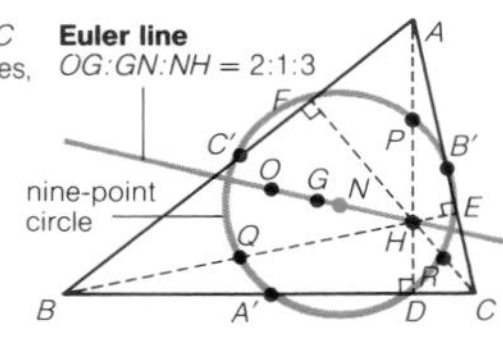

nine-point circle of a triangle: the circle which passes through the mid-points of its sides, the feet (p. 175) of its altitudes (↑) and its Euler points (↑). Also known as **Euler circle**, **Feuerbach circle**. *Do not mistake for* Euler circles (p. 248).

nine-point centre the centre of the nine-point circle (↑).

Euler line of a triangle: the straight line (p. 167) passing through its circumcentre (↑) *O*, its centroid (↑) *G*, its nine-point centre (↑) *N* and its orthocentre (↑) *H* such that $OG : GN : NH = 2:1:3$.

Miquel point
ADB, *BCF*, *AEC*, *DEF* are the sides of a complete quadrilateral: the circles *ABC*, *ADE*, *BFD*, *CFE* meet at the Miquel point *M*

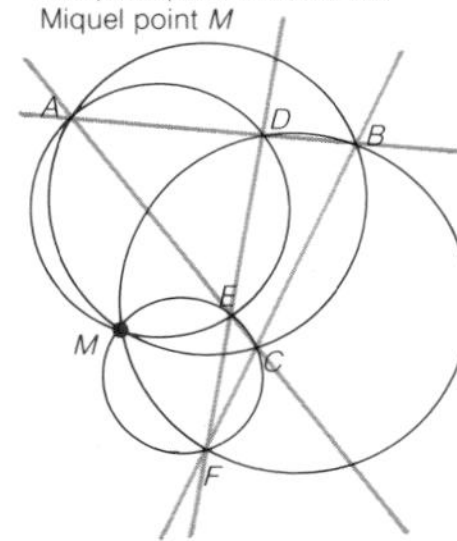

Miquel point if the circumcircles (↑) are drawn to the four triangles which make up the complete quadrilateral (p. 181), then they all meet at the Miquel point.

inscribe (*v*) to draw a closed curve (p. 190) (especially a circle) inside a polygon touching each of its sides, or to take a closed surface (p. 209) (especially a sphere) inside a polyhedron (p. 200) touching each of its faces.

incircle (*n*) an inscribed (↑) circle, often of a triangle. Similarly **incentre**, its centre, and **inradius**, its radius.

interior angle bisector the angle bisector (p. 171) of the interior angle of a polygon; for a triangle these meet at the incentre (↑). Also known as **internal bisector**.

interior angle bisectors
in a triangle these meet at the incentre *I*

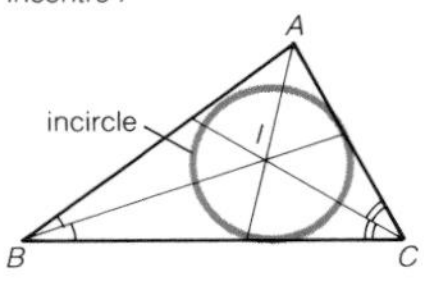

escribe (*v*) to draw one of the three circles touching the outside of one side of a triangle between the vertices and the other two sides produced (p. 167). Not often used for other curves and/or polygons. Also written **exscribe**.

excircle (*n*) an escribed (↑) circle. Similarly **excentre**, its centre, and **exradius**, its radius. Also known as **e-circle**. Any given triangle is the pedal triangle (p. 196) of its three excentres.

excircle
the incircle and three excircles of a triangle

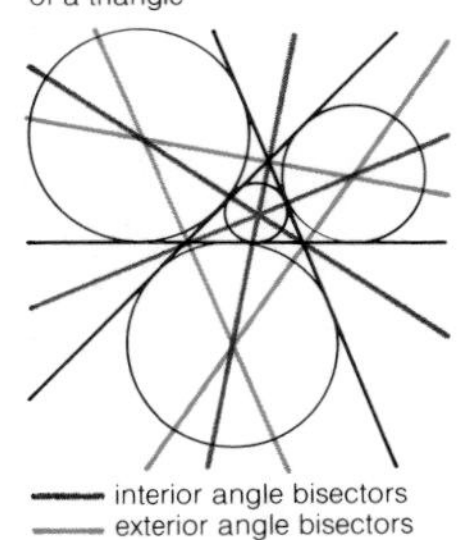

exterior angle bisector angle bisector (p. 171) of the exterior angle (p. 177) of a polygon; for a triangle any two of these meet on the interior angle bisector (↑) of the third angle at one of the three excentres (↑). Also known as **external bisector**.

angle bisector theorem if *AD* is an angle bisector (interior, ↑, or exterior, ↑) of a triangle *ABC* which meets *BC* (produced, p. 167, if necessary) at *D*, then $BD/DC = AB/AC$.

angle bisector theorem
$\frac{BD}{DC} = \frac{AB}{AC}$

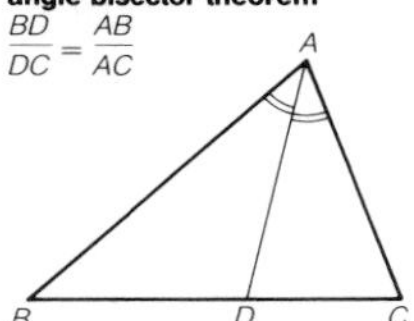

Feuerbach's theorem in any triangle the nine-point circle (p. 197) touches all three excircles (↑).

isogonal lines lines which make equal angles on either side of a given line at a given point on that line. A pair of isogonal lines are the reflections (p. 234) of one another in the given line.

isogonal lines
the red lines are isogonal with respect to the black line at the point *P*

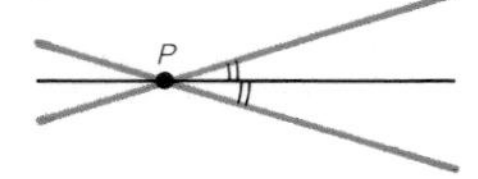

symmedian (*n*) a line which is isogonal (↑) to the median (p. 196) of a triangle at its vertex with respect to the interior angle bisector (↑).

Lemoine point that point of a triangle at which the symmedians (↑) meet.

Tucker circle if a triangle *PQR* lies inside a triangle *XYZ* and is homothetic (p. 236) to it with respect to the Lemoine point (↑), then the six points in which *PQ, QR* and *RP* produced (p. 167) meet the sides of *ABC* lie on the Tucker circle.

Lemoine circle the three lines through the Lemoine point (↑) of a triangle which are parallel to the sides of the triangle meet its sides in six points which lie on the Lemoine circle (a special case of the Tucker circle, ↑).

Brocard points

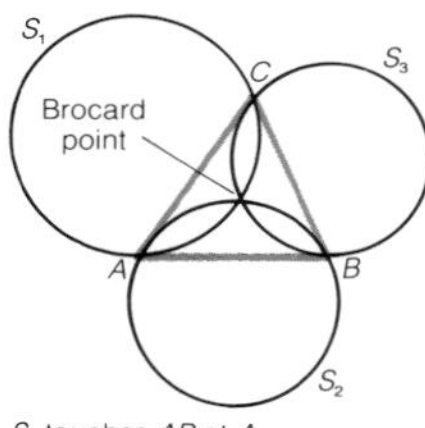

S_1 touches *AB* at *A*
S_2 touches *BC* at *B*
S_3 touches *CA* at *C*

Brocard points if, in a triangle *ABC*, circles are drawn touching *AB* at *A* and passing through *C*, touching *BC* at *B* and passing through *A* and touching *CA* at *C* and passing through *B*, then these three circles meet at one of the two Brocard points. The other Brocard point can be found by drawing circles touching *AB* at *B* and passing through *C* and so on.

isogonic point
P is the isogonic point where $AP + BP + CP$ is a minimum

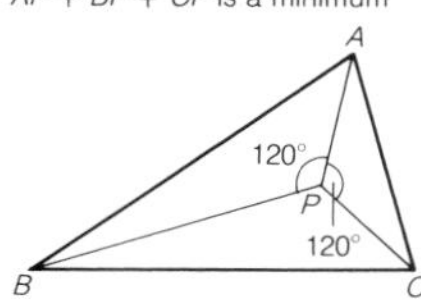

isogonic point that point *P* in any triangle *ABC* for which $AP + BP + CP$ is a minimum. If all angles of the triangle are less than 120°, the point lies inside the triangle, otherwise it lies on a side. It is important in finding a network (p. 239) of minimum length linking three points.

Fermat's problem the problem of finding the isogonic point (↑). Also known as **Steiner's problem**.

isoperimetric point that point *Q* in a triangle *ABC* for which the triangles *AQB, BQC* and *CQA* have equal perimeters (p. 191). Also known as **Kimberling point**.

orthopole
K is the orthopole of line *PQR* with respect to triangle *ABC*

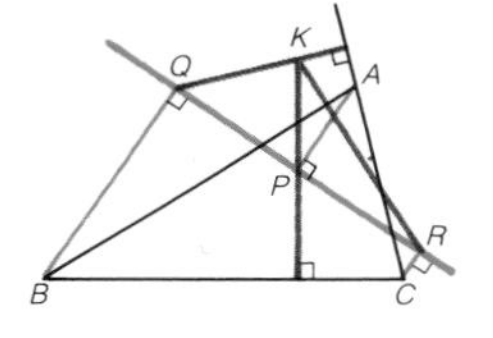

orthopole (*n*) if *P, Q* and *R* are the feet (p. 175) of the perpendiculars from the vertices of a triangle *ABC* to a fixed line, then the feet of the perpendiculars from *P* to *BC*, from *Q* to *CA* and from *R* to *AB* all pass through a point which is called the orthopole. If the fixed line passes through the circumcentre (p. 197) of the triangle, then the orthopole lies on the nine-point circle (p. 197).

solid geometry the geometry of space with three dimensions and especially of solids.

cross-section (*n*) the plane figure obtained from the intersection of a solid or surface with a plane.

dihedral angle the angle between two intersecting planes. It is the angle between perpendiculars in each plane through a point on the line of intersection of the planes.

solid angle the three dimensional concept similar to angle in two dimensions; the space shaped like a cone (p. 210) subtended (p. 175) by a region of a surface at a point not on the surface.

steradian (*n*) a measure of solid angle (↑). That area of the surface of a unit sphere with centre at the vertex of the solid angle which is covered by the cone (p. 210) defining the solid angle. A whole sphere subtends (p. 175) an angle of 4π steradians at its centre. **sr** (*abbr*). Symbol: Ω.

trihedral angle a solid angle (↑) formed by three intersecting planes.

spherical degree the area of a unit sphere formed by great circles (p. 204) joining the ends of four radii each at 1° to the next one. No longer used.

skew (*adj*) not lying in a single plane, used especially of lines and polygons. *See also* p. 122.

skew lines straight lines (p. 167) in three or higher dimensional space which are neither parallel nor intersecting, and which do not therefore have a common plane.

volume (*n*) the measure of the amount of space inside a solid obtained by approximating to the sums of volumes of cuboids (↓) which are defined to have
volume = length × breadth × height.

polyhedron (*n*) a solid figure which is bounded by plane surfaces. **polyhedra** (*pl*), **polyhedral** (*adj*).

regular polyhedron a polyhedron (↑) having every face the same regular polygon (p. 177). Also known as **regular solid**, **Platonic solid**. *See also* tetrahedron (↓) etc.

dihedral angle
θ between two planes

regular polyhedra

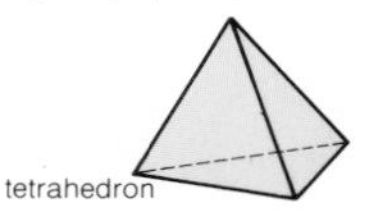

tetrahedron

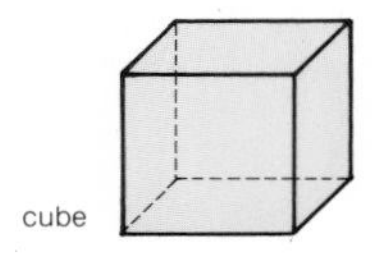

cube

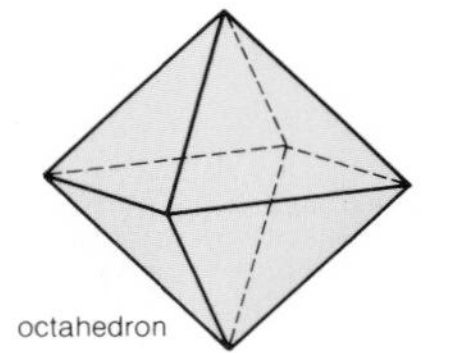

octahedron

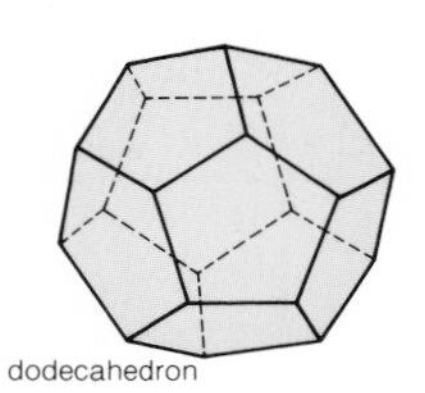

dodecahedron

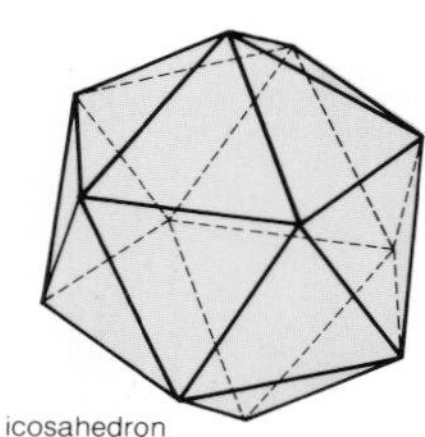

icosahedron

parallelepiped

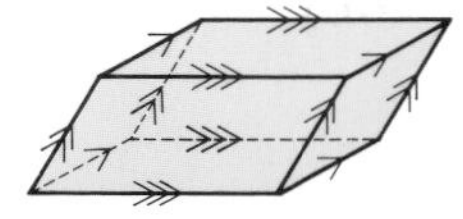

cuboid

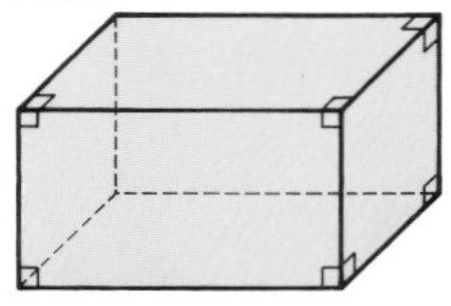

stellated
small stellated dodecahedron (one of the Kepler-Poinsot polyhedra)

faceted
great dodecahedron (one of the Kepler-Poinsot polyhedra)

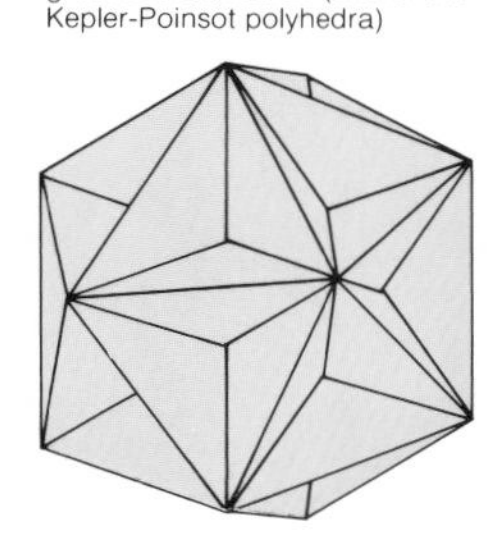

face (*n*) the plane or curved surface (often a polygon) which forms part of the boundary surface of a polyhedron (↑) or other solid.

edge (*n*) the line in which two faces of a polyhedron (↑) meet. Often used loosely for boundary, and unwisely for the side of a polygon. *See also* p. 177, arc (p. 239).

tetrahedron a polyhedron (↑) with four faces (which must be triangles). It can be a regular polyhedron. Similarly **hexahedron** (*see also* cube, ↓), **octahedron**, **dodecahedron**, **icosahedron** have six, eight, 12 and 20 faces respectively. All these can be regular, in which case they have triangular faces except for the hexahedron (squares) and the dodecahedron (pentagons, p. 177). Names for polyhedra with other numbers of faces can be formed using the Greek prefixes (*see* Appendix 2, p. 294), e.g. a **decahedron** has ten faces, but cannot be regular.

cube[2] (*n*) a regular hexahedron (↑) having six square faces. *See also* p. 34.

cubic (*adj*) related to a cube, also used of measurements of volume, e.g. cubic metres.

parallelepiped (*n*) a hexahedron (↑) whose faces are six parallelograms (p. 181).

rhombohedron (*n*) usually a hexahedron (↑) whose faces are six rhombuses (p. 181), but other definitions are found.

cuboid (*n*) a hexahedron (↑) whose faces are six rectangles. Also known as **rectangular block**.

boundary (*n*) loosely the edges, faces and vertices of polyhedra (↑), sides and vertices of polygons, closed curves (p. 190) round a region, and the surfaces of solids. Also known as **bound**.

stellated (*adj*) (of solids) star-shaped, formed by producing non-adjacent faces of a polyhedron (↑). **stellate** (*v*).

faceted (*adj*)(of solids) formed by drawing planes through non-adjacent (p. 171) vertices of a polyhedron (↑). **facet** (*n*).

Kepler-Poinsot polyhedron one of the four regular stellated (↑) or faceted (↑) polyhedra (↑) which have equal faces and angles, but in which the faces intersect (*compare* star polygons, p. 178, in two dimensions).

semi-regular polyhedron a polyhedron (p. 200) with the same arrangement of regular polygons around each vertex. Also known as **Archimedean polyhedron**, **facially regular polyhedron**.

semi-regular dual polyhedron a polyhedron (p. 200) with every face having the same non-regular polygon which is the dual (p. 256) of a semi-regular polyhedron (↑) in which each vertex is replaced by a face and each face by a vertex. Also known as **Archimedean dual polyhedron**, **vertically regular polyhedron**.

simple closed polyhedron a polyhedron (p. 200) with no missing faces, holes or intersecting faces.

Euler's formula[2] $V + F = E + 2$, where V is the number of vertices, F is the number of faces and E is the number of edges, which holds for any simple closed polyhedron (↑) as well as for any planar network (p. 239) or any network on the surface of a sphere. *Do not mistake for other formulae known as* Euler's formula (pp. 49, 149). *See also* Euler characteristic (p. 242).

Schlegel diagram a plane diagram of a polyhedron (p. 200) formed by making a hole in one face and stretching the surface into a plane, the face with the hole becoming the outer part of the diagram formed.

net[2] (*n*) of a solid: a plane configuration of lines drawn on paper which when cut out and folded forms the solid. *See also* pp. 40, 239.

Schäfli number an ordered pair (p. 19) of numbers (p, q) or p^q which describes a polyhedron (p. 200) having at each vertex q faces, each face having p sides, e.g. a dodecahedron is (5, 3).

dihedron (*n*) a plane regarded as a polyhedron (p. 200) with two faces and an axis which is perpendicular to it. **dihedral** (*adj*). *See also* p. 200.

prism (*n*) a polyhedron (p. 200) with two congruent (p. 179) parallel faces and a constant cross-section (p. 200) parallel to them. Prisms are usually described by the shape of the two parallel faces, which are known as the *bases*. A prism of planes is a set of any number of planes whose lines of intersection in pairs are all parallel but not coplanar (p. 171).

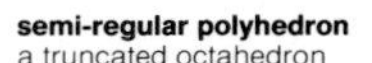

semi-regular polyhedron
a truncated octahedron

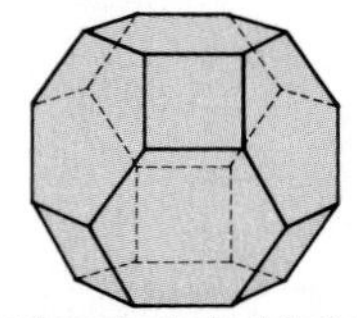

semi-regular dual polyhedron
a rhombic dodecahedron

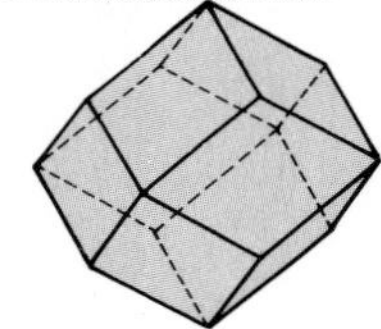

Schlegel diagram
a dodecahedron

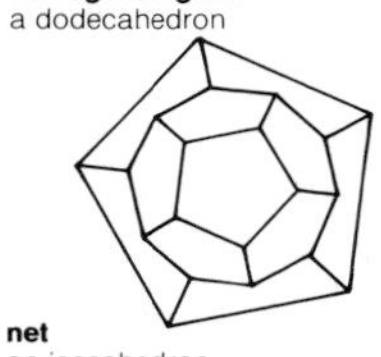

net
an icosahedron

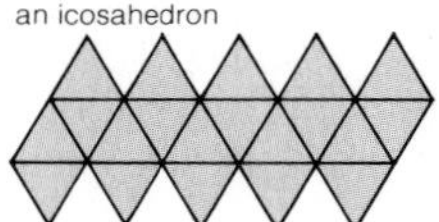

prism

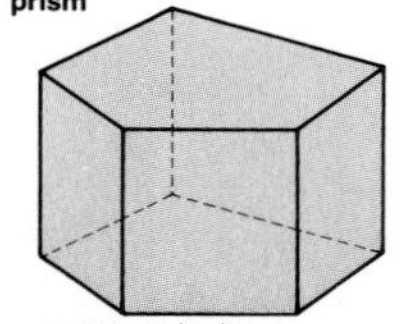

a pentagonal prism

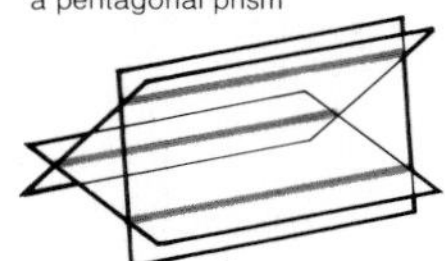

a prism of three planes meeting in the three red parallel lines

antiprism
a pentagonal antiprism

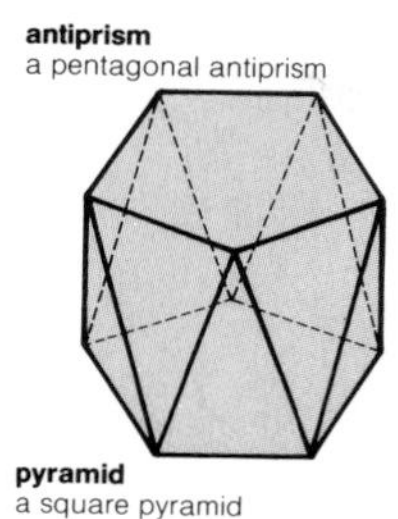

pyramid
a square pyramid

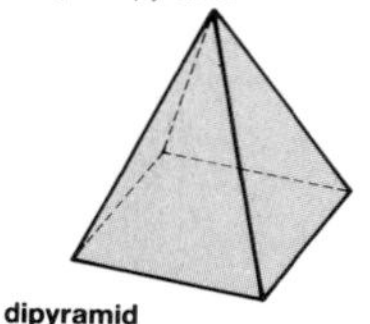

dipyramid
a square dipyramid

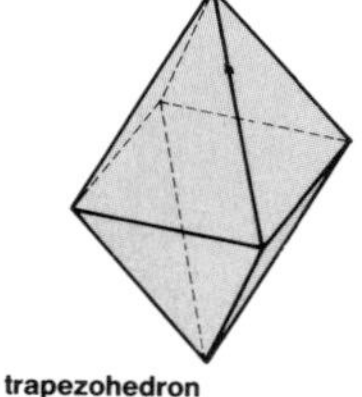

trapezohedron
the dual of a square dipyramid

truncated
a truncated pyramid

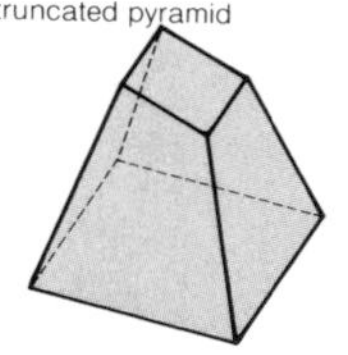

lateral (*adj*) used of a face or side, e.g. of a face of a prism (↑) which is not a base.

antiprism (*n*) a polyhedron (p. 200) similar to a prism (↑), usually with bases which are regular polygons, but with one base turned through an angle in its plane and the lateral (↑) faces replaced by double the number of triangles.

pyramid (*n*) a polyhedron (p. 200) with one face called the *base* having each vertex joined to a point not in its plane which forms the remaining vertex (often known as *the* vertex), every face other than the base being a triangle. Usually described by the shape of the base.

dipyramid (*n*) a polyhedron (p. 200) formed of two pyramids (↑) with equal bases joined base to base. It is the dual (p. 256) of a prism (↑).

trapezohedron (*n*) a sort of twisted dipyramid (↑) in which every face is a kite (p. 181). It is the dual (p. 256) of an antiprism (↑).

right (*adj*) perpendicular; in particular of cylinders (p. 209) and prisms (↑), having the lines joining corresponding points on the two bases perpendicular to the bases, and of pyramids (↑) and cones (p. 210) where the base has a centre of symmetry (p. 233) and the line joining this to the vertex is perpendicular to the base.

oblique (*adj*) not at right angles, as in oblique cartesian coordinates (p. 214); and of cylinders (p. 209), pyramids (↑) etc, which are not right (↑).

truncated (*adj*) used of a solid (often a pyramid, ↑, or cone, p. 210) part of which has been cut off by a plane. *See also* truncate (p. 44).

desmic (*adj*) used of pairs of tetrahedra (p. 201) which are such that each edge of either meets two opposite edges of the other.

polytope (*n*) a configuration in *n* dimensions where *n* is greater than three, corresponding to a polygon in two dimensions and a polyhedron (p. 200) in three dimensions.

hypercube (*n*) the polytope (↑) corresponding to a square and a cube in more than three dimensions; in four dimensions it is a **tesseract**.

hypersphere (*n*) the polytope (↑) corresponding to a circle and a sphere in more than three (and often just four) dimensions.

sphere (*n*) the surface formed by the locus (p. 187) of a point in three dimensions moving at a fixed distance from a fixed point. *See also* radius (p. 192). **spherical** (*adj*).

sphere

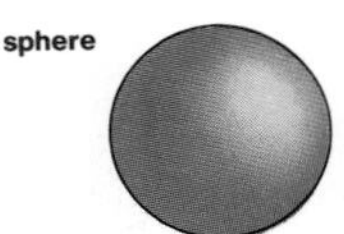

ball (*n*) the region contained by a sphere.

spherical shell a thin hollow sphere which has inside and outside surfaces but no thickness.

hemisphere (*n*) half a sphere, bounded by a plane through the centre.

great circle a circle lying on the surface of a sphere whose centre is the centre of the sphere.

small circle a circle lying on the surface of a sphere whose centre is not the centre of the sphere and whose radius is therefore less than that of the sphere.

circles on a sphere

great circle small circle

lune (*n*) a part of a sphere's surface bounded by two half great circles (↑). A spherical wedge is bounded by a lune and two semicircles (p. 192).

lune

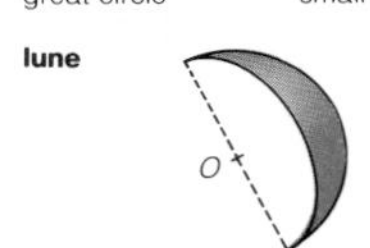

spherical segment the solid formed by cutting off part of a sphere by a plane or by two parallel planes. Also known as **zone**, **spherical zone**.

spherical segment

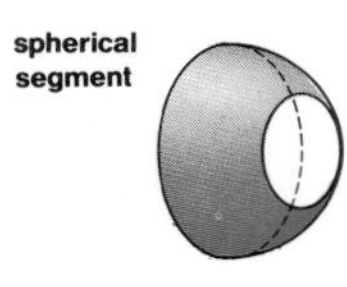

cap^2 (*n*) a spherical segment (↑) cut off by one plane which is smaller than a hemisphere. *See also* p. 13. Also known as **spherical cap**.

cap

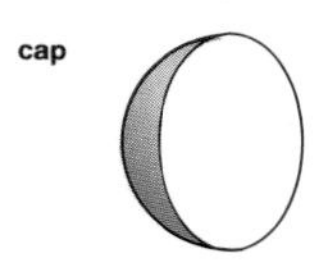

spherical triangle a configuration on the surface of a sphere bounded by the minor arcs (p. 192) of three great circles (↑). A spherical triangle is oblique (p. 203), right (p. 203), birectangular or trirectangular according to whether it has none, one, two or three right angles respectively.

spherical excess the amount by which the sum of the angles of a spherical triangle (↑) is greater than two right angles.

spherical triangle
the three arcs have centre at the sphere's centre, *O*

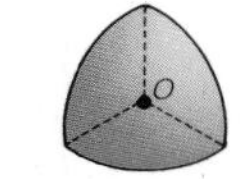

pseudosphere (*n*) a surface of constant curvature (p. 213) which is in the opposite sense to the curvature of a sphere so that triangles formed by geodesic (p. 212) lines have an angle sum less than two right angles, not greater than two right angles as in a spherical triangle (↑). It is the surface obtained by rotating a tractrix (p. 228) about its asymptote (p. 188).

equator (*n*) the great circle (↑) of a sphere which is perpendicular to the axis of rotation (p. 234). Used of the earth and heavenly bodies. **equatorial** (*adj*).

pseudosphere
the solid of revolution of a tractrix about its asymptote

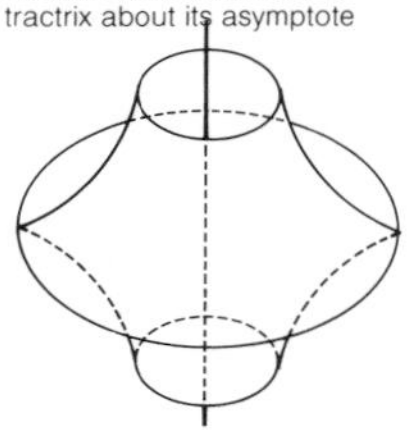

pole

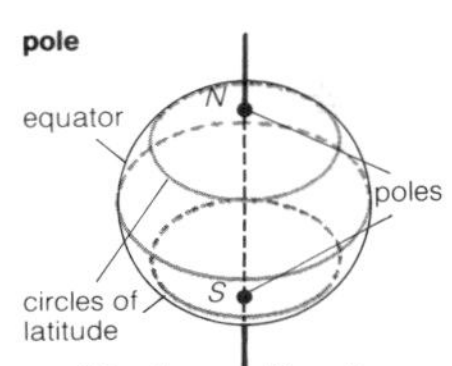

meridian through P meeting the equator at E: the latitude of P is θ = angle POE where O is the centre of the sphere

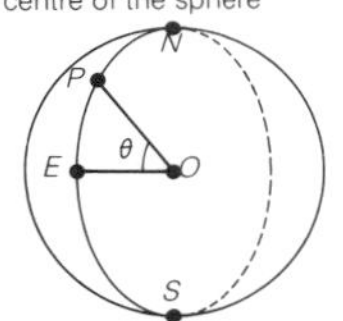

longitude
if G is a fixed point on the equator, then φ = angle EOG is the longitude of P

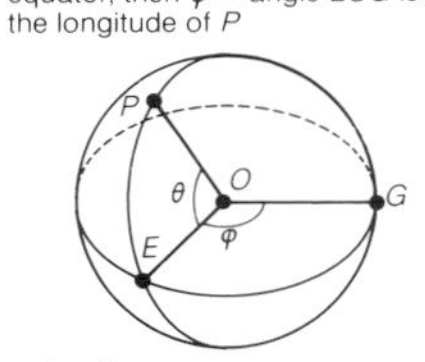

azimuth
θ is the azimuth of B from A where arc AB is part of a great circle

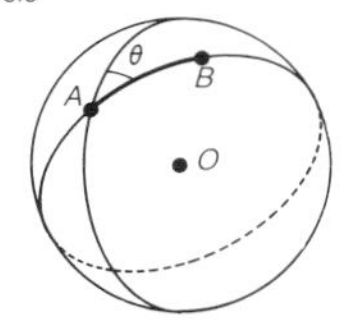

bearing and backbearing
θ is the bearing of B from A, φ is the backbearing of A from B; $\varphi = \theta \pm 180°$

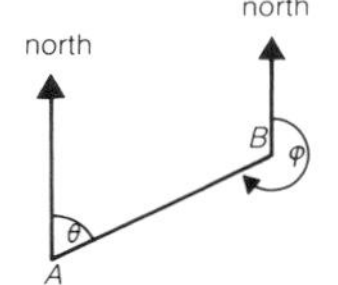

pole[1] (*n*) a point on the surface of a sphere which is also on the axis of rotation (p. 234). *See also* pp. 216, 223. **polar** (*adj*).

circle of latitude a small circle (↑) on the surface of a sphere which is parallel to the equator (↑).

meridian (*n*) a great circle (↑) of a sphere which passes through the two poles (↑); sometimes the half of such a circle between the poles. Also known as **circle of longitude**.

latitude (*n*) if P is a point on the surface of a sphere, O is the centre of the sphere and E is the point on the equator (↑) and on the same meridian (↑) as P so that angle POE is acute (p. 173), this angle is called the latitude of P. Conventionally (p. 9) P is given as north or south of the equator.

colatitude (*n*) the complementary angle (p. 175) of the latitude, found by subtracting it from 90°.

longitude (*n*) if P, O and E are defined as under latitude (↑) and G is a fixed point on the equator (↑), then the angle EOG is the longitude of P. Conventionally (p. 9) on the surface of the earth G lies on the meridian (↑) through Greenwich (London) and angle EOG is given as an angle between 0° and 180° east or west of this.

azimuth (*n*) the azimuth of a point B from a point A on a sphere is the angle between the great circle (↑) through A and B and the meridian (↑) through A.

bearing (*n*) the azimuth (↑) expressed as an angle between 0° and 360° measured clockwise (p. 174) from north using the convention (p. 9) in geography (the study of the earth) and not that in mathematics, and giving the direction of one point on the earth's surface from another.

backbearing (*n*) the bearing (↑) in the opposite direction to the one given, found by adding or subtracting 180° to or from the bearing.

zenith (*n*) the direction in the heavens immediately overhead at any given time. **zenithal** (*adj*).

nadir (*n*) the direction in the heavens exactly opposite the zenith (↑).

celestial sphere an imaginary sphere of infinite radius on which heavenly bodies are taken to lie and whose centre lies at the point of the person looking at the heavens.

equinoctial (*n*) a circle on the celestial sphere (↑) lying in the same plane as the equator (p. 204).

declination (*n*) the angle of a point in the celestial sphere (↑) corresponding to latitude (p. 205) for points on the surface of the earth, i.e. the angle above or below the equinoctial (↑) of the celestial sphere. *See also* variation (↓).

right ascension the angle of a point in the celestial sphere (↑) corresponding to longitude (p. 205) for points on the surface of the earth. Conventionally (p. 9) the fixed point in this case is the point known as the *First Point in Aries*.

ecliptic (*n*) the great circle (p. 204) on the celestial sphere (↑) which is the path on which the sun appears to move.

true bearing the bearing (p. 205) measured from true north, i.e. the bearing from the direction of the north pole (p. 205).

magnetic bearing the bearing (p. 205) measured by a compass from the direction of the north-pointing end of the needle.

grid bearing the bearing (p. 205) measured on a map using the grid (p. 207) lines on it which are approximately but not usually exactly true north.

variation (*n*) (1) the angle between the true bearing (↑) and the magnetic bearing (↑) at a point on the earth's surface. The variation changes with the position of the point, and also with time. Also known as **magnetic variation**, **declination**. (2) also frequently used in mathematics for change.

nautical mile a distance equal to the distance along a meridian (p. 205) of the earth which subtends (p. 175) an angle of one minute (p. 173) at the centre of the earth. The length of a nautical mile varies slightly depending on the part of the earth but is approximately 2000 metres.

equinoctial
α is the declination of P
β is the right ascension of P

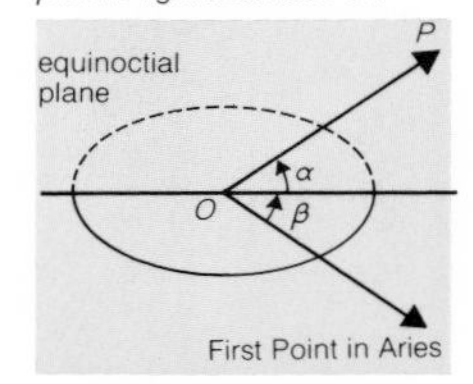

variation
variation at a point P on the earth's surface

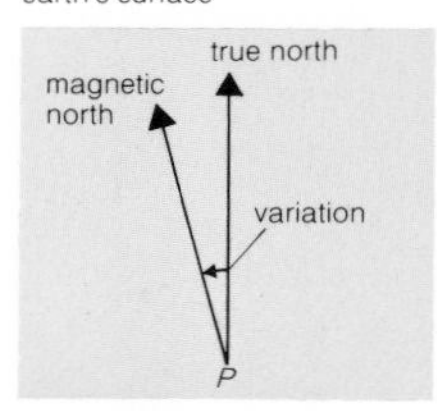

triangulation
R and *S* are fixed by measuring the angles at *P* and *Q*. *T* can then be fixed by measuring the angles at *R* and *S*

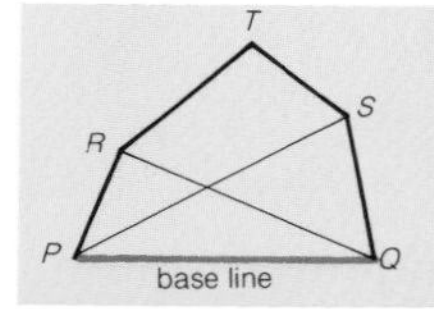

survey (*n*) a map or plan of a part of the earth's surface. **survey** (*v*).

triangulation (*n*) a method of surveying (↑) a map by measuring angles from the ends of a line of known length to draw triangles and calculate the lengths of their sides, and then using these sides as starting points for more triangles and so on. **triangulate** (*v*).

base line[1] the line of known length in a triangulation (↑). *See also* p. 216.

offset
P, Q, R, S, T are fixed by measuring offsets from a traverse *AB*

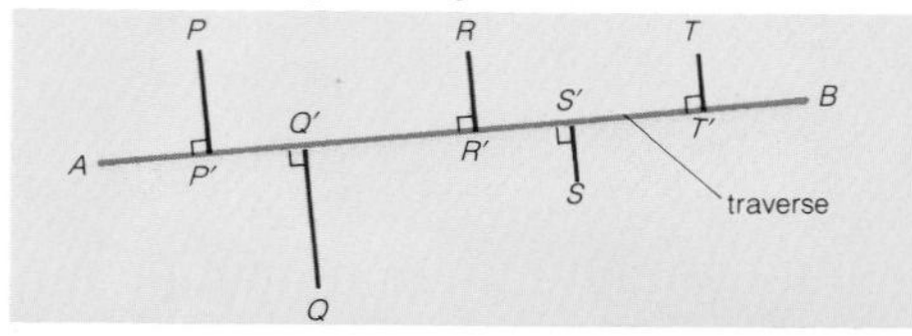

offset (*n*) a method of surveying (↑) a map of a small area by measuring distances or offsets at right angles to a fixed line from points at known distances along the line.

traverse (*n*) the fixed line in an offset (↑) survey (↑).

representative fraction the scale of a map given as a fraction or as a ratio of two numbers.

grid (*n*) intersecting sets of lines, usually those forming a coordinate system on a map or plan.

graticule (*n*) the grid (↑) on a map formed by the lines of latitude (p. 205) and longitude (p. 205), sometimes used more widely with the same meaning as grid.

easting (*n*) the distance of a point measured eastwards on the grid (↑) of a map from some origin (p. 215), corresponding to the *x*-coordinate (p. 215). Similarly **northing** corresponding to the *y*-coordinate (p. 215).

grid reference a number giving the coordinates of a point on a map measured on a grid (↑) from a conventional (p. 9) origin (p. 215) and axes, e.g. a grid reference 361058 might show a point with easting (↑) 361 units or 36.1 km. and northing 058 units or 5.8 km.

grid reference
if *P* has easting 36.1 km and northing 5.8 km its grid reference with respect to origin *O* is 361058. (Coordinates (36.1, 5.8))

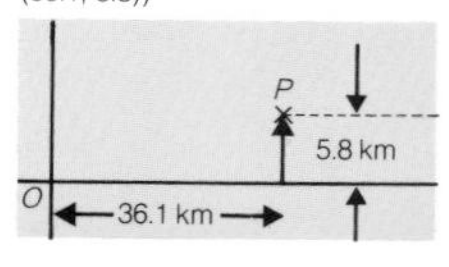

isogram (*n*) a line on a map joining points which have a constant value of some quantity, e.g. *isotherm*, a line of equal temperature.

contour (*n*) an isogram (↑), often a line on a map joining points of equal height above sea level.

map projection a transformation (p. 54) of the surface of the earth or part of it onto a plane, allowing a map to be drawn. Conical (p. 210) and cylindrical (p. 209) projections (p. 237) are obtained by first imagining the map on the surface of a cone or cylinder and then unrolling it. Equidistant (p. 169) and equal area projections are those in which equal distances (p. 169) or areas (p. 191) on the earth become equal distances or areas on the map (this is possible only for some of the distances). Also known as **projection**. *See also* projection (p. 237).

gnomonic (*adj*) used of a map projection (↑) from the centre of the earth onto a tangent plane (↓) in which all great circles (p. 204) become straight lines (p. 167). *See also* gnomon (p. 172).

stereographic (*adj*) used of a map projection (↑) from a point at one end of a diameter of the earth to the tangent plane (↓) at the opposite end of the diameter. It has the property that angles remain unaltered.

orthographic (*adj*) used of a map projection (↑) from a point at infinity to a perpendicular tangent plane (↓) at a point on the surface of the earth which gives a view of the earth as seen from outer space.

Mercator's projection a common cylindrical (↓) projection (↑) much used in maps for seamen and which gives correct angles.

transverse Mercator's projection Mercator's projection (↑) based on a meridian (p. 205) rather than on a parallel of latitude (p. 205), and used on some British government maps.

rhumb line a line of constant bearing (p. 205) on a map. In Mercator's projection (↑) it is a straight line (p. 167); on the earth's surface it is a spiral (p. 230) with centre at a pole (p. 205). Also known as **loxodrome**.

gnomonic
if *AB* is a diameter of the earth and *O* is its centre, the point *P* projects along *OP* to meet the tangent plane at *B* in *Q*

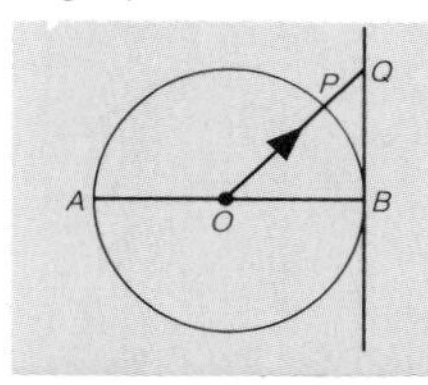

stereographic
if *AB* is a diameter of the earth, the point *P* projects along line *AP* to meet the tangent plane at *B* in *Q*

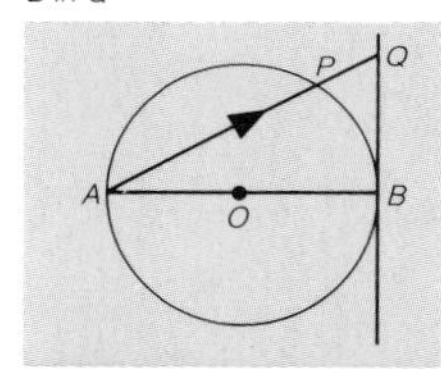

orthographic
if *AB* is a diameter of the earth, the point *P* projects parallel to *AB* to meet the tangent plane at *B* in *Q*

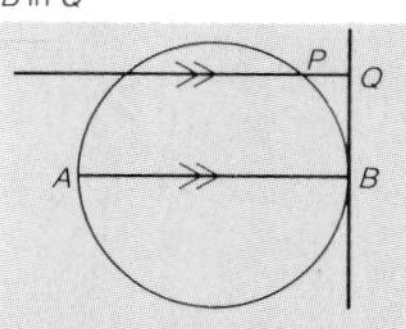

Mercator's projection

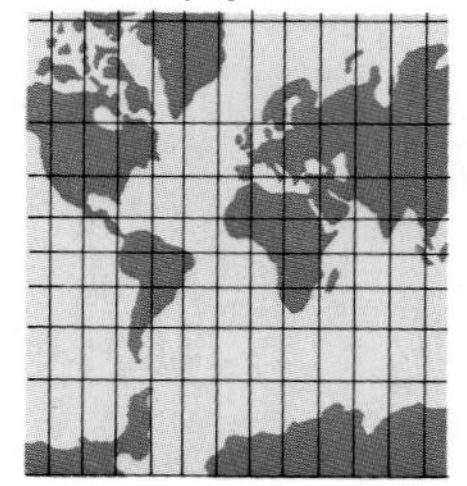

synclastic
a synclastic tangent plane

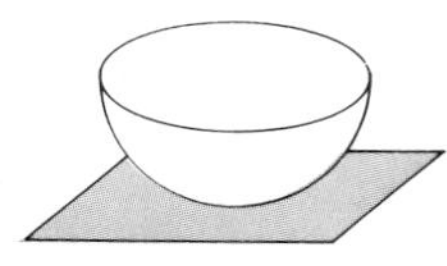

anticlastic
an anticlastic tangent plane

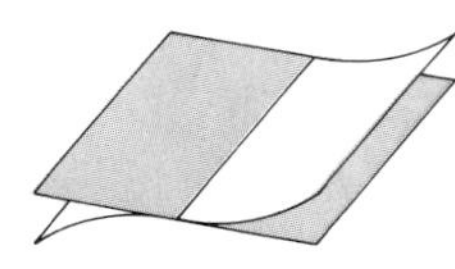

generator
a hyperboloid of one sheet showing the two sets of generators: this is an example of a ruled surface, it is not a developable surface but a skew surface

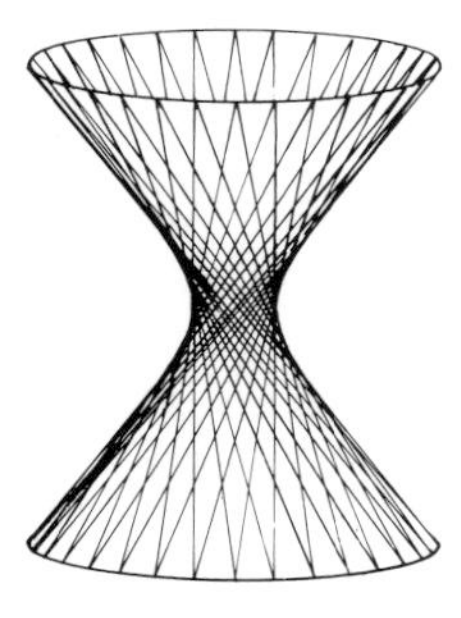

cylinder
right and oblique circular cylinders

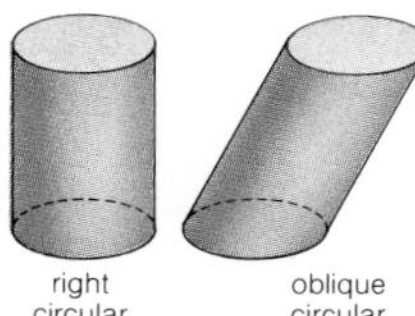

tangent plane the tangent plane to a surface at a point is that plane for which every line through the point in the plane touches the surface or lies on the surface.

synclastic (*adj*) used of a surface which lies wholly on one side of the tangent plane (↑) near the point of contact (p. 187).

anticlastic (*adj*) used of a surface which lies on both sides of the tangent plane (↑) near the point of contact (p. 187). Also known as **antisynclastic**.

generator[2] (*n*) that which produces a mathematical entity (p. 288); one of a set of straight lines (p. 167) or curves lying on certain surfaces which make up the surface; a base vector (p. 80). *See also* p. 74.

generate (*v*) to determine or make by means of a rule, e.g. to generate the terms of a sequence (p. 91). *See also* span (p. 80). **generation** (*n*).

ruled surface a surface which has through any point one or more straight lines (p. 167) or generators (↑) lying on that surface, e.g. a cone (p. 210) and a hyperboloid of one sheet (p. 211). It may, but need not, be a developable surface (↓).

developable surface a ruled surface (↑) in which consecutive generators (↑) intersect and which can be bent without stretching or tearing into a plane, e.g. a cone (p. 210). A cylinder (↓) is a special case where the intersections are at infinity.

skew surface a ruled surface (↑) in which consecutive generators (↑) of the same set do not intersect, e.g. a hyperboloid of one sheet (p. 211).

closed surface a finite surface with no boundary curves. (Edges, p. 201, as in polyhedra, p. 200, are not boundary curves in this sense.)

cylinder (*n*) a surface formed by drawing a set of parallel lines through every point of a curve (often a plane closed curve, p. 190, especially a circle). Most often used for the solid formed by two circles in parallel planes joined by a set of line segments (p. 167). If these are perpendicular to the circles it is a right (p. 203) circular cylinder, if not it is an oblique (p. 203) circular cylinder. Other cylinders are similarly called by the shapes of the curves. **cylindrical** (*adj*).

cone (*n*) a solid with one face called the *base* which is bounded by a plane closed curve (p. 190), every point of which is joined by a straight line to another point called the *vertex* outside its plane. Used most often to mean a right (p. 203) circular cone where the curve is a circle such that the line joining the vertex to its centre is perpendicular to it. Also often used with the same meaning as nappe (↓). **conical** (*adj*).

slant height the slant height of a right (p. 203) circular cone (↑) is the length of a straight line (p. 167) from the boundary of the base to the vertex. Also used similarly of other solids.

nappe (*n*) the infinite surface formed by producing (p. 167) every straight line (p. 167) on the curved surface of a right (p. 203) circular cone (↑) to infinity in both directions. *See also* cone (↑). Also known as **double cone**.

zone (*n*) a part of a surface (often of a sphere) lying between two parallel planes. *See also* spherical segment (p. 204), frustrum (↓).

Archimedes' theorem the surface area (p. 191) of a spherical (p. 204) zone (↑) is equal to the area of the curved surface of the cylinder (p. 209) of equal radius to the sphere and whose bases are the two planes of the boundaries of the zone.

frustrum (*n*) usually a truncated (p. 203) cone (↑) between circles in two parallel planes; also part of a solid between two parallel planes. *See also* zone (↑).

torus (*n*) the solid of revolution (p. 155) formed by rotating a circle about an axis in its plane and lying outside it. Also known as **ring**, **anchor ring**.

minimal surface a surface of minimum area which is bounded by closed curves (p. 190), e.g. *see* catenoid (↓). It is important in the study of soap films.

catenoid (*n*) the solid of revolution (p. 155) formed by rotating a catenary (p. 228) $y = a \cosh (x/a) + c$ about the axis $y = 0$. It is the minimal surface (↑) formed by a soap film joining two parallel circles with a common axis.

cone

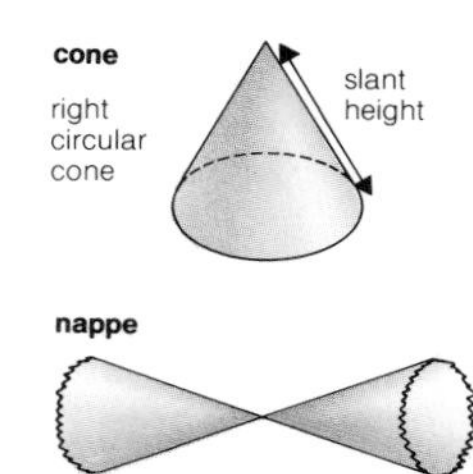

nappe

part of a nappe or double cone

Archimedes' theorem
the surface areas of the spherical zone and the cylinder are equal when they lie between the same two parallel planes

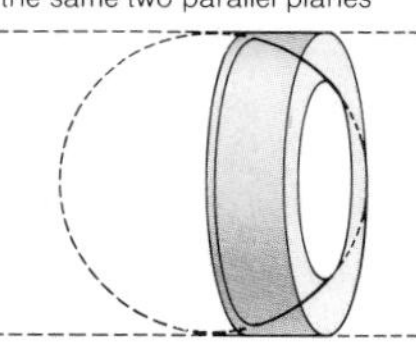

frustrum
frustrum of a cone

torus

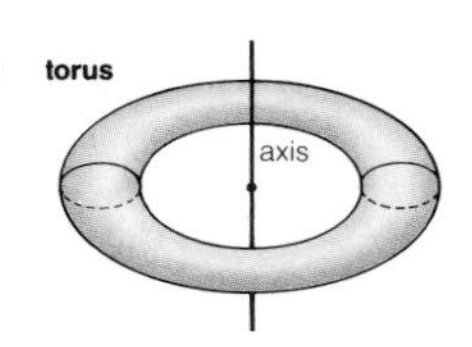

catenoid

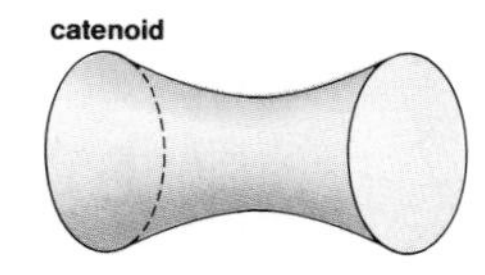

quadric (*n*) a surface in three dimensions corresponding to a conic (p. 220) in two dimensions and given by the general equation of the second degree in three variables. Also known as **quadric surface**, **conicoid**.

central quadric a quadric (↑) which has a centre of symmetry (p. 233).

ellipsoid (*n*) a central quadric (↑) whose equation may be written in the form

$$x^2/a^2 + y^2/b^2 + z^2/c^2 = 1,$$

where *a*, *b* and *c* are the non-zero lengths of the semi-axes (p. 222).

ellipsoid

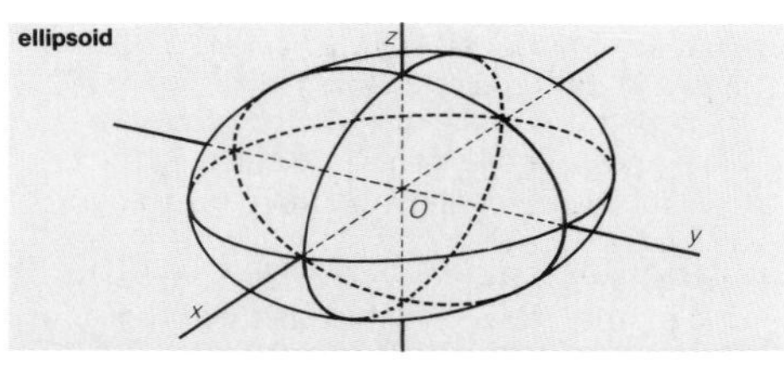

hyperboloid of one sheet

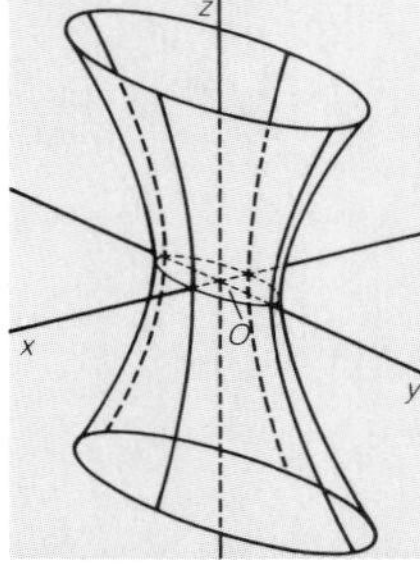

hyperboloid of two sheets

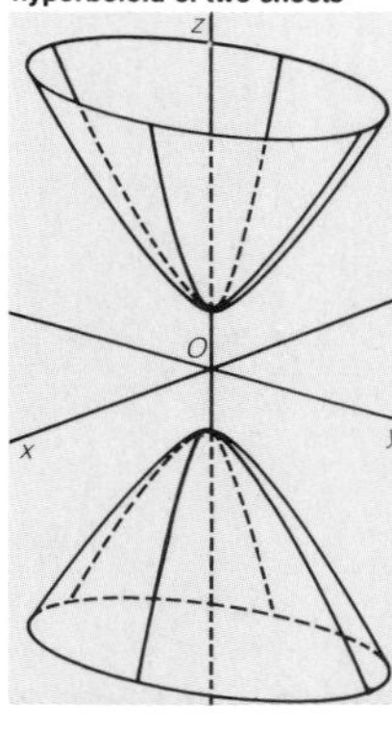

spheroid (*n*) an ellipsoid (↑) which has two equal semi-axes (p. 222) and is formed by the rotation of an ellipse (p. 220) about one of its axes.

oblate spheroid a spheroid (↑) formed by rotating an ellipse (p. 220) about its major axis (p. 221).

prolate spheroid a spheroid (↑) formed by rotating an ellipse (p. 220) about its minor axis (p. 221); it is approximately the shape of the earth.

geoid (*n*) a prolate spheroid (↑) which is the shape and size of the earth, forgetting relatively small irregularities (p. 289) such as mountains.

hyperboloid of one sheet a central quadric (↑) whose equation may be written in the form

$$x^2/a^2 + y^2/b^2 - z^2/c^2 = 1,$$

where *a*, *b* and *c* are the non-zero lengths of the semi-axes (p. 222).

hyperboloid of two sheets a central quadric (↑) whose equation may be written in the form

$$x^2/a^2 - y^2/b^2 - z^2/c^2 = 1,$$

where *a*, *b* and *c* are the non-zero lengths of the semi-axes (p. 222). The surface has two separate parts.

paraboloid (*n*) a non-central quadric (p. 211) whose equation may be written in the form $x^2/a^2 \pm y^2/b^2 - z/c = 0$, where *a, b* and *c* are non-zero constants. The positive sign gives an elliptic (p. 220) paraboloid, which meets $z = k$ (k with the same sign as c) in an ellipse. The negative sign gives a hyperbolic (p. 221) paraboloid, which meets $z \mp k$ in a hyperbola (p. 221) for all k; this is a ruled surface (p. 209) sometimes used in building roofs.

regulus (*n*) a set of generators (p. 209) of a quadric (p. 211) which is a ruled surface (p. 209), no two of which intersect.

umbilic (*n*) if a set of parallel circular sections is drawn to a quadric (p. 211), the locus (p. 187) of their centres is a diameter of the quadric. If there are point circles (p. 195) at the ends of this diameter, then these points are called umbilics.

space-curve (*n*) a curve which does not lie in a single plane. Also known as **twisted curve**, **skew curve**.

geodesic (*n*) a curve lying in a surface which is the line of minimum length joining two points of that surface. The geodesics on a sphere are minor arcs (p. 192) of great circles (p. 204)

helix (*n*) a space-curve (↑) on the surface of a right (p. 203) circular cylinder (p. 209) which cuts the generators (p. 209) at a constant angle, often loosely used for other similar space-curves. Also often incorrectly known as a spiral (p. 230) in everyday life. **helical** (*adj*).

differential geometry the application of the methods of analysis to geometry, especially to space-curves (↑) and surfaces in three dimensions, and extensions of these ideas to higher dimensions.

osculating plane if *Q, P* and *R* are three points on a space-curve (↑), the limiting position of the plane *QPR* as *Q* and *R* tend to *P* is the osculating plane at *P*.

principal normal of a space-curve (↑) at a point *P* on the curve: the line through *P* perpendicular to the tangent and lying in the osculating plane (↑) at *P*. Its positive direction is arbitrary (p. 287).

paraboloid

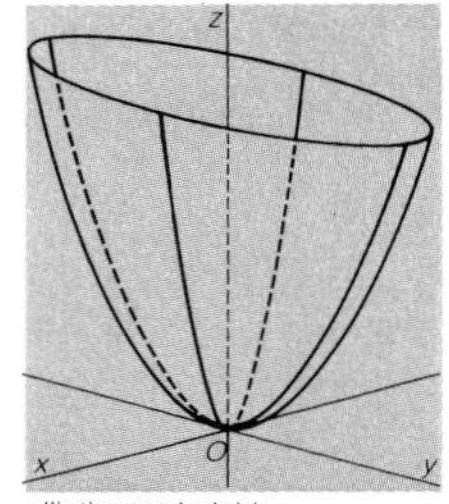

elliptic paraboloid

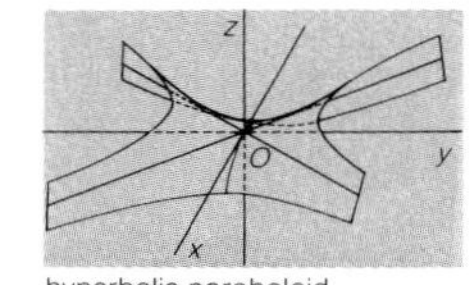

hyperbolic paraboloid

umbilic
PP′ is a diameter of centres of circular cross-sections of the quadric, *P* and *P′* are umbilics

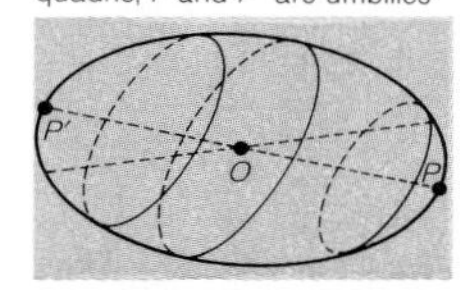

helix

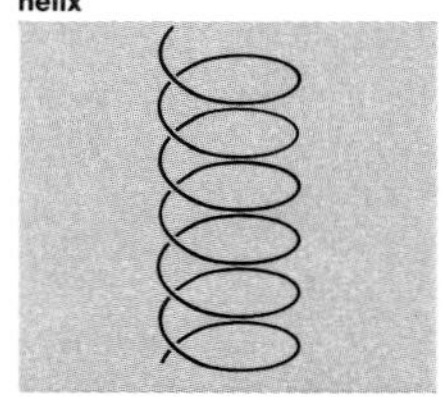

binormal
the osculating plane contains the principal normal and the tangent and is perpendicular to the binormal

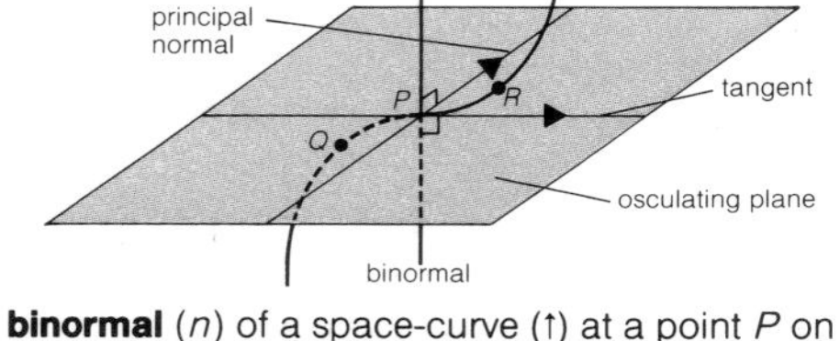

normal and rectifying planes
the space-curve itself is omitted

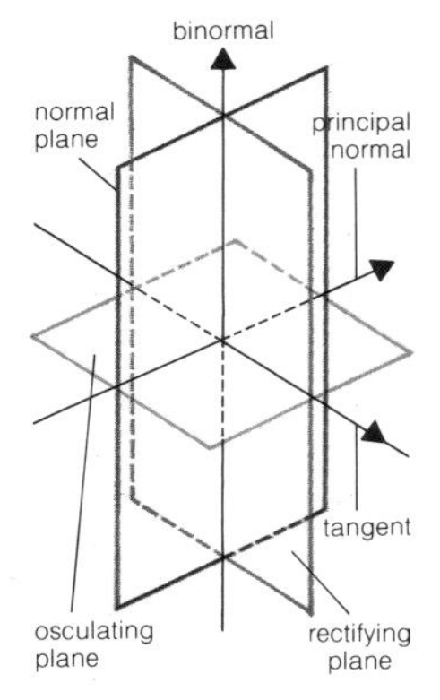

binormal (*n*) of a space-curve (↑) at a point *P* on the curve: a line through *P* perpendicular to the tangent and principal normal (↑) through *P* (and to the osculating plane, ↑). Its positive direction is chosen so that the unit tangent vector $\mathbf{t}$, the unit principal normal (↑) vector $\mathbf{n}$, and the unit binormal vector $\mathbf{b}$ form a right-handed set (p. 82).

normal plane of a space-curve (↑) at a point *P*: the plane through *P* containing the principal normal (↑) and binormal (↑).

rectifying plane of a space-curve (↑) at a point *P*: the plane through *P* containing the tangent and binormal (↑).

curvature vector if $\mathbf{r}$ is the position vector of a point *P* on a space-curve (↑), the curvature vector at *P* is the rate of change of the unit tangent vector at *P* with respect to arc length given by $d\mathbf{t}/ds = \mathbf{t}' = d^2\mathbf{r}/ds^2 = \mathbf{r}''$ and is equal to $\kappa\mathbf{n}$ where κ is the curvature (↓) at *P* and $\mathbf{n}$ is the unit principal normal (↑) at *P*.

curvature[2] (*n*) the quantity κ under curvature vector (↑). *See also* p. 189.

torsion (*n*) the quantity τ which is the rate of turning with respect to arc length of the osculating plane (↑) about the tangent of a space-curve (↑) at any given point. It is given by $\mathbf{b}' = -\tau\mathbf{n}$ where $\mathbf{b}'$ is the rate of change of the unit binormal (↑) with respect to arc length and $\mathbf{n}$ is the unit principal normal (↑).

Serret-Frenet formulae the formulae

$\mathbf{t}' = \kappa\mathbf{n}$ (*see* curvature, ↑),

$\mathbf{b}' = -\tau\mathbf{n}$ (*see* torsion, ↑) and

$\mathbf{n}' = \tau\mathbf{b} - \kappa\mathbf{t}$,

where $\mathbf{n}'$ is the rate of change of the unit principal normal (↑) with respect to arc length.

analytical geometry the study of geometry by using ordered sets (p. 19) of numbers to define the positions of points in space. The number of elements in the ordered set is the dimension of the geometry. Also known as **coordinate geometry**, **cartesian geometry**.

position (*n*) place in space, the place defined by a set of coordinates or by other means.

coordinate (*n*) one of the elements in an ordered set (p. 19) which defines the position of a point in space. Symbol: $(x, y, \ldots)$.

cartesian (*adj*) related to a coordinate system based on a set of as many concurrent (p. 171) straight lines as the dimension of the space (usually two or three), and distances measured either parallel or perpendicular to these lines. The base lines must be placed in such a way that points on any one cannot be defined from the rest, i.e. they must be independent. *See especially* rectangular cartesian coordinates (↓).

rectangular cartesian coordinates the commonest coordinate system which uses lines which are all perpendicular to each other where it does not matter whether measurements are parallel or perpendicular to the base lines. *See also* cartesian (↑).

oblique cartesian coordinates cartesian (↑) coordinates which are not rectangular cartesian coordinates (↑).

covariant (*adj*) used of oblique cartesian coordinates (↑) using measurements taken from the origin (↓) to the feet (p. 175) of the perpendiculars from the point to the axes. *See also* contravariant (↓).

contravariant (*adj*) used of oblique cartesian coordinates (↑) using measurements parallel to the axes. This is the usual system. *See also* covariant (↑).

real Euclidean plane the plane obtained by taking all ordered pairs (p. 19) of real numbers (p. 46) as pairs of cartesian (↑) coordinates defining points. Similarly **real Euclidean space** which may have more than two dimensions.

rectangular cartesian coordinates
in three dimensions
forming a right-handed set: *P* is the point (*a*, *b*, *c*)

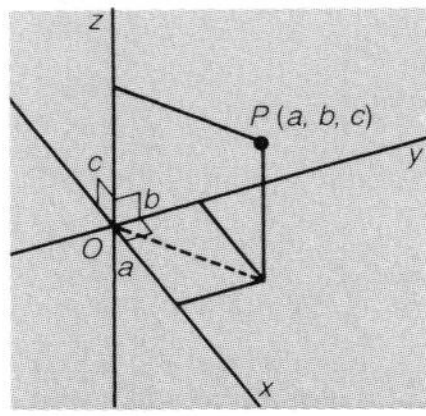

covariant
covariant oblique cartesian coordinates in two dimensions

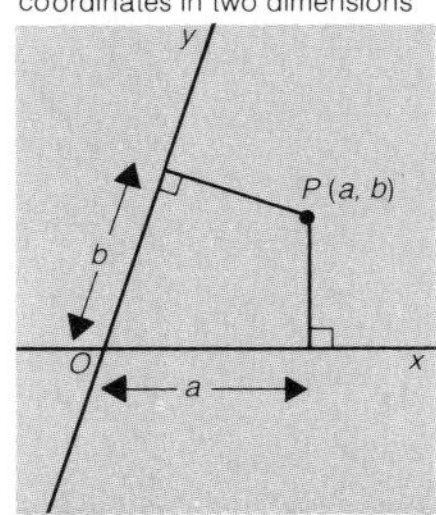

contravariant
contravariant oblique cartesian coordinates in two dimensions

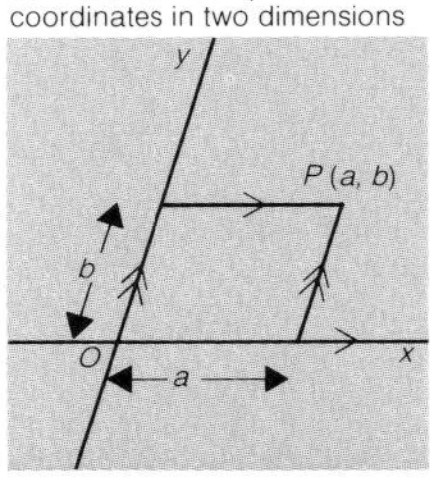

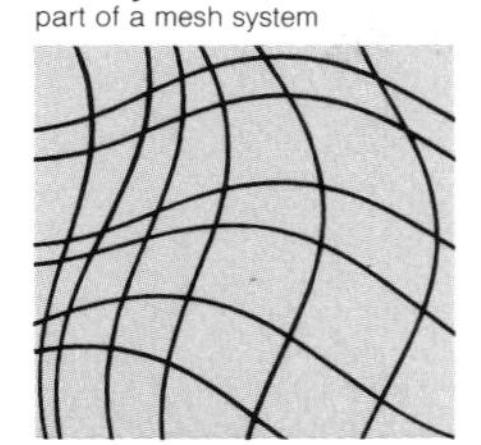
mesh system
part of a mesh system

cartesian graph a graph representing sets of points in cartesian (↑) coordinates, e.g. $y^2 = 4ax$ is represented by a parabola (p. 220).

mesh system two sets of straight lines (p. 167) or curves in a plane such that no two lines of any one set intersect, but all lines of one set intersect each line of the other. The mesh system can be used to define the position of a point in a plane; rectangular plane cartesian (↑) coordinates are an example of such a system, as are the lines of latitude (p. 205) and longitude (p. 205) on a map, except at the poles (p. 205).

axis (*n*) (1) a straight line (p. 167) about which rotation takes place; (2) a straight line about which a configuration has rotational (p. 232) or reflective symmetry (p. 232); (3) a straight line which is a base line (p. 216) in cartesian (p. 214) coordinates and from which distances are measured. **axes** (*pl*), **axial** (*adj*).

axial plane in cartesian (p. 214) coordinates of more than two dimensions, a plane containing a pair of axes.

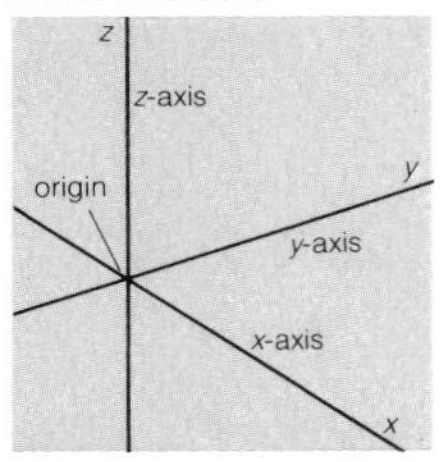

origin and axes
in three dimensions

origin (*n*) the starting point from which a coordinate system begins, in cartesian (p. 214) coordinates the intersection of the axes. *See also* p. 167.

***x*-axis** (*n*) the first axis in cartesian (p. 214) coordinates. It is the one along or parallel to which the first measurement is made in giving the coordinates of a point. Similarly ***y*-axis**, ***z*-axis**, which are the second and third axes.

***x*-coordinate** (*n*) the first coordinate in cartesian (p. 214) coordinates which is measured along or parallel to the *x*-axis (↑). Similarly ***y*-coordinate**, ***z*-coordinate**, which are the second and third coordinates.

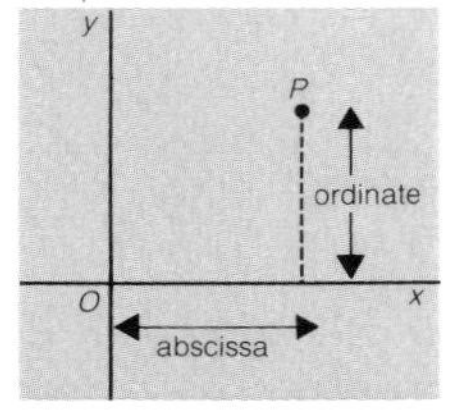

abscissa and ordinate
of a point *P* in two dimensions

abscissa (*n*) the *x*-coordinate (↑) of a point in plane rectangular cartesian coordinates (↑).

ordinate (*n*) (1) the *y*-coordinate (↑) of a point in plane rectangular cartesian coordinates (↑); (2) a line drawn perpendicular to a given line from a given point.

polar coordinates (r, θ) of a point P

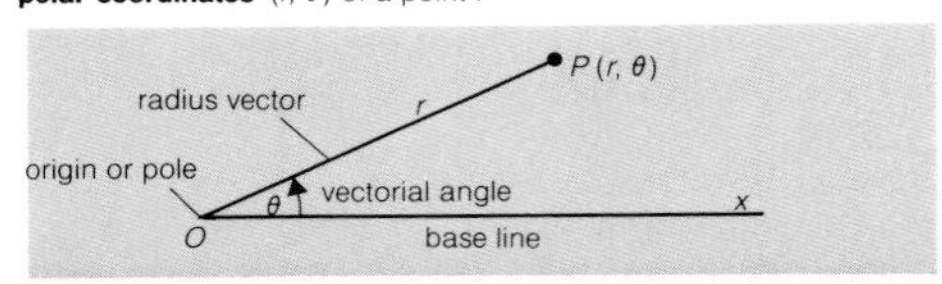

polar coordinates plane coordinates in which the position of a point P is described by its distance from a fixed point (the pole, ↓) O and the angle that OP makes with a fixed ray (p. 167) or half-line (the base line, ↓) through O, the angle being measured in the positive or anticlockwise (p. 174) direction from the base line.

pole[2] (*n*) the origin or fixed point in polar coordinates (↑). *See also* pp. 205, 223. **polar** (*adj*).

base line[2] the ray (p. 167) or half-line from which angles are measured in polar coordinates (↑). *See also* p. 207. Also known as **initial line**.

radius vector the position vector (p. 79) of a point in polar coordinates (↑). **radius vectors** (*pl*), **radii vectores** (*pl*).

vectorial angle the angle measured anticlockwise (p. 174) from the base line (↑) to the radius vector (↑) in polar coordinates (↑).

spherical polar coordinates three dimensional coordinates in which the first coordinate is the distance from a fixed point or pole (↑) and the other two coordinates are angles. These correspond respectively either to latitude (p. 205) and longitude (p. 205) on the earth, or alternatively to colatitude (p. 205) and longitude. In older books other definitions of the angles may be found.

cylindrical polar coordinates three dimensional coordinates similar to three dimensional cartesian (p. 214) coordinates except that the x- and y-coordinates (p. 215) are replaced by polar coordinates (↑), the z-coordinate remaining the same.

bipolar coordinates plane coordinates in which the position of a point is given by the distances (p. 169) r_1 and r_2 from two fixed points. In general, each value of r_1 and r_2 defines two points.

spherical polar coordinates
two ways of giving (r, θ, φ) for a point P

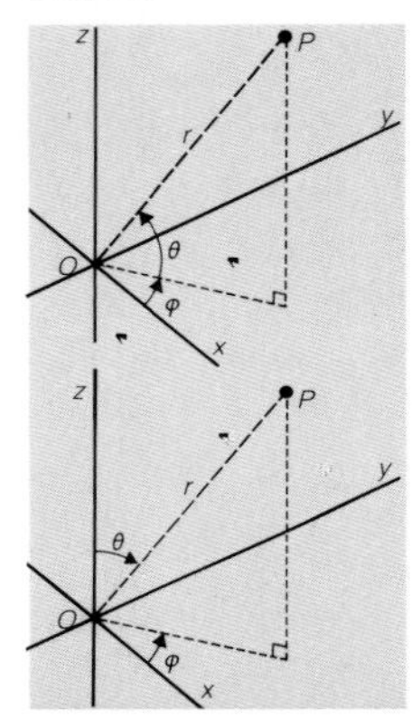

cylindrical polar coordinates
(r, θ, z) of a point P

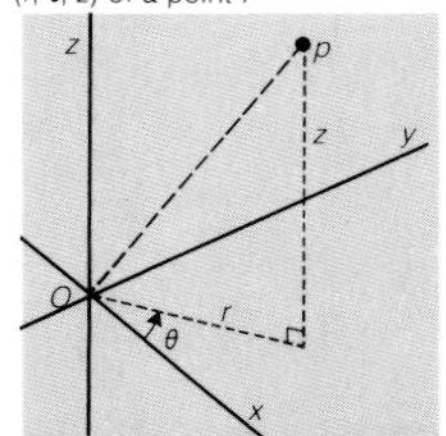

bipolar coordinates
the two points P_1 and P_2 with bipolar coordinates (r_1, r_2) using poles A and B

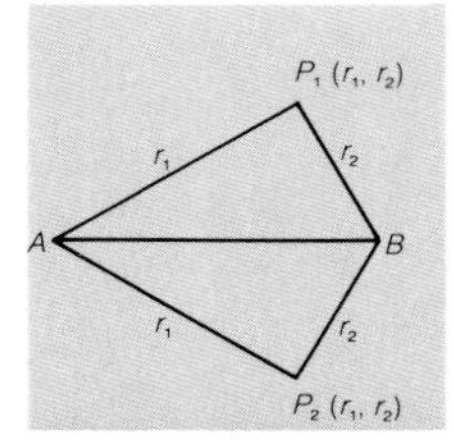

transformation of coordinates changing from one type of coordinate system to another; changing the origin (p. 215) and/or axes of a coordinate system. *See also* transformation (p. 54).

homogeneous coordinates cartesian (p. 214) coordinates to which another coordinate is added and points are then defined by ratios. In two dimensions a point is defined by (x, y, z) which can be regarded as corresponding to $(x/z, y/z)$ in cartesian coordinates. This allows points at infinity to be defined by putting $z = 0$.

accessible point a point in homogeneous coordinates (↑) for which z is non-zero. Similarly **inaccessible point**.

areal coordinates

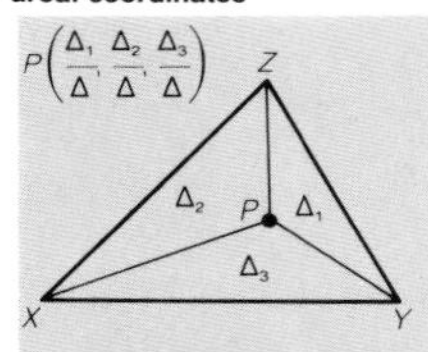

areal coordinates if P is a point inside a triangle XYZ, then P may be defined as $(\triangle_1/\triangle, \triangle_2/\triangle, \triangle_3/\triangle)$ where $\triangle_1$, $\triangle_2$, $\triangle_3$ and $\triangle$ are the areas of triangles (p. 135) PYZ, PZX, PXY and XYZ respectively. The sum of the three coordinates is always unity. (The definition may be extended, p. 262, to include cases where P is outside the triangle by using negative areas.) Also known as **barycentric coordinates**.

trilinear coordinates

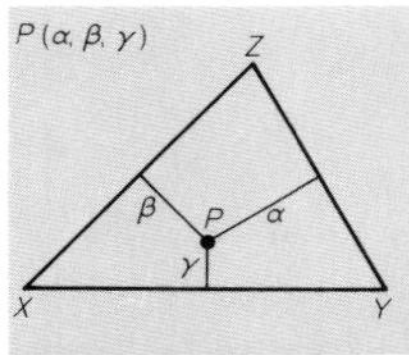

trilinear coordinates coordinates similar to areal coordinates (↑) but in which P is defined by the ratios of its distances from the three sides of the triangle XYZ. The first coordinate is taken to be positive if P is on the same side of YZ as X and so on. The sum of the coordinates is not necessarily unity.

triangle of reference the triangle whose coordinates are (1, 0, 0), (0, 1, 0) and (0, 0, 1) in the definitions of homogeneous coordinates (↑) and areal coordinates (↑); the triangle which is used in the definition of trilinear coordinates (↑).

intrinsic equation
if the arc P_0P is length s, then $s = f(\psi)$

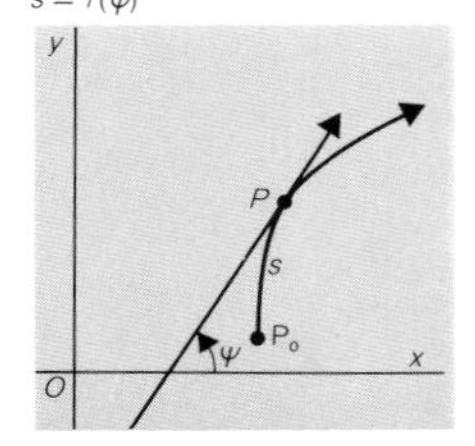

intrinsic equation the equation of a plane curve given in terms of s and ψ where s is the distance along the curve from a fixed point on the curve to a given point P and ψ is the angle made by the tangent at P (in the direction in which s increases) with the positive x-axis (p. 215).

pedal equation the equation of a plane curve given in terms of p and r where r is the radius vector (p. 216) of a point P on the curve in polar coordinates (p. 216) and p is the perpendicular distance of the tangent at P from the origin.

tangential polar coordinates the coordinates p and r which give the pedal equation (↑).

parametric coordinates coordinates of any type given in terms of a parameter (p. 67), e.g. the parabola (p. 220) $y^2 = 4ax$ may be given with a parameter t as $x = at^2$, $y = 2at$.

ideal point a point in a geometry which is not found in the real world, e.g. the circular points at infinity (↓), the ideal complex point (p. 49). Similarly **ideal line**.

circular points at infinity two ideal points (↑) whose homogeneous coordinates (p. 217) are $(1, \pm i, 0)$ (*see* square root, p. 47, of -1) through which all circles in the plane pass, but no other conics (p. 220). Also known as the **isotropic points**.

isotropic lines the ideal lines (↑) $x^2 + y^2 = 0$ or $x \pm iy = 0$, each of which passes through one of the circular points at infinity (↑) and through the origin.

line at infinity the ideal line (↑) whose homogeneous (p. 61) equation is $z = 0$ and which passes through the circular points at infinity.

point at infinity (1) the ideal complex point (p. 49); (2) a point in homogeneous coordinates (p. 217) whose z-coordinate (p. 215) is zero.

principle of continuity any theorem which can be proved true for all real (p. 49) points or lines can be generalized to imaginary (p. 49) points or lines.

syzygy (n) an identity (p. 57) between the homogeneous coordinates (p. 217) of n points of the type $\lambda_1\mathbf{P}_1 + \lambda_2\mathbf{P}_2 + \ldots + \lambda_n\mathbf{P}_n = 0$, where $\mathbf{P}_i = (x_i, y_i, z_i, \ldots)$. Points for which such an identity exists are said to be in syzygy.

general equation of a straight line the equation $lx + my + n = 0$ where l and m are not both zero. This gives all straight lines (p. 167) in plane cartesian (p. 214) coordinates.

pedal equation
if PT is the tangent at P and $OT = p$ is perpendicular to PT, this is the equation $r = f(p)$

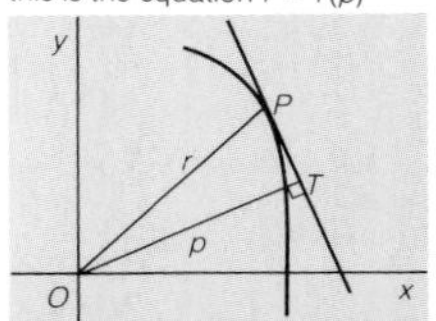

gradient-intercept form
$y = mx + c$, $m = \tan\theta$, and when $x = 0$, $y = c$

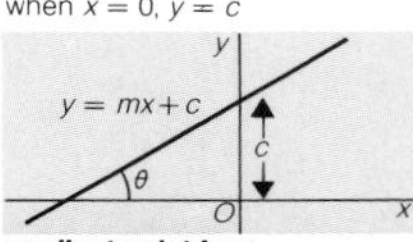

gradient-point form
$y - y_1 = m(x - x_1)$, $m = \tan\theta$, and the line passes through $P(x_1, y_1)$

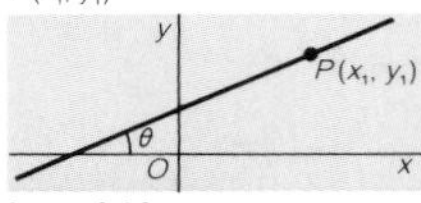

two-point form
$\frac{y - y_1}{y_2 - y_1} = \frac{x - x_1}{x_2 - x_1}$ passes through $P_1(x_1, y_1)$ and $P_2(x_2, y_2)$

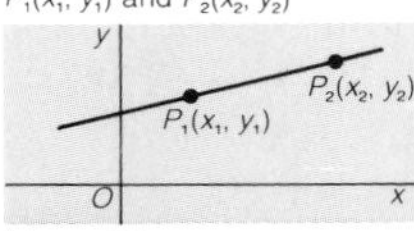

intercept form
$\frac{x}{a} + \frac{y}{b} = 1$ passes through $(a, 0)$ and $(0, b)$, having intercepts a, b on the axes

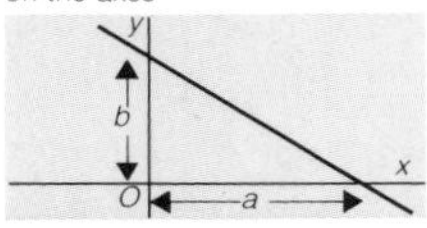

gradient of the line P_1P_2 is
$\tan\theta = \frac{y_2 - y_1}{x_2 - x_1}$ in both diagrams

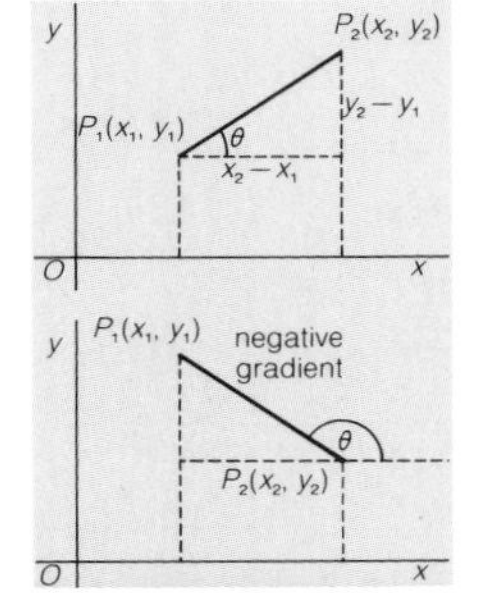

gradient (*n*) a measure of steepness; for any segment of a straight line it is the ratio of the difference between the *y*-coordinates (p. 215) at the ends of the segment divided by the difference between the *x*-coordinates (p. 215), so that if the line passes through (x_1, y_1) and (x_2, y_2) it is the quantity $(y_2 - y_1)/(x_2 - x_1)$. The gradient of a curve at any point is the gradient of the tangent at that point. If two lines have gradients m_1 and m_2 they are parallel if $m_1 = m_2$ and perpendicular if $m_1m_2 + 1 = 0$. Also known as **slope**.

gradient-intercept form the equation of a straight line in the form $y = mx + c$ where m is the gradient (↑) and c is the intercept (p. 182) on the *y*-axis (p. 215).

gradient-point form the equation of a straight line in the form $y - y_1 = m(x - x_1)$, where m is the gradient (↑) of the line and (x_1, y_1) is a point on it. Also known as **point-slope form**.

two-point form the equation of a straight line (p. 167) in the form $(y - y_1)/(y_2 - y_1) = (x - x_1)/(x_2 - x_1)$, where (x_1, y_1) and (x_2, y_2) are two points on the line.

intercept form the intercept form of the equation of a straight line (p. 167) is the form $(x/a) + (y/b) = 1$ where a and b are the intercepts (p. 182) on the *x*- and *y*-axes (p. 215) respectively.

Plücker coordinates the coordinates $(-1/a, -1/b)$ of a line where a and b are defined as in the intercept form (↑). From this the concept of lines and points as duals (p. 256) can be studied.

line coordinates if $lx + my + 1 = 0$ is a line not through the origin (p. 215), $[l, m]$ are the coordinates of the line. If the line is $lx + my + n = 0$, then the line has homogeneous coordinates (p. 217) $[l, m, n]$. This allows lines through the origin to be defined, and the duality (p. 256) of points and lines to be studied.

envelope equation when the coordinates of a line (↑) $[l, m]$ satisfy an equation $\varphi(l, m) = 0$, this equation defines an envelope (p. 190) of lines. Similarly with homogeneous line coordinates (p. 217), a homogeneous (p. 61) equation $\varphi(l, m, n) = 0$ defines an envelope of lines.

conic (*n*) a curve which is obtained from the intersection of a plane and a double cone or nappe (p. 210); a parabola (↓), ellipse (↓), hyperbola (↓), circle, pair of straight lines (p. 167) or other degenerate conic (↓). The first three are often defined as the locus (p. 187) of a point *P* whose distance *PS* from a fixed point *S* (the focus, ↓) is in a constant ratio to its perpendicular distance *PN* from a fixed line (the directrix, ↓) (the focus-directrix property). All second degree equations $ax^2 + 2hxy + by^2 + 2gx + 2fy + c = 0$ form conics. Also known as **conic sections**.

focus (*n*) the fixed point *S* in the definition of a conic (↑). A central conic (↓) has two foci, a parabola (↓) only one. **foci** (*pl*), **focal** (*adj*).

confocal (*adj*) having the same foci (↑).

directrix (*n*) (1) the fixed line through *N* in the definition of conic (↑). A central conic (↓) has two directrices, a parabola (↓) only one; (2) the line $y = 0$ in the catenary (p. 228) $y = a \cosh (x/a)$. **directrices** (*pl*).

focus-directrix property *see* conic (↑). Also known as **Pappus' theorem** (*do not mistake for* Pappus' theorem, p. 183, and Pappus-Guldin theorems, p. 156).

eccentricity (*n*) the ratio *e* which is equal to *PS/PN* in the definition of conic (↑).

parabola (*n*) a conic (↑) of eccentricity (↑) 1; the locus (p. 187) of a point in a plane equidistant (p. 169) from a fixed point and a fixed line. The cartesian (p. 214) equation can be written $y^2 = 4ax$, with parametric equations (p. 68) $x = at^2$, $y = 2at$ where *t* is the parameter (p. 67). It is the path of a projectile (p. 268), ignoring air resistance (p. 265). **parabolae** (*pl*), **parabolas** (*pl*), **parabolic** (*adj*).

ellipse (*n*) a conic (↑) of eccentricity (↑) less than 1, shaped like a flattened circle. The cartesian (p. 214) equation can be written $x^2/a^2 + y^2/b^2 = 1$, with parametric equations (p. 68) $x = a \cos \theta$, $y = b \sin \theta$ where θ is the parameter (p. 67). **elliptic** (*adj*), **elliptical** (*adj*).

conic
the conics as cross-sections of the double cone or nappe

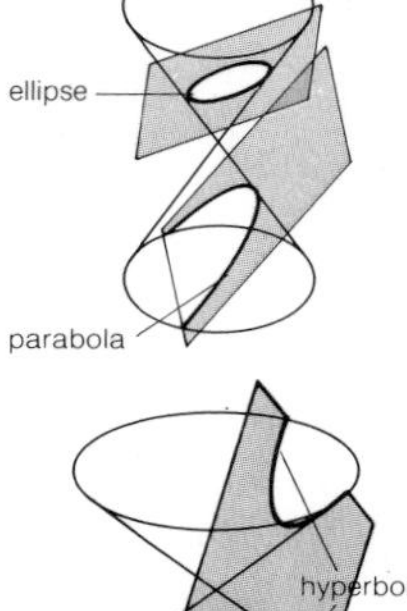

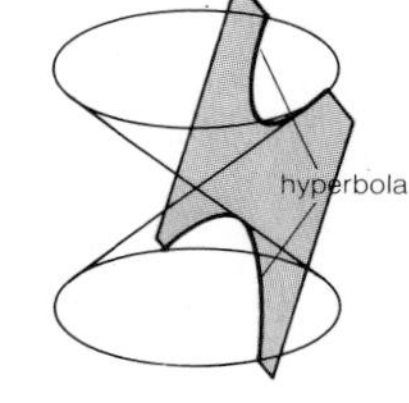

parabola $y^2 = 4ax$

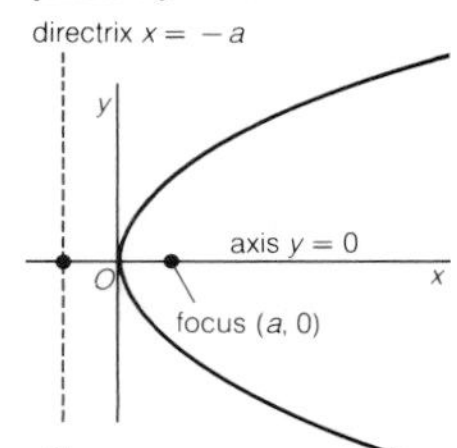

ellipse
$\frac{x^2}{a^2} + \frac{y^2}{b^2} = 1$ where $b^2 = a^2(1 - e^2)$ and *e* is the eccentricity

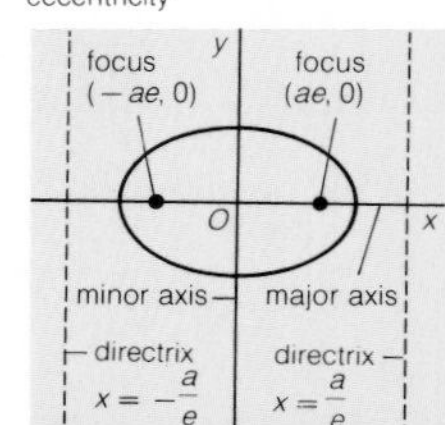

hyperbola
$\frac{x^2}{a^2} - \frac{y^2}{b^2} = 1$ where
$b^2 = a^2(e^2 - 1)$ and e is the eccentricity

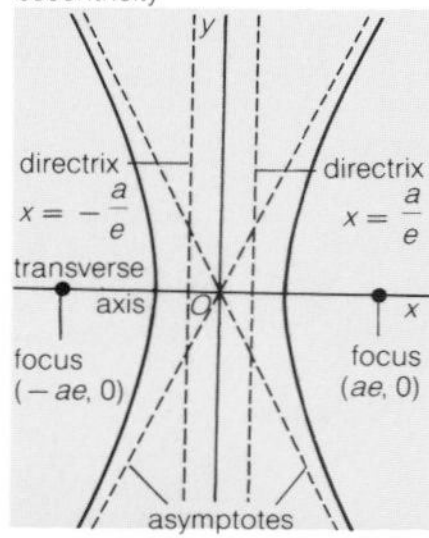

rectangular hyperbola
$xy = c^2$

directrix
$x + y = -\sqrt{2}c$
focus
$(\sqrt{2}c, \sqrt{2}c)$
asymptote
conjugate axis
$x + y = 0$
$x - y = 0$
y
O
asymptote
x
transverse axis
focus
$(-\sqrt{2}c, -\sqrt{2}c)$
directrix
$x + y = \sqrt{2}c$

degenerate conic
the pair of lines $x + y = 2$ and $x - y = 1$ which form the degenerate conic
$(x + y - 2)(x - y - 1) = 0$ or
$x^2 - y^2 - 3x + y + 2 = 0$

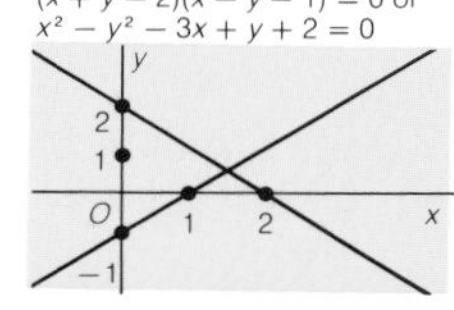

latus rectum
one latus rectum of an ellipse

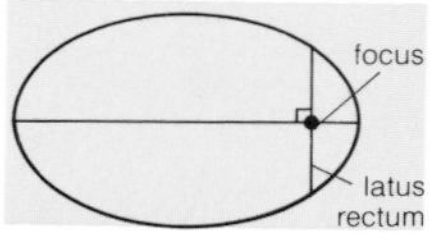

hyperbola (*n*) a conic (↑) of eccentricity (↑) greater than 1, having two branches (p. 189) and two asymptotes (p. 188). The cartesian (p. 214) equation can be written $x^2/a^2 - y^2/b^2 = 1$, with parametric equation (p. 68) $x = a \sec\theta$, $y = b \tan\theta$ where θ is the parameter (p. 67). **hyperbolae** (*pl*), **hyperbolas** (*pl*), **hyperbolic** (*adj*).

rectangular hyperbola a hyperbola (↑) with perpendicular asymptotes (p. 188) whose equation may be written as for that of the hyperbola with $a = b$, but which is more often written as $xy = c^2$, with parametric equations (p. 68) $x = ct$, $y = c/t$ and the axes as asymptotes.

central conic a conic (↑) with a centre of symmetry (p. 233), i.e. an ellipse (↑), a hyperbola (↑) or a degenerate conic (↓).

degenerate conic a conic (↑) whose second degree equation factorizes (p. 27) and thus forms a pair of straight lines (p. 167) (which may coincide, p. 171, or result in only a single real, p. 49, point). Also known as **reducible conic**.

proper conic a conic (↑) which is not degenerate (↑).

focal chord a chord (p. 187) of a conic (↑) which passes through a focus (↑).

latus rectum a focal chord (↑) of a conic (↑) which is perpendicular to the axis and passes through the focus (↑).

semi-latus rectum the length along the latus rectum (↑) from the focus (↑) to the curve.

principal axes the axes of symmetry (p. 232) of an ellipse (↑) or hyperbola (↑).

major axis the axis of symmetry (p. 232) of an ellipse (↑) which has the larger diameter and passes though the foci (↑); the length of this axis cut off by the curve. If $a > b$ (with the usual equation) then it is the x-axis (p. 215).

minor axis the axis of symmetry (p. 232) of an ellipse (↑) which has the smaller diameter and does not pass through the foci (↑); the length of this axis cut off by the curve. If $a > b$ (with the usual equation), then it is the y-axis (p. 215).

transverse axis the axis of symmetry (p. 232) of a hyperbola (p. 221) which intersects both branches (p. 189) of the curve and passes through the foci (p. 220); the length of this axis cut off by the curve.

conjugate axis the axis of symmetry (p. 232) of a hyperbola (p. 221) which does not intersect the curve or pass through the foci (p. 220) and which is perpendicular to the transverse axis (↑).

semi-axis (*n*) half the length of the major axis (↑) of an ellipse (p. 220), or of the transverse axis (↑) of a hyperbola (p. 221), lying between the two intersections of the axis with the curve; the corresponding length for a central quadric (p. 211).

bifocal property if S, S' are the foci (p. 220) of an ellipse (p. 220) or hyperbola (p. 221), and P is a point on it, then for an ellipse $SP + S'P$ is twice the semi-axis (↑), and for a hyperbola the difference between SP and $S'P$ is twice the semi-axis.

auxiliary circle the circle whose diameter is the major axis (p. 221) of an ellipse (p. 220) or the transverse axis (↑) of a hyperbola (p. 221).

eccentric angle the eccentric angle of a point P on an ellipse (p. 220) is the angle QON where O is the origin (p. 215), and the ordinate (p. 215) PN meets the auxiliary circle (↑) at Q.

auxiliary rectangle the rectangle whose vertices are formed by the points of intersection with the asymptotes (p. 188) of the tangents to a hyperbola (p. 221) which are parallel to the conjugate axis (↑).

director circle the circle which is the locus (p. 187) of perpendicular tangents to an ellipse (p. 220) or hyperbola (p. 221). Also known as **orthoptic circle**.

conormal points a set of four points on a central conic (p. 221) for which the four normals (p. 188) are concurrent (p. 171).

Appolonius' hyperbola the rectangular hyperbola (p. 221) on which the four conormal points (↑) lie for any given set of concurrent (p. 171) normals (p. 188).

bifocal property

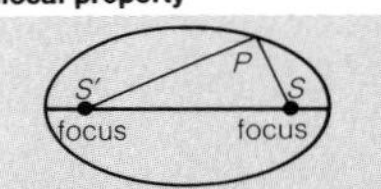

if P is a point on an ellipse, $SP + S'P$ = constant

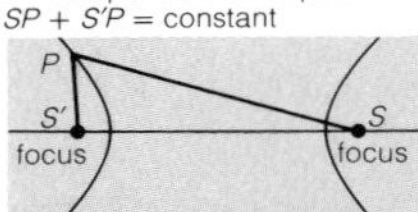

if P is a point on a hyperbola, $|SP - S'P|$ = constant

auxiliary circle

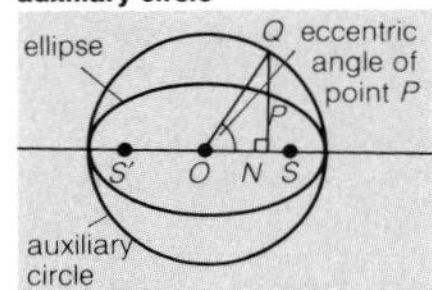

auxiliary circle

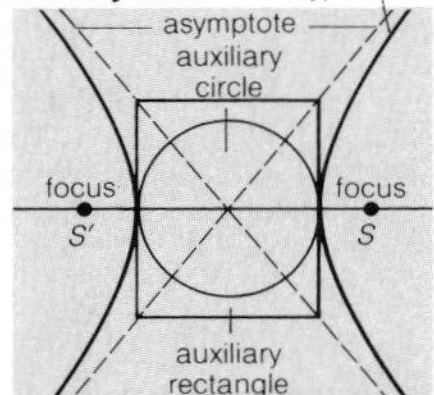

director circle

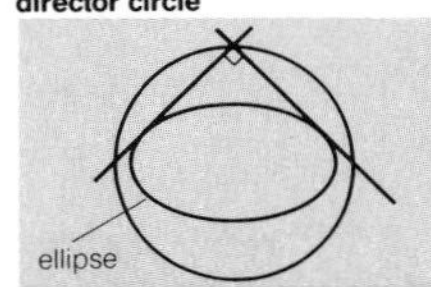

conormal points

the four conormal points P_1, P_2, P_3, P_4 of a point N with respect to a conic

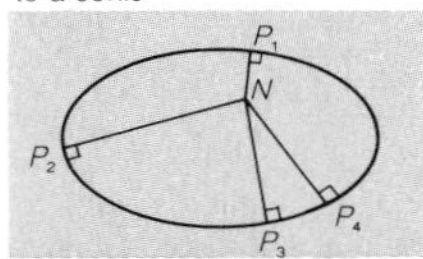

Joachimstal's ratio equation
if P divides P_1P_2 in the ratio $k_1 : k_2$ and lies on $S = 0$, then
$S_{11}k_1^2 + 2S_{12}k_1k_2 + S_{22}k_2^2 = 0$

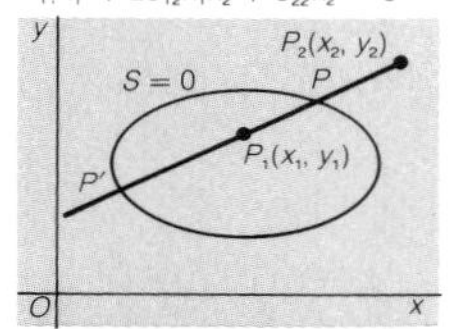

the second solution to the ratio $k_1 : k_2$ gives the point P'

conjugate diameters
the mid-points of chords parallel to diameter AB lie on diameter CD, and conversely

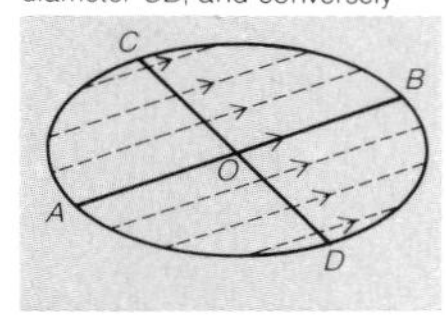

conjugate hyperbolae
$9x^2 - y^2 = 1$ and $y^2 - 9x^2 = 1$ (in red) which have the same asymptotes $y \pm 3x = 0$

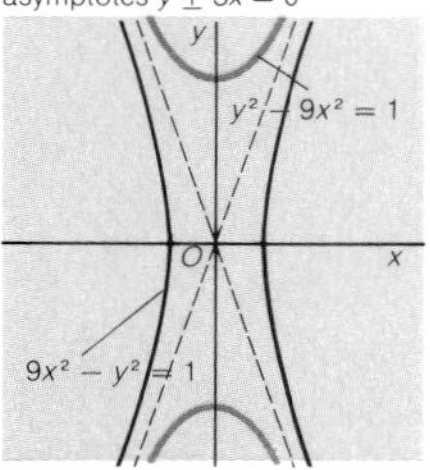

pole and polar
line p is the polar of P, point P is the pole of p

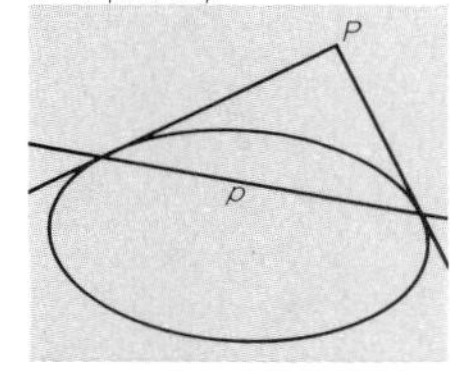

Frégier point all chords (p. 187) of a parabola (p. 220) which subtend (p. 175) a right angle at a given point on the parabola pass through the Frégier point of the given point.

Joachimstal's ratio equation if a point P lies on a conic (p. 220) $S = 0$ and divides the line joining $P_1\,(x_1, y_1)$ and $P_2\,(x_2, y_2)$ so that

$$P_1P : PP_2 = k_1 : k_2, \text{ then}$$

$$S_{11}k_1^2 + 2S_{12}k_1k_2 + S_{22}k_2^2 = 0$$

where

$$S \equiv ax^2 + 2hxy + by^2 + 2gx + 2fy + c = 0,$$

$$S_{12} \equiv ax_1x_2 + h(x_1y_2 + x_2y_1) + by_1y_2 + g(x_1 + x_2) + f(y_1 + y_2) + c,$$

and S_{11} and S_{22} are the results of substituting (p. 69) (x_1, y_1) and (x_2, y_2) respectively in S. *See also* rule of alternate suffices (p. 224).

conjugate diameters pairs of diameters of central conics (p. 221), each of which forms the locus (p. 187) of the mid-points of chords (p. 187) lying parallel to the other.

equiconjugate diameters pairs of conjugate diameters (↑) having the same lengths.

conjugate hyperbolae pairs of hyperbolae (p. 221) with common asymptotes (p. 188) but with transverse (↑) and conjugate axes (↑) interchanged and which can be written in the form $x^2/a^2 - y^2/b^2 = \pm 1$.

polar (*n*) the polar of a point P with respect to a conic (p. 220) (often a circle) is the line joining the points of contact (p. 187) of the tangents from P to the curve. If the conic is

$$S \equiv ax^2 + 2hxy + by^2 + 2gx + 2fy + c = 0,$$

then its equation is

$$S_1 \equiv axx_1 + h(xy_1 + yx_1) + byy_1 + g(x + x_1) + f(y + y_1) + c = 0,$$

where P is (x_1, y_1). This gives the polar of P even when there are no real (p. 49) tangents, and the tangent when P lies on the conic. Also known as **chord of contact of tangents** of P. *See also* pp. 205, 216, and rule of alternate suffices (p. 224).

pole[3] the pole of a line with respect to a conic (p. 220) (often a circle) is the point whose polar (↑) is the given line. *See also* pp. 205, 216.

rule of alternate suffices a rule for remembering the equations $S_1 = 0$ of tangents and polars (p. 223) at $P(x_1, y_1)$ with respect to a conic (p. 220) $S = 0$. Replace x^2, $2xy$, y^2, $2x$ and $2y$ by xx, $xy + yx$, yy, $x + x$ and $y + y$ respectively, and place the suffix 1 on alternate (p. 287) variables. *See also* S_{12} in Joachimstal's ratio equation (p. 223) which may be remembered in the same way.

interior point of a conic a point which does not have a real (p. 49) polar (p. 223).

self-polar triangle a triangle in which each side is the polar (p. 223) of the opposite vertex with respect to a given conic (p. 220).

conjugate points with respect to a conic (p. 220), points such that each lies on the polar (p. 223) of the other.

conjugate lines with respect to a conic (p. 220), lines such that each passes through the pole (p. 223) of the other.

reciprocation (*n*) the process of replacing a pole (p. 223) *P* by its polar (p. 223) *p* with respect to a given conic (p. 221) (often a circle) and the other way round. The locus (p. 187) of *P* then becomes the envelope (p. 190) of *p* and similarly envelopes become loci. **reciprocate** (*v*).

reciprocal curves corresponding loci (p. 187) and envelopes (p. 190) under reciprocation (↑).

point reciprocation reciprocation (↑) with respect to a circle is often called point reciprocation with respect to the centre of the circle.

Pascal's theorem if a hexagon (p. 177) is inscribed (p. 198) in a conic (p. 220), the three pairs of opposite sides meet in three collinear (p. 171) points. *See also* Brianchon's theorem (↓).

Brianchon's theorem if a hexagon (p. 177) is circumscribed (p. 197) about a conic (p. 220), the three lines joining pairs of opposite vertices are concurrent (p. 171). This is the dual (p. 256) of Pascal's theorem (↑).

Chasles' theorem if *A*, *B*, *C* and *D* are four fixed points on a conic (p. 220) and *P* is a variable point also on it, the cross-ratio (p. 184) of the pencil (p. 167) $P\{AB, CD\}$ is constant.

self-polar triangle
triangle *PQR*: *p*, *q*, *r* are the polars of *P*, *Q*, *R* with respect to the conic

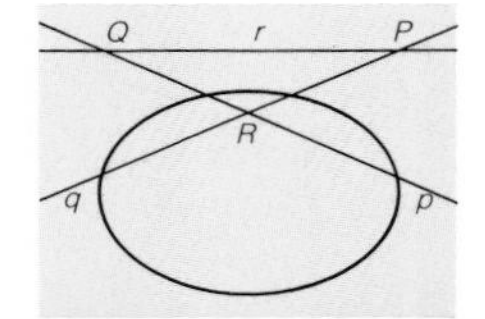

Pascal's theorem
AB, *ED* meet at *L*; *BC*, *FE* meet at *M*; *CD*, *FA* meet at *N*; then *L*, *M*, *N* are collinear

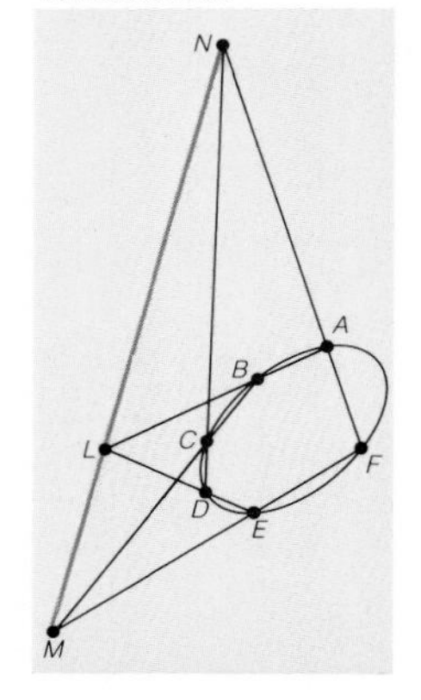

Brianchon's theorem
if *A B C D E F* is a hexagon circumscribed about a conic, *AD*, *BE* and CF are concurrent

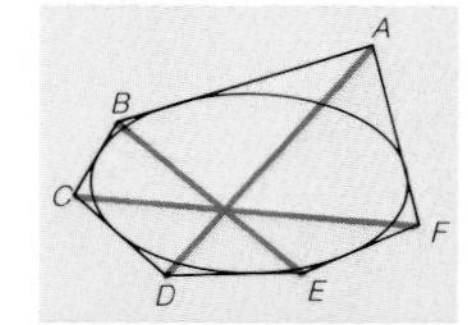

cycloid

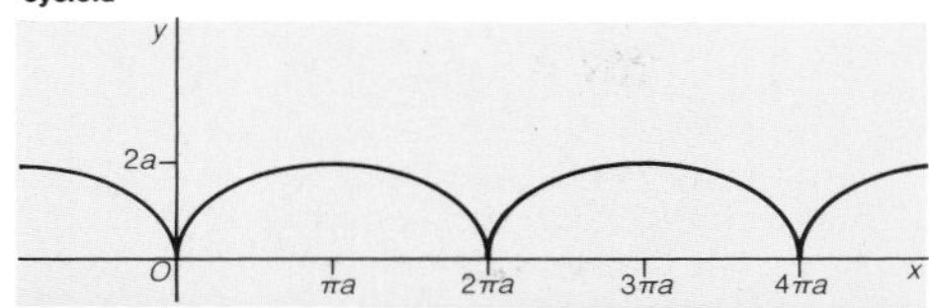

cycloid (*n*) the locus (p. 187) of a point on the circumference (p. 192) of a circle of radius *a* as it rolls along a straight line. Its parametric equation (p. 68) can be written $x = a(\theta + \sin\theta)$, $y = a(1 - \cos\theta)$.

Torricelli's theorem the area (p. 191) of one arch of a cycloid (↑) between two consecutive (p. 288) cusps (p. 189) and the fixed line is three times the area of the rolling circle.

curtate and prolate cycloids

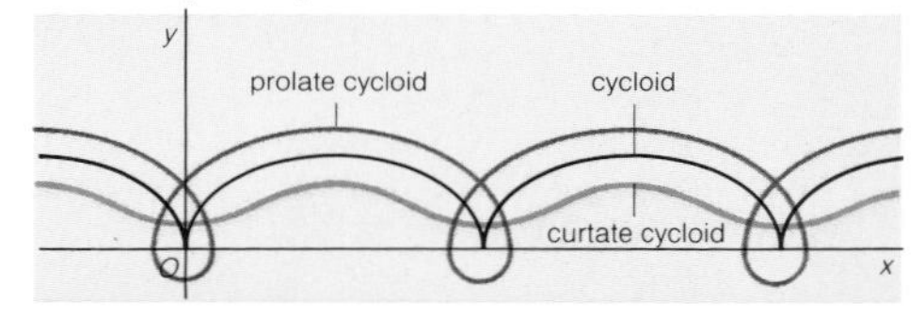

curtate cycloid the locus (p. 187) of a point inside a circle as it rolls with the circle along a straight line.

prolate cycloid the locus (p. 187) of a point outside a circle as it rolls with the circle along a straight line (p. 167), e.g. a point on the flange or edge of a railway wheel.

brachistochrone (*n*) the curved path down which a particle will slide fastest between two points if there is no friction (p. 265). It is a cycloid (↑) whose fixed line is horizontal through the upper point, which is a cusp (p. 189), the curve being drawn below the line.

tautochrone (*n*) the curved path such that a particle sliding down it without friction (p. 265) will reach a fixed point in the same time no matter which point on the curve it starts at. The fixed point is the mid-point of an arch of a cycloid (↑) drawn below the fixed line.

brachistochrone
the curve of fastest descent between two points

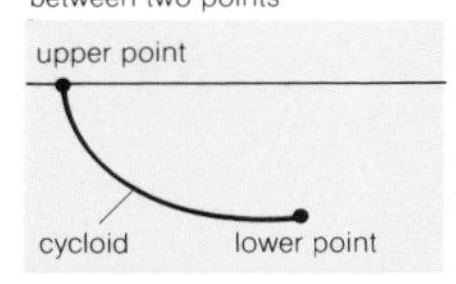

tautochrone
a particle takes the same time to slide to *P* whatever point on the arch it starts

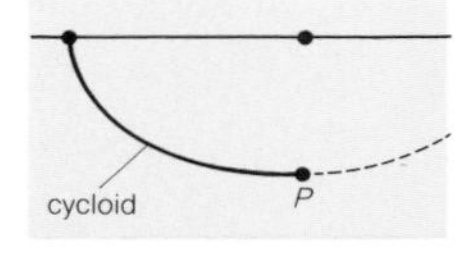

isochrone (*n*) (1) an isogram (p. 208) on a map joining points of equal time; (2) the property of a cycloid (p. 225) which, when it is drawn below the fixed line, means that a pendulum (p. 275) with a string which bends freely and is hung from a cusp (p. 189) between two arches always has the same period (p. 276) no matter what the swing's amplitude (p. 276). *See also* isochronic curve (↓). **isochronic** (*adj*).

isochrone

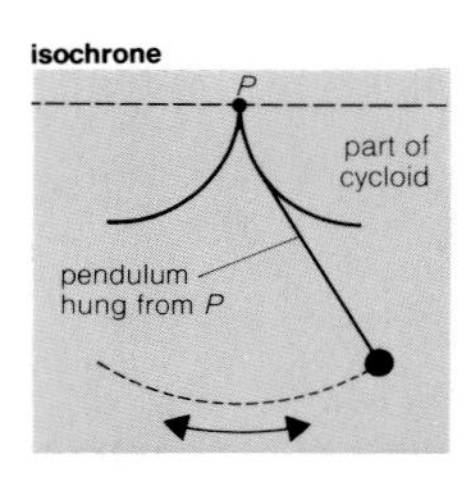

semi-cubic parabola a curve whose equation can be $y^2 = x^3$. It is the approach curve (↓).

approach curve the curve along which the rate of change of acceleration (p. 258) is constant for a given speed; it is the semi-cubic parabola (↑) and is used in laying railway track.

semi-cubic parabola $y^2 = x^3$

isochronic curve the curve along which a particle falls under gravity (p. 266) through equal distances in equal times if there is no friction (p. 265); it is the semi-cubic parabola (↑) in the form $x^2 + y^3 = 0$, where the particle falls from the cusp (p. 189).

isochronic curve $x^2 + y^3 = 0$: a particle takes equal times to travel between each pair of red lines

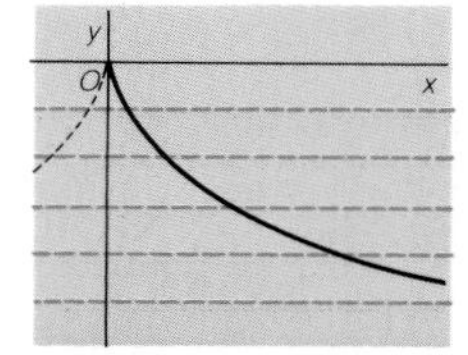

roulette (*n*) a curve formed by a point or line moving with a curve which rolls along another fixed curve, e.g. the cycloid (p. 225).

point roulette a roulette (↑) formed by the locus (p. 187) of a point.

line roulette a roulette (↑) formed by the envelope (p. 190) of a line.

trochoid (*n*) a roulette (↑), often used only when the fixed curve is a circle or straight line (p. 167) and the rolling curve is a circle.

epicycloid (*n*) a roulette (↑) formed by a point on the circumference (p. 192) of a circle rolling round the outside of a fixed circle, e.g. cardioid (↓), nephroid (↓).

hypocycloid (*n*) a roulette (↑) formed by a point on the circumference (p. 192) of a circle rolling round the inside of a fixed circle, e.g. deltoid (↓), astroid (↓).

deltoid (*n*) a hypocycloid (↑) with three cusps (p. 189) whose parametric equations (p. 68) can be written $x = 2a \cos t + a \cos 2t$, $y = 2a \sin t - a \sin 2t$. It is the envelope (p. 190) of the Simson's line (p. 197) of a triangle for the set of points on the circumcircle (p. 197).

deltoid
three-cusped hypocycloid

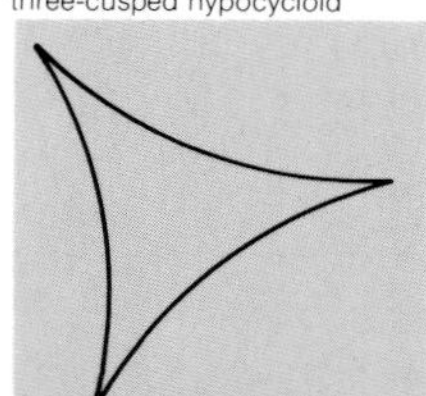

astroid (*n*) a hypocycloid (↑) with four cusps (p. 189) whose parametric equations (p. 68) can be written $x = a\cos^3 t$, $y = a\sin^3 t$. It is the envelope (p. 190) of a line segment (p. 167) of fixed length with its ends on two perpendicular axes.

astroid
$x = a\cos^3 t$, $y = a\sin^3 t$:
the first quadrant is shown as the envelope of a line-segment

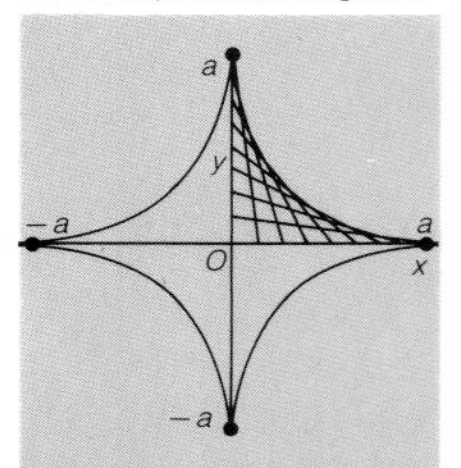

cardioid (*n*) an epicycloid (↑) with one cusp (p. 189) whose polar (p. 216) equation can be written $r = 2a(1 - \cos\theta)$.

cardioid
$r = 2a(1 - \cos\theta)$

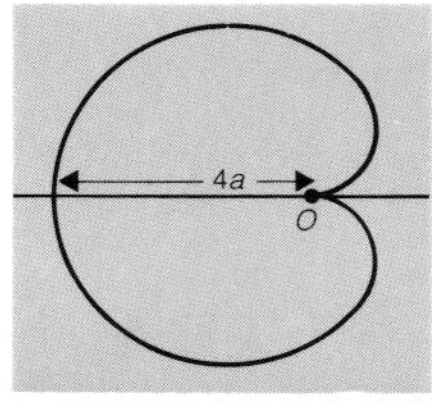

nephroid (*n*) an epicycloid (↑) with two cusps (p. 189) whose parametric equations (p. 68) may be written $x = a(3\cos t - \cos 3t)$, $y = a(3\sin t - \sin 3t)$. It is the envelope (p. 190) or caustic (p. 190) formed by light from a distant point reflected (p. 234) in the inside of a semicircle (p. 192) and may sometimes be seen on the surface of a cup of liquid.

nephroid
$x = a(3\cos t - \cos 3t)$
$y = a(3\sin t - \sin 3t)$

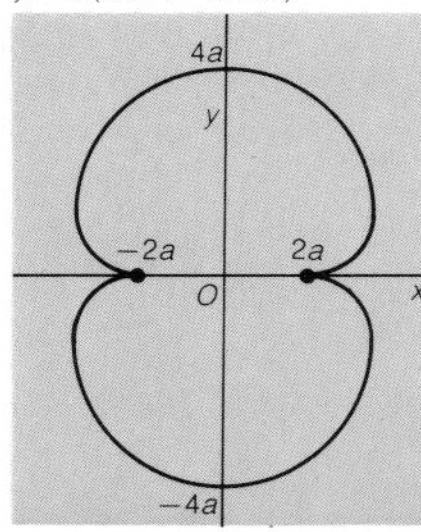

epitrochoid (*n*) a curve formed like an epicycloid (↑) except that the moving point need not lie on the circumference (p. 192) of the rolling circle, but may lie inside or outside it, though it still moves with it, e.g. limaçon (↓). Similarly **hypotrochoid** corresponds to hypocycloid (↑).

limaçon (*n*) an epitrochoid (↑) formed from two equal circles of radius *a*, the point generating (p. 209) the locus (p. 187) being a distance *k* from the centre of the rolling circle. Its polar (p. 216) equation can be written $r = k - 2a\cos\theta$. If $k = 2a$, the curve is a cardioid (↑), if $k = a$ it is a trisectrix (↓).

trisectrix (*n*) (1) usually the limaçon (↑) in which $k = a$ and whose polar (p. 216) equation can be written $r = a(1 - 2\cos\theta)$; (2) any other curve which may be used to trisect (p. 171) angles, e.g. the trisectrix of Maclaurin (p. 229) or the trisectrix of Catalan (p. 229).

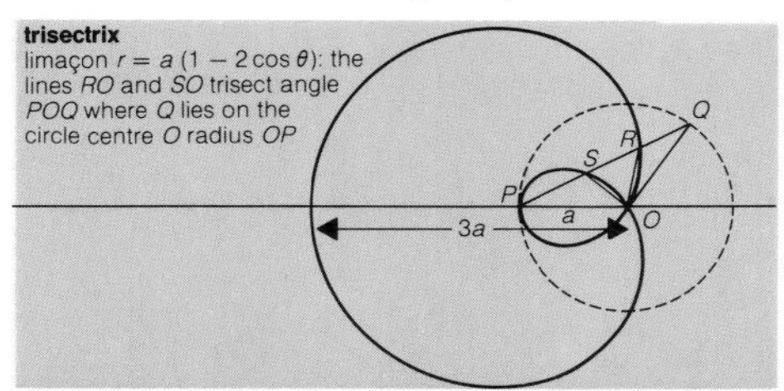

trisectrix
limaçon $r = a(1 - 2\cos\theta)$: the lines *RO* and *SO* trisect angle *POQ* where *Q* lies on the circle centre *O* radius *OP*

rose-petal curves curves with polar (p. 216) equations $r = a \sin n\theta$, which have n loops (p. 189) if n is odd, e.g. trifolium (↓), and $2n$ loops if n is even, e.g. quadrifolium (↓). Also known as **rose curves**.

trifolium (n) the rose-petal curve (↑) $r = a \sin 3\theta$ which has three loops (p. 189).

quadrifolium (n) the rose-petal curve (↑) $r = a \sin 2\theta$ which has four loops (p. 189).

folium of Descartes a curve with the equation $x^3 + y^3 - 2axy = 0$ which has a folium or loop (p. 189).

witch of Agnesi a curve with the equation $xy^2 = 4a^2(2a - x)$.

quadratrix (n) the curve $y = x \cot(\pi x/2a)$ which when $0 \leqslant x \leqslant a$ may be used to trisect (p. 171) angles.

catenary (n) the curve taken by a heavy string hanging between two points whose equation can be written $y = a \cosh(x/a)$ (*see* hyperbolic cosine, p. 141).

tractrix (n) the involute (p. 190) of a catenary (↑), it is the curve of pursuit (↓) formed by pulling a tight string with an object tied to one end in a straight line at an angle to the string. Its equation may be written

$$\pm x = c \cosh^{-1}(c/y) - (c^2 - y^2).$$

See also pseudosphere (p. 204).

curve of pursuit a path of a given curve such as that which one animal takes in running towards another animal which is itself running along the given curve, both animals running at constant speeds. Also known as **pursuit curve**. *See also* tractrix (↑).

curve of pursuit
the tractrix as a curve of pursuit: the line segments are tangents to the curve and all have the same length

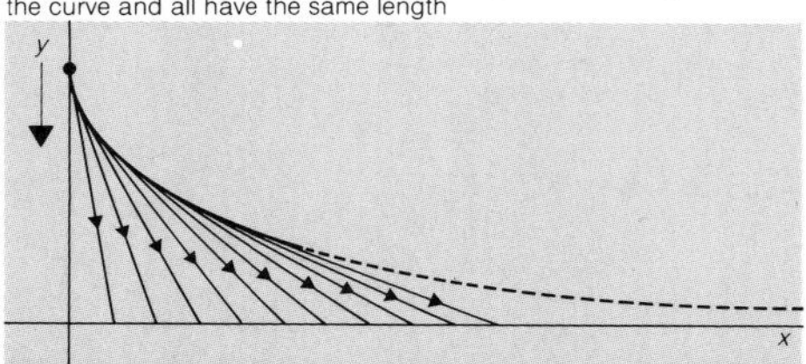

trifolium $r = a \sin 3\theta$

quadrifolium $r = a \sin 2\theta$

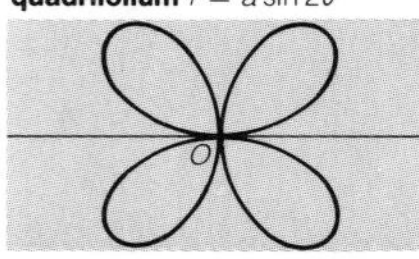

folium of Descartes

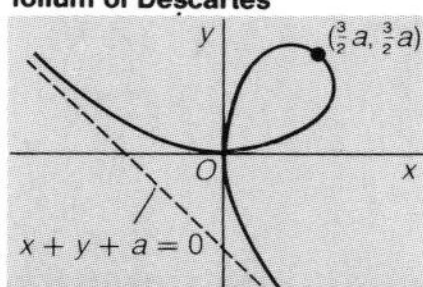

witch of Agnesi
$x = 0$ is an asymptote

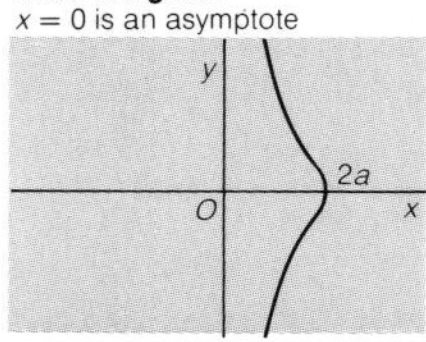

quadratrix
$y = x \cot(\pi x/2a)$ for $0 \leqslant x \leqslant a$

catenary $y = c \cosh \frac{x}{c}$
and its involute, the tractrix

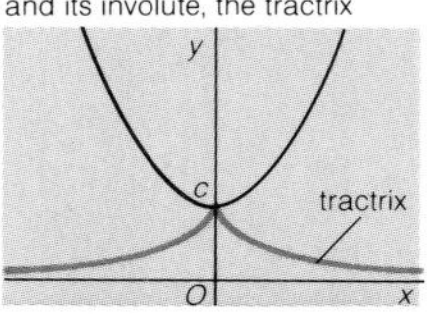

trisectrix of Maclaurin
$r = a \sec(\theta/3)$ which is the pedal curve of the parabola $y^2 = 4a(x - a)$ with respect to the point S'

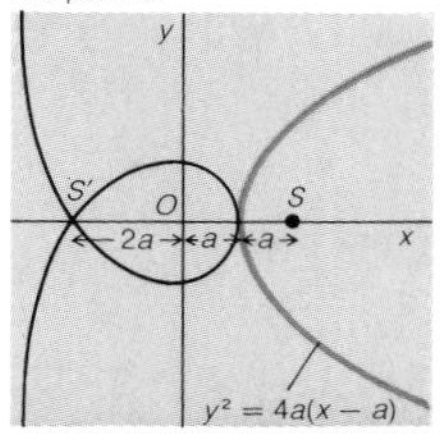

strophoid
the right strophoid
$y^2(a - x) = x^2(a + x)$

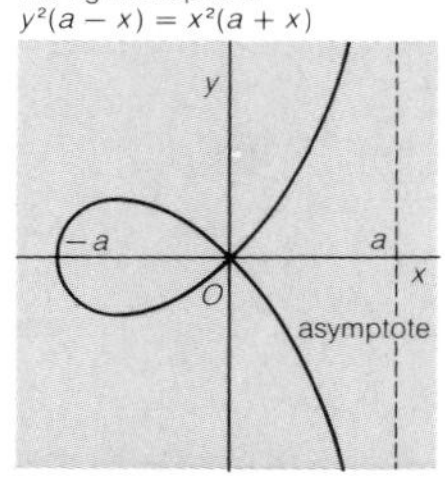

pedal curve the pedal curve of a given curve from a given point is the locus (p. 187) of the feet (p. 175) of the perpendiculars drawn from that point to tangents to the given curve, *see* trisectrix of Maclaurin (↓).

trisectrix of Maclaurin the curve which can be written with the polar (p. 216) equation $r = a \sec(\theta/3)$ which is the pedal curve (↑) of a parabola (p. 220) with respect to the point which is the reflection (p. 234) of the focus (p. 220) in the directrix (p. 220). *See also* trisectrix (p. 227).

strophoid (n) if l is a variable line through a point K, meeting a curve at Q, and A is a fixed point, then the locus (p. 187) of P and P' on l such that $QP = QP' = QA$ is the strophoid of the curve with respect to pole (p. 216) K and fixed point A, e.g. the curve called a *right strophoid* whose equation can be written $y^2(a - x) = x^2(a + x)$ for $a > 0$, which has asymptote (p. 188) $x = a$ and is the pedal curve (↑) of a parabola (p. 220) with respect to the point of intersection of its axis and its directrix (p. 220).

negative pedal if Q is a point on a curve and O is a fixed point, the envelope (p. 190) of a line through Q perpendicular to OQ is the negative pedal of the curve with respect to O. The parabola (p. 220) is the negative pedal of a straight line, *see also* trisectrix of Catalan (↓).

trisectrix of Catalan the curve whose polar (p. 216) equation may be written $r = a \sec^3(\theta/3)$ which is the negative pedal (↑) of a parabola (p. 220) with respect to its focus (p. 220). Also known as **Tschirnhausen's cubic**, **l'Hôpital's cubic**.

trisectrix of Catalan
$r = a \sec^3(\theta/3)$: it is the negative pedal of the parabola with focus at the pole O shown in red

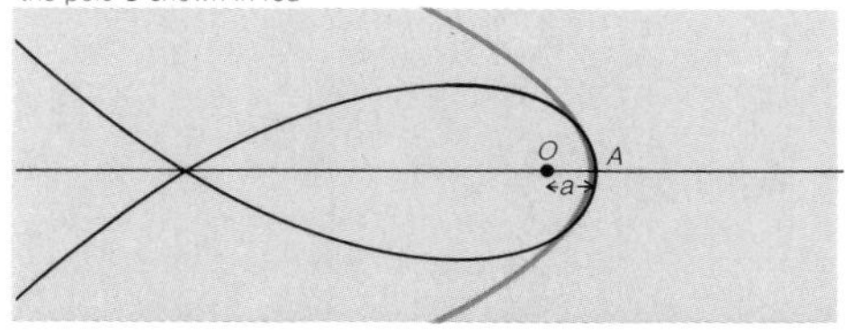

equiangular spiral
the angle θ is a constant

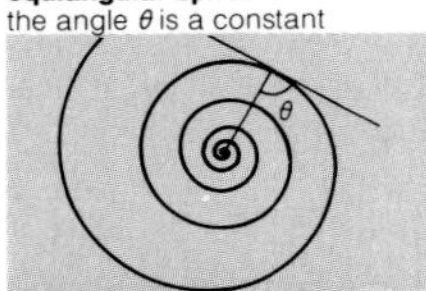

Archimedean spiral
$r = a\theta$

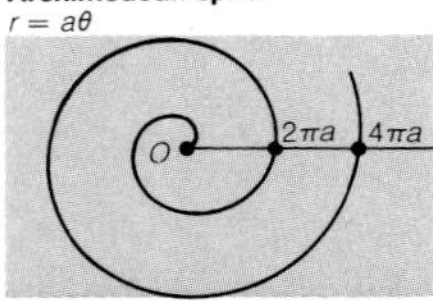

parabolic spiral

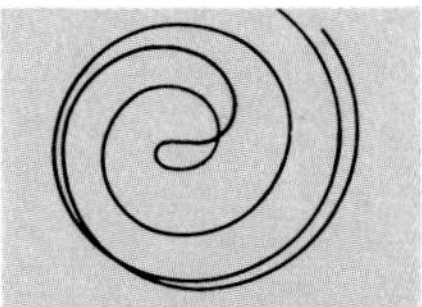

spiral (*n*) one of a number of plane curves shaped like the cross-section (p. 200) of a roll of cloth. *Do not mistake for* helix (p. 212).

equiangular spiral the spiral (↑) with the polar (p. 216) equation $r = ae^{(\theta \cot \alpha)}$ in which the angle between the tangent and the radius vector (p. 216) is a constant α. Many seashells have this shape. Also known as **logarithmic spiral**.

Archimedean spiral the curve with polar (p. 216) equation $r = a\theta$.

parabolic spiral a curve with polar (p. 216) equation $(r - b)^2 = a^2\theta$ of which Fermat's spiral (↓) is an example.

Fermat's spiral the parabolic spiral (↑) with polar (p. 216) equation $r^2 = a^2\theta$.

reciprocal spiral the curve with polar (p. 216) equation $r\theta = a$. Also known as **hyperbolic spiral**.

lituus (*n*) the curve with polar (p. 216) equation $r^2\theta = a^2$. If P is a point on the curve and Q is a point on the base line (p. 216) such that $OP = OQ$, the sector (p. 193) POQ has constant area (p. 191). The base line is also an asymptote (p. 188).

sinusoidal spirals curves with polar (p. 216) equations of the form $r^n = a^n \cos n\theta$, e.g. lemniscate of Bernoulli (↓).

lemniscate of Bernoulli the sinusoidal spiral (↑) $r^2 = a^2 \cos 2\theta$ which is both the inverse (p. 195) and the pedal (p. 229) curve of a rectangular hyperbola (p. 221) with respect to its centre. *See also* Cassini ovals (↓). Sometimes written **leminiscate**.

oval (*n*) a closed curve (p. 190) of approximately elliptical (p. 220) shape, a flattened circle. The word has no exact mathematical meaning.

reciprocal spiral
part of $r\theta = a$ with asymptote distance a from the pole O

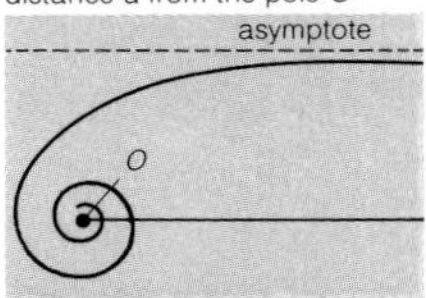

lituus
part of $r^2\theta = a^2$: sector POQ has constant area

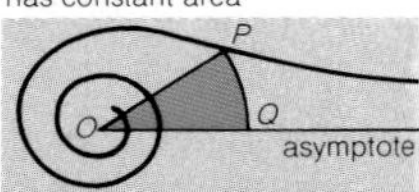

lemniscate of Bernoulli
$r^2 = a^2 \cos 2\theta$

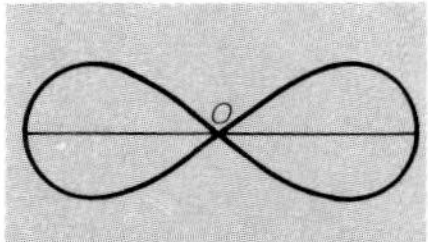

conchoid of Nicomedes
$PQ = QP' =$ constant

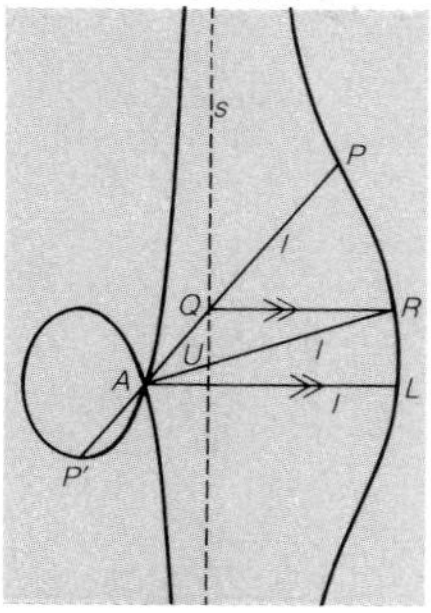

cissoid of Diocles
$AP = QR$

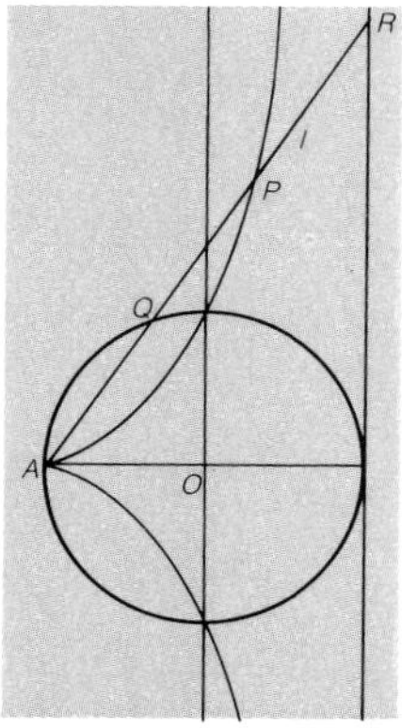

glissette
the circle as a glissette

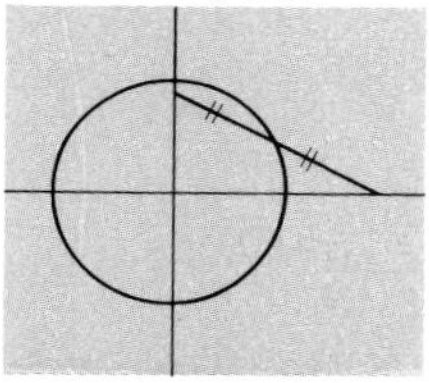

Cassini ovals the set of curves for which the products of the distances from two fixed points is a constant, in bipolar coordinates (p. 216) $r_1r_2 = b^2$. The lemniscate of Bernouilli (↑) is the special case for which the product is half the square of the distance between the fixed points.

cartesian ovals the set of curves for which in bipolar coordinates (p. 216) $lr_1 \pm mr_2 = k$ where k, l and m are constants, e.g. the ellipse (p. 220). Also known as **ovals of Descartes**.

conchoid (*n*) if a line l is drawn through a fixed point A, while P, Q and P' are points on l such that $PQ = QP' =$ constant, and if Q moves on a given curve, the locus (p. 187) of P and P' with respect to the given curve is a conchoid, e.g. conchoid of Nicomedes (↓).

conchoid of Nicomedes a conchoid (↑) in which the point Q lies on a straight line (p. 167) s, and whose polar (p. 216) equation is $r = a \sec \theta + k$. It can be used to trisect (p. 171) an angle or to find the side of a cube having double (p. 288) the volume of a given cube. If $AQ = \frac{1}{2}QP$ in the diagram, and QR is drawn parallel to AL so that $UR = QP = l$, then angle $RAL = (1/3)$ angle QAL.

cissoid (*n*) if a line l is drawn through a fixed point A to cut two given curves at Q and R, and if P is the point on l such that $AP = QR$, then the locus (p. 187) of P is the cissoid of the two curves with respect to A, e.g. cissoid of Diocles (↓).

cissoid of Diocles the cissoid (↑) of a circle and its tangent with respect to a point of the circle at the opposite end of the diameter through the point of contact (p. 187). Its polar (p. 216) equation can be written $r = 2a(\sec \theta - \cos \theta)$. It can be used to find the side of a cube having double (p. 288) the volume of a given cube.

glissette (*n*) if a moving curve S slides against two fixed curves, the locus (p. 187) of any point (or the envelope, p. 190, of any line) moving with S is called a glissette, e.g. the locus of the mid-point of a line segment (p. 167) sliding between two perpendicular axes which is a circle.

symmetry (*n*) the one : one correspondence (p. 52) of parts of a configuration either on opposite sides of a straight line (p. 167) (*see* reflective symmetry, ↓) or by rotation about an axis (*see* rotational symmetry, ↓). *See also* commutative (p. 15), p. 18. **symmetric** (*adj*), **symmetrical** (*adj*).

asymmetry (*n*) lack of symmetry. **asymmetrical** (*adj*).

line of symmetry a line about which a plane configuration has symmetry. If the configuration is drawn on a piece of paper, then the paper may be folded at the line of symmetry so that the two halves of the configuration lie on top of each other, e.g. the two tangents from an external (p. 289) point *P* to a circle are equal in length as there is a line of symmetry which is the diameter of the circle through *P*. Also known as **mirror line**.

line of symmetry
an isosceles trapezium

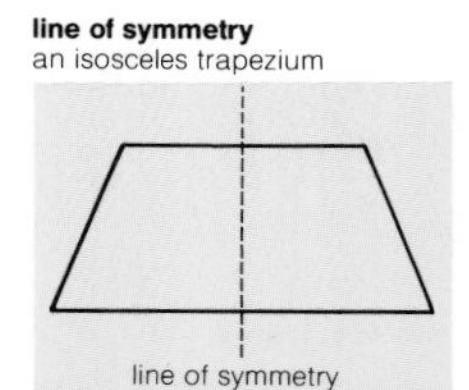

plane of symmetry a plane about which a solid configuration has symmetry, corresponding in three dimensions to a line of symmetry (↑) in two dimensions.

reflective symmetry symmetry about either a line or a plane. Also called **bilateral symmetry**, or in the case of a line **line symmetry**.

rotational symmetry symmetry of rotation or turning about an axis. Also known as **central symmetry** or **point symmetry** for plane configurations and **axial symmetry** for three dimensional configurations.

order of rotational symmetry the number of ways a configuration may be placed when rotating about an axis so that it looks the same. Thus a square has rotational symmetry (↑) of order 4, and a regular tetrahedron (p. 201) has order 3 about each of four different axes of symmetry (↓). All configurations have order of symmetry at least one about any axis.

axis of symmetry a line about which a configuration can be rotated to become coincident (p. 171) with itself. For a plane configuration the axis is perpendicular to the plane and through the centre of symmetry (↓).

centre of symmetry

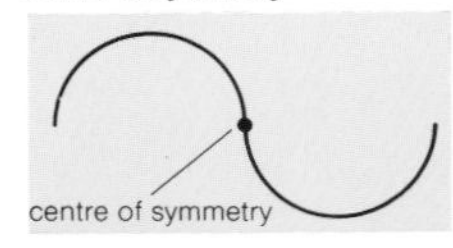

triskelion

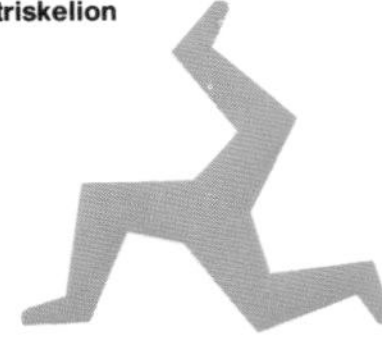

swastika

translation
translation of a configuration described by the column vector $\begin{pmatrix}2\\1\end{pmatrix}$

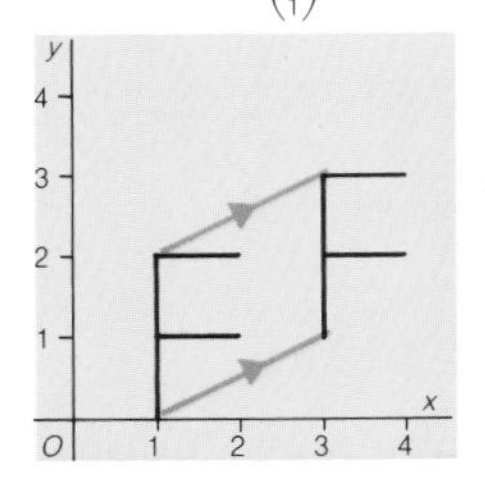

centre of symmetry a point about which a plane configuration can be rotated to become coincident (p. 171) with itself, e.g. the centre of a square. Also known as **point of symmetry**.

triskelion (*n*) a configuration having rotational symmetry (↑) of order 3 but no other symmetry, showing three bent legs joined at the top and sometimes known as the 'legs of Man'.

swastika (*n*) a configuration having rotational symmetry (↑) of order 4 but no other symmetry, shaped like a cross but with bent arms.

transformation geometry geometry using transformations (p. 54) of elements such as points, lines, planes and other plane and solid configurations. Such transformations are in fact transformations or mappings (p. 51) of the whole plane or space into itself, although only certain configurations which are moved will be of interest in any given example.

isometry (*n*) a geometrical transformation (p. 54) in which lengths and angles remain unchanged. The four types are translation (↓), rotation, reflection (p. 234) and glide reflection (p. 234). These each map (p. 51) a configuration onto a congruent (p. 179) configuration. *See also* isometric graph paper (p. 9). **isometric** (*adj*).

direct isometry an isometry (↑) in which no reflection (p. 234) or turning over takes place. The word **direct** is used in the same way with other geometrical transformations (p. 54).

indirect isometry an isometry (↑) in which a reflection (p. 234) is needed so that a configuration must be moved into a higher dimension to carry out the transformation (p. 54). In a plane this needs a third dimension, for solids the transformation cannot really take place, e.g. transforming a left shoe into a right one. The word **indirect** is used in the same way with other geometrical transformations.

translation (*n*) a direct isometry (↑) in which any line becomes another parallel line and any point is moved in a fixed direction by a fixed amount which can be described by a vector. **translate** (*v*), **translational** (*adj*), **translatory** (*adj*).

rotation (*n*) turning about an axis; a direct isometry (p. 233) in which in a plane one point (or in three dimensions one line) remains unchanged. **rotate** (*v*), **rotational** (*adj*), **rotatory** (*adj*).

reflection (*n*) the transformation (p. 54) which is produced by a looking glass or by a mirror except that it takes place in both directions; an indirect isometry (p. 233) in which in a plane one line (the line of symmetry, p. 232) and in three dimensions one plane (the plane of symmetry, p. 232) remains fixed and in which every other point is transformed perpendicular to it by an amount which is equal to twice its distance from it. *See also* central inversion (↓) for reflection in a point. **reflect** (*v*), **reflective** (*adj*).

glide reflection an indirect isometry (p. 233) which is produced by a translation (p. 233) and a reflection (↑) in a line parallel to the direction of translation. It is one of the four types of isometry (p. 233).

enantiomorph (*n*) a configuration which is a reflection (↑) of a given configuration, e.g. as a left shoe is of a right shoe. **enantiomorphic** (*adj*).

amphicheiral (*adj*) able to be deformed (p. 242) continuously without reflection into its enantiomorph (↑).

rotary reflection an indirect isometry (p. 233) which is produced by a rotation and a reflection (↑). It has the same result as a glide reflection (↑).

centre of rotation the fixed point in a rotation in a plane.

axis of rotation the fixed line in a rotation in three dimensions.

central inversion a transformation (p. 54) of a configuration that happens in such a way that the centre of symmetry (p. 232) of the configuration is the mid-point of the line which joins each point to its image (p. 51); it has the same result as a half-turn. *Do not mistake for* inversion (p. 195). Also known as **reflection in a point**.

rotation rotation of a configuration with centre (0, 1) through an angle of 135°

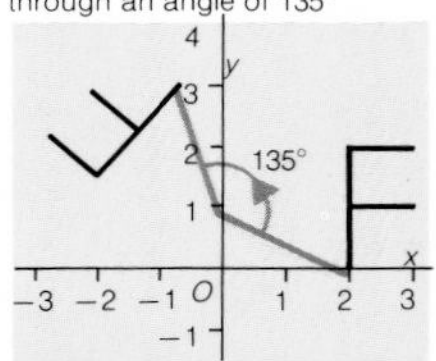

reflection
reflection of a configuration in the line $y = x$

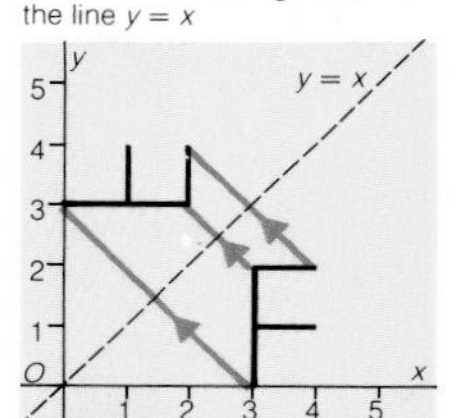

glide reflection
a translation parallel to a line followed by a reflection in it

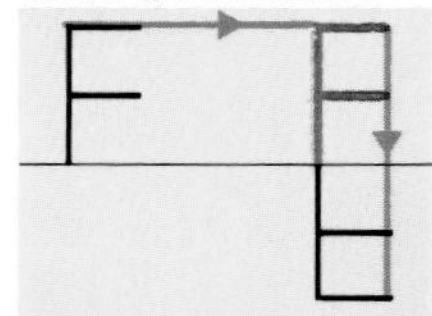

rotary reflection
a rotation and a reflection giving the same result as a glide reflection

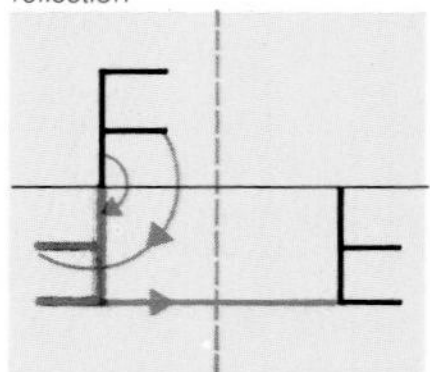

enlargement with centre (1, 0) and scale factor 3

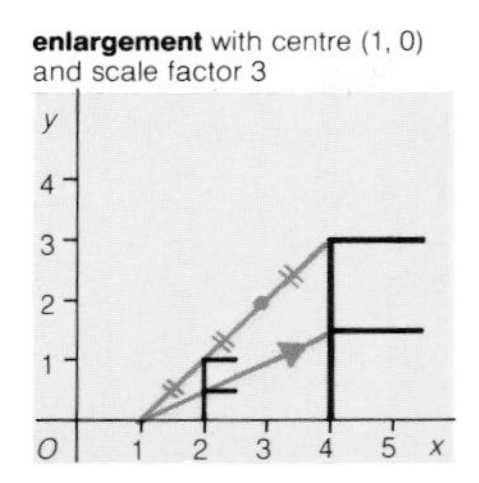

scale factor
fractional enlargement or contraction with scale factor $\frac{1}{2}$

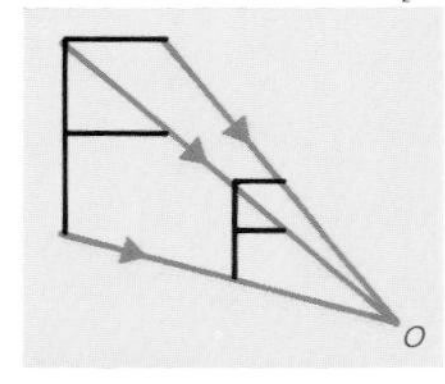

scale factor
negative enlargement with centre *O* and scale factor −3

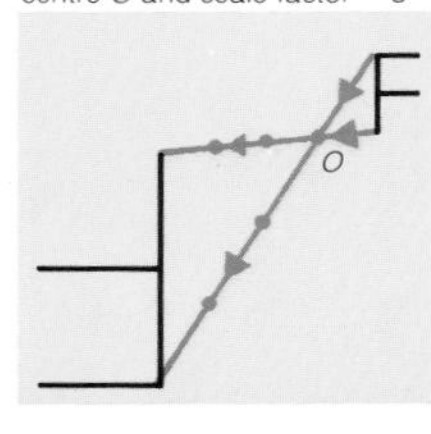

spiral similarity with angle of rotation 90° and scale factor 2

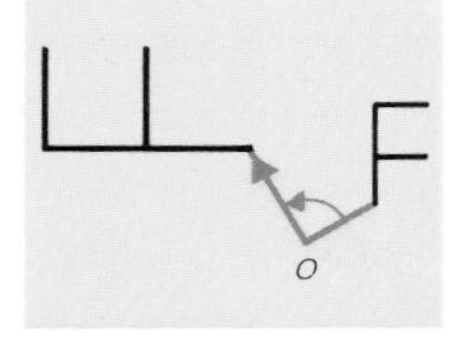

half turn a rotation through two right angles.

quarter turn a rotation through one right angle, usually in the positive sense (p. 174).

deflect (*v*) to turn, usually through a small angle. **deflection** (*n*).

conformal (*adj*) angle preserving (apart from sense, p. 174). Conformal is usually used of transformations (p. 54) in which angles remain unchanged, e.g. isometries (p. 233) and inversions (p. 195).

collineation (*n*) a plane transformation (p. 54) in which a set of collinear (p. 171) points becomes another set of collinear points.

axial collineation a collineation (↑) in which the two sets of points are reflections (↑) of one another in a line.

Erlangen programme a classification (p. 110) by Klein of geometries of different types by the properties which remain invariant (p. 55) under different transformations (p. 54).

enlargement (*n*) a geometrical transformation (p. 54) in which all lines become parallel lines but their lengths change in a constant ratio. Also known as **dilatation**, **similitude**, **magnification**. *See also* scale factor (↓). **enlarge** (*v*).

centre of enlargement the one point in an enlargement (↑) which does not change position. Also known as **centre of similitude**, **homothetic centre**.

scale factor the constant ratio in an enlargement (↑). A scale factor between −1 and 1 makes the configuration smaller (in spite of the name). Negative scale factors in plane isometries (p. 233) are the same as the corresponding positive ones with a half-turn (↑) added. **local scale factor** is sometimes used for derivative (p. 143). *See also* scale (p. 185). Also known as **magnification ratio**.

contraction (*n*) an enlargement (↑) whose scale factor (↑) is between −1 and 1. **contract** (*v*).

spiral similarity a transformation (p. 54) produced by a rotation and an enlargement (↑) with the same centre.

homothetic (*adj*) used of configurations which are enlargements (p. 235) of each other (as with isometries, p. 233, they can be direct, p. 233, or indirect, p. 233).

squashing transformation a transformation (p. 54) in which points in *n* dimensions become points in a smaller number of dimensions, e.g. points in a plane become points in a line.

perspective affinity a plane geometrical transformation (p. 54) in which straight lines (p. 167) remain straight lines and parallel lines and only parallel lines remain parallel. It is defined by a line *l*, a direction *d* and a non-zero ratio *r*. A point *P* is transformed into *P′* by drawing a line *PN* in the given direction *d* to meet the given line *l* at *N* and then making $P'N/PN = r$. It includes as special cases isometries (p. 233), similarities (p. 185), shears (↓) and stretches (↓).

perspective affinity
P transforms into *P′*, in this case $r = -3$ since *PN* and *P′N* are in opposite directions

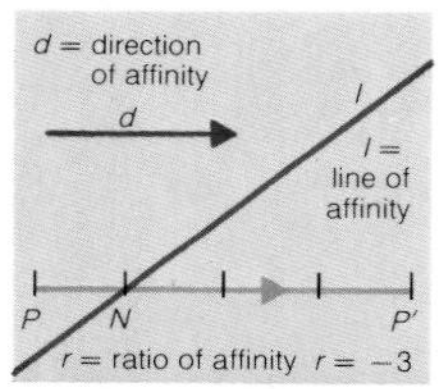

general affinity the result of one or more perspective affinities (↑). Also known as **affinity**, **affine transformation**.

affine geometry geometry using affinities (↑).

affine plane the plane whose points undergo affinities (↑).

axis of affinity the straight line *l* in the definition of perspective affinity (↑).

direction of affinity the direction *d* in the definition of perspective affinity (↑).

ratio of affinity the ratio *r* in the definition of perspective affinity (↑).

normal affinity a perspective affinity (↑) in which the direction *d* is perpendicular to the axis of affinity (↑) *l*. Reflection (p. 234) is a special case when the ratio of affinity (↑) *r* is −1.

oblique reflection a perspective affinity (↑) in which the ratio *r* is −1. It has the same result as a glide reflection (p. 234). Reflection is a special case in which the direction *d* is perpendicular to the axis of affinity (↑).

shear (*n*) a perspective affinity (↑) whose direction *d* and axis *l* are parallel. Every point moves parallel to the axis a distance proportional to its distance from the axis.

shear with direction and axis of affinity parallel

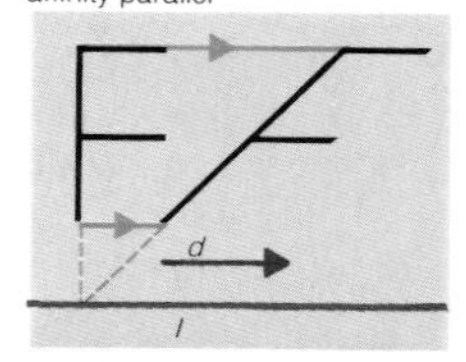

central projection from line *l* to line *l′*

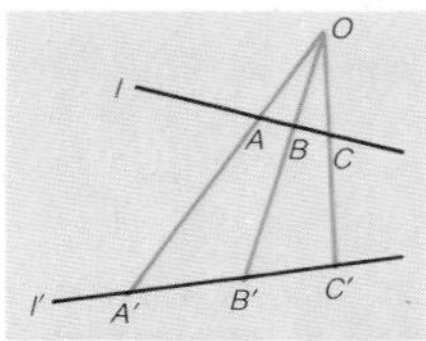

orthogonal projection from line *l* onto line *l′*

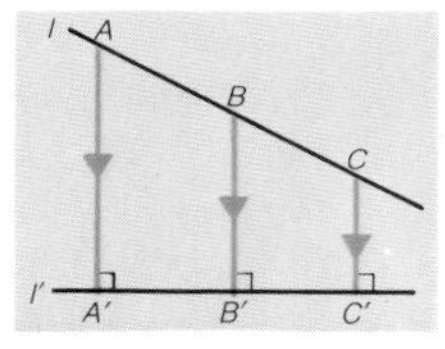

oblique projection from line *l* onto line *l′*

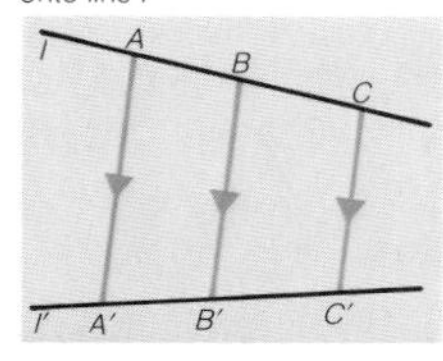

strip pattern
two of the seven types

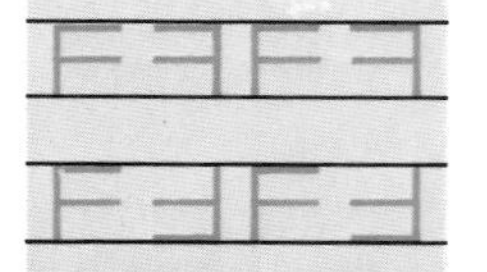

wallpaper pattern
parts of two of seventeen types

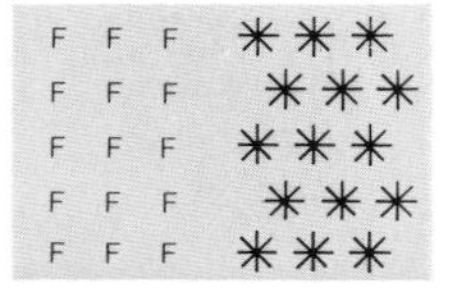

stretch[1] (*n*) a normal affinity (↑); sometimes one in which the ratio *r* is positive, so that it is direct. Also *one-way stretch*, in which case a *two-way stretch* is the result of two one-way stretches, usually perpendicular to each other. *See also* isometry (p. 233), p. 262.

squeeze (*n*) the same as a stretch (↑) but with the ratio positive and less than 1.

projection (*n*) a geometrical transformation (p. 54) of points; the word **transformation** is now commoner. *See also* component (p. 80), map projection (p. 208). **project** (*v*).

projective geometry especially used of geometry with one : one mappings (p. 51) and of the properties which remain unchanged under them, e.g. the ratios of corresponding lengths.

central projection a projection (↑) in which points on one plane (or line) are projected into corresponding points on another plane (or line) by lines which all pass through a fixed point outside both planes. Also known as **conical projection**.

centre of projection the fixed point in a central projection (↑).

orthogonal projection a projection (↑) in which points on one plane (or line) are projected onto points on a second plane (or line) by parallel lines drawn perpendicular to the second plane (or line).

oblique projection (1) a projection (↑) in which points on one plane (or line) are projected onto points in another plane (or line) by parallel lines not drawn perpendicular to the second plane (or line); (2) a map projection (p. 208) which is not in a usual position, i.e. it is neither zenithal (p. 205) nor equatorial (p. 204).

strip pattern a pattern which is repeated along a line by a set of translations (p. 233). There are seven ways of making such patterns. Also known as **frieze pattern**.

wallpaper pattern a pattern which is repeated over a plane by translations (p. 233) in two different directions. There are 17 ways of making such patterns.

topology (*n*) the part of mathematics which studies the properties of geometrical configurations which are unchanged under continuous transformations (p. 54); often thought of as 'rubber sheet' geometry, where the configuration may be stretched in any way possible without tearing. In algebra, if a non-empty (p. 11) set S has a system of subsets (p. 12) T, then T is called a topology if the null set and S itself are members of T and the intersection and union (p. 12) of any subsets of T are also subsets of T. Once known as **analysis situs**. **topological** (*adj*).

topological space a space defined by S and T as defined under topology (↑) and written (S, T).

topological equivalence used of topological spaces (↑) between whose elements there is a bijection (p. 52); in geometry, those configurations which can be transformed (p. 54) continuously into each other.

topological invariant any property which remains invariant (p. 55) or unchanged under a topological (↑) transformation (p. 54).

deleted neighbourhood a neighbourhood (p. 95) of a point without the point itself. Also called **punctured neighbourhood**.

fixed-point theorems theorems in topology (↑) about points which remain fixed or unchanged under transformations (p. 54), e.g. the hairy ball theorem (↓).

Brouwer's fixed-point theorem in any continuous transformation (p. 54) of a disc (p. 192) (or any topologically equivalent, ↑, surface) into itself, there is at least one point which is mapped (p. 51) onto itself.

hairy ball theorem a fixed-point theorem (↑) for a sphere or other simple closed surface (p. 209), so called because it means that a hairy sphere must, if brushed, have at least one point to which no hair can be brushed.

hairy ball theorem
there is a point to which hairs cannot be brushed

frontier (*n*) the frontier of a set A in a topological space (↑) X is the set of all points x such that every neighbourhood (p. 95) of x meets A and the relative complement (p. 13) of A in X. It gives a more exact definition of the idea of boundary.

Jordan curve
simple closed curve on a plane

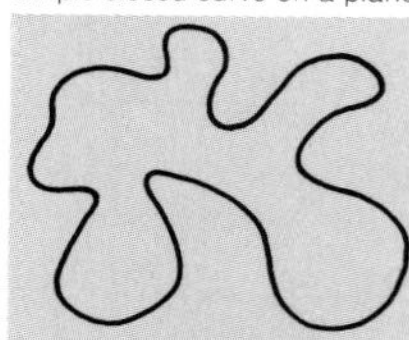

Jordan curve a simple closed curve (p. 191) on a plane which is topologically equivalent (↑) to a circle, and thus divides the plane into two regions.

network (*n*) a set of points joined by lines, often called arcs. Only the ways in which the points are joined is important, and not either the positions of the points or the lines; two networks which are topologically equivalent (↑) are taken as having the same properties. Also known as **net**, **graph**.

graph theory the part of topology (↑) which studies networks (↑). Also known as **network theory**.

planar network a network (↑) on a surface which is topologically equivalent (↑) to a plane.

node² (*n*) one of the set of points in a network (↑). Also known as **point**, **vertex**, **junction**, **intersection**. *See also* pp. 189, 279.

arc² (*n*) a line joining two nodes (↑) in a network (↑). Also known as **edge**, **link**, **route**, **branch**. *See also* p. 187.

order of a node the number of arcs leading from that node.

region² (*n*) an area into which a planar network (↑) divides the plane. Also known as **face** if the network (↑) is a net (p. 202) of a polyhedron (p. 200). *See also* p. 168.

network
five points joined by six lines

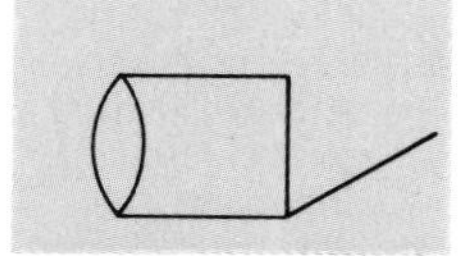

node
five nodes *A*, *B*, *C*, *D*, *E* of a network

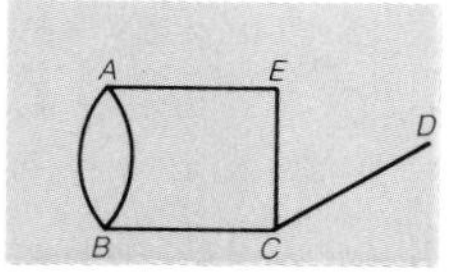

arc
six arcs *a*, *b*, *c*, *d*, *e*, *f* of a network

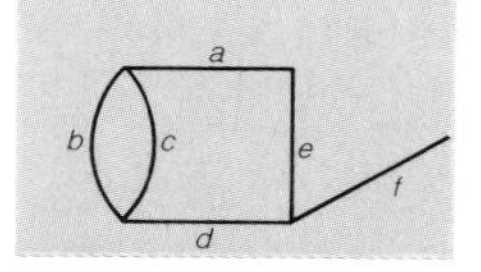

region
three regions of a plane network: one region is unbounded

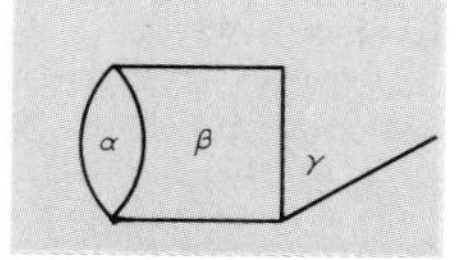

circuit (*n*) the set of arcs around a given region. Also known as **loop**.
collapse (*v*) to remove one by one the arcs from a network (p. 239).
unicursal (*adj*) used of networks (p. 239) having a continuous path which can be followed along every edge once and only once. Such a network must have not more than two nodes (p. 239) with an odd number of arcs leading to them.
Königsberg bridge problem a famous topological (p. 238) problem which has no solution. A path had to be found which crossed each of the bridges in the town once and once only. This was impossible since the network (p. 239) with a bridge on each arc was not unicursal (↑).

unicursal
the path *A C D A B C* covers each arc once and only once

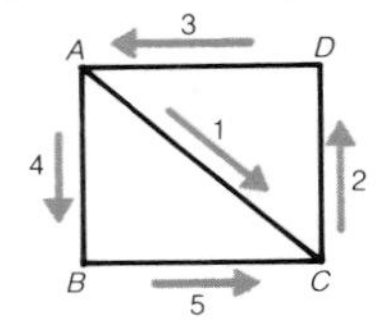

Königsberg bridge problem
the network is not unicursal

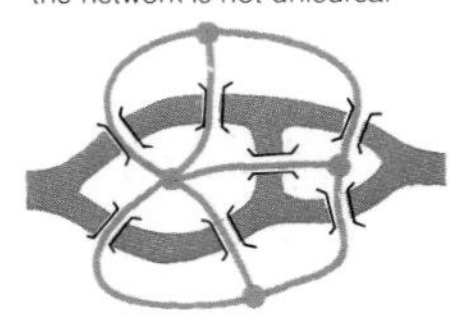

connected

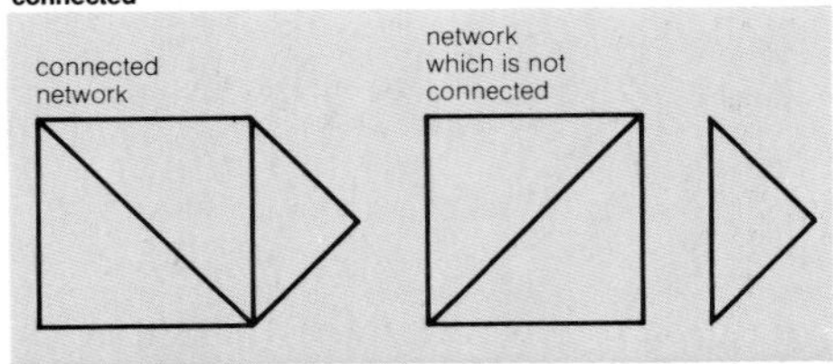

connected (*adj*) joined; used in networks (p. 239) where every node (p. 239) is joined to every other node by a set of arcs and in algebraic topology (p. 238) of a region whose points cannot be partitioned (p. 13) into two disjoint (p. 13) open (p. 65) sets (p. 10), any points on the partition must make one of the sets closed. (The corresponding word for functions is *continuous*, p. 63.) **connect** (*v*), **connection** (*n*).
singly connected used of a network (p. 239) where the removal of any one arc separates it into two parts. *Do not mistake for* simply connected (↓). *See also* multiply connected (↓).
tree (*n*) a network (p. 239) which is singly connected (↑) like a tree with branches.
co-tree (*n*) the set of arcs removed from a connected (↑) network (p. 239) to turn it into a given tree (↑).

singly connected
network or tree

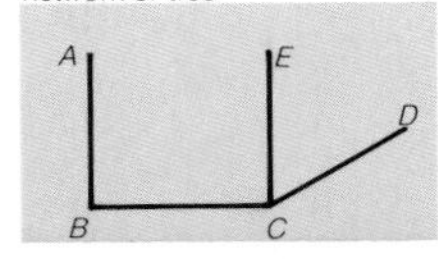

tree
a network (both colours). When the co-tree (in red) is removed a tree (in black) is obtained

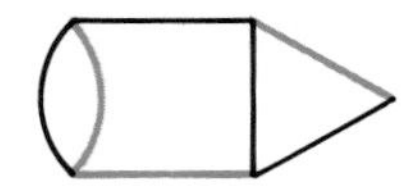

incidence matrix
for a network

A a E
b α c β e γ D
f
B d C

network

incidence matrix for arcs

	a	b	c	d	e	f
a	0	1	1	0	1	0
b	1	0	1	1	0	0
c	1	1	0	1	0	0
d	0	1	1	0	1	0
e	1	0	0	1	0	1
f	0	0	0	0	1	0

incidence matrix for nodes

	A	B	C	D	E
A	0	2	0	0	1
B	2	0	1	0	0
C	0	1	0	1	1
D	0	0	1	1	0
E	1	0	1	0	0

incidence matrix for regions. Note the 2 in the principal diagonal

	α	β	γ
α	0	1	1
β	1	0	3
γ	1	3	2

incidence matrix a matrix (p. 83) whose rows and columns correspond to arcs, nodes (p. 239) or regions of a network showing which are adjacent (p. 171) or joined, the number in each case being shown by the number in the corresponding element of the matrix (*see figure*). A zero shows that the two nodes are not joined or that the two regions or arcs are not next to one another.

knot in topology

knot (*n*) a closed curve (p. 190) formed by threading string with itself and then joining the two ends. This is a topological (p. 238) concept and the word has a slightly different meaning from that in everyday life.

reducible loop
the region in red has a hole, so that the black curve is not reducible in the region, which is not therefore simply connected

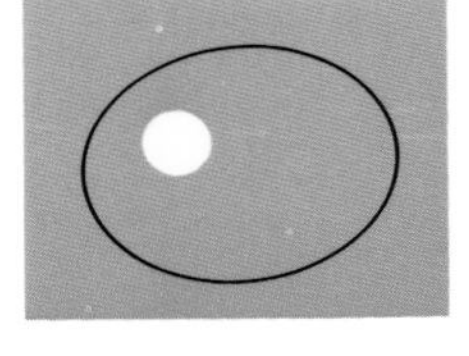

reducible loop a closed curve (p. 190) which can be reduced (p. 291) continuously to a point while remaining within the region concerned.

simply connected used of a region in which every closed curve (p. 190) forms a reducible loop (↑). *Do not mistake for* singly connected (↑).

multiply connected (1) (of regions) not simply connected (↑); (2) (of networks) not singly connected (↑).

connectivity
a surface (in red) with three holes requires three cuts to make it homeomorphic to a disc. It has connectivity 4

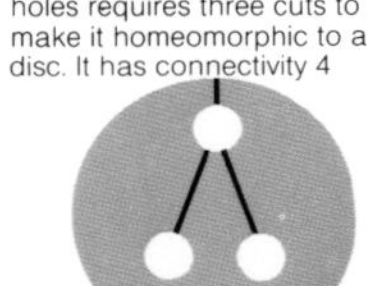

connectivity (*n*) if the smallest number of cuts which a surface needs to make it homeomorphic (p. 54) to a disc (p. 192) is $n - 1$, then the connectivity of the surface is n.

genus (*n*) (of a surface) the largest number of distinct (p. 60) non-intersecting closed curves (p. 190) which may be drawn on the surface without dividing it into two regions.

deformation (*n*) a continuous transformation (p. 54) which stretches or twists a shape in space (or a line on a surface) but gives no new intersections or boundaries. **deform** (*v*).

homotopic (*adj*) used of curves which can be deformed (↑) continuously into each other. **homotopy** (*n*).

fundamental group the group (p. 72) whose elements are the set of all homotopic (↑) closed curves (p. 190) in a given space.

winding number the number of times a closed curve (p. 190) lying on a surface with a discontinuity (p. 64) goes round the discontinuity.

Euler characteristic of a surface: the quantity $V - E + F$ for a network (p. 239) drawn on the surface, where V is the number of nodes (p. 239), E is the number of arcs and F is the number of regions in the network. It is a constant for a given surface, and for a plane or a sphere it is 2. *See also* Euler's formula (p. 202).

Möbius strip a surface with one side and one edge. It can be made from a long, thin rectangle of paper which is given a half twist, the opposite short edges then being joined. Also known as **Möbius band**.

cross-cap (*n*) a topological (p. 238) shape which is a sphere with a cap cut off and replaced with a Möbius strip (↑) joined edge to edge.

Klein bottle a topological (p. 238) shape which has only one surface, no edges and no outside or inside.

four-colour problem once a conjecture (p. 254), now proved, it is the theorem that in colouring the regions on a plane or spherical (p. 204) map (p. 51) so that no two regions with a common edge have the same colour, only four different colours are needed. For a torus (p. 210) as many as seven colours are needed.

chromatic number of a surface: the least number of different colours required to colour any given map (p. 51) on that surface.

genus
the genus of a torus is one: the curve *C* is drawn on a torus, but does not divide it in two

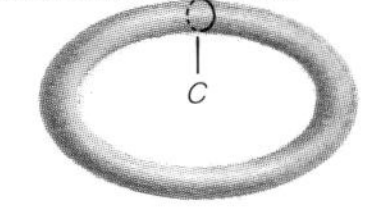

winding number
the closed curve (in black) has winding number 2 in the region (in red)

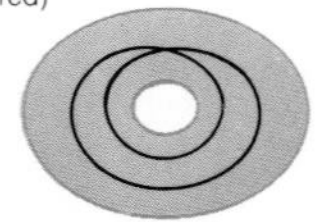

Möbius strip
the different colours are in fact the same side since the strip has only one side

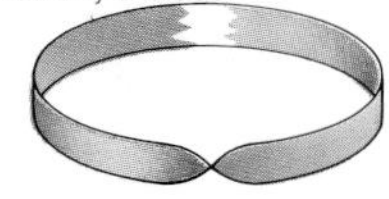

cross-cap
formed by joining *AB and CD* in the directions shown and similarly joining *BC* and *DA*

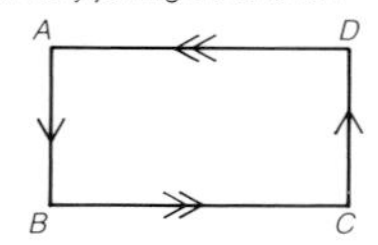

Klein bottle

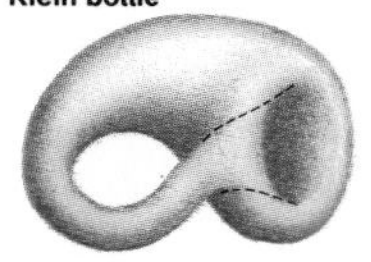

four-colour problem
a plane map requiring four colours

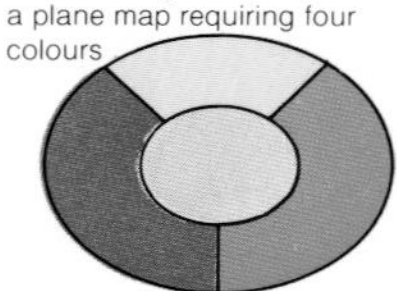

logic (*n*) the study of reasoning, and of methods of obtaining correct (p. 288) proofs. **logical** (*adj*).

argument[2] (*n*) a set of reasons given as part of a proof. *See also* p. 49. **argue** (*v*).

statement (*n*) a set of words giving a mathematical idea or concept or forming part of a proof or argument (↑). **state** (*v*). *See also* p.292.

true (*adj*) correct, especially of an argument (↑), agreeing with fact. *See also* false (↓). **truth** (*n*).

false (*adj*) incorrect, not agreeing with fact. Most mathematical statements (↑) are either true or false, but some are undecidable (↓). **falsity** (*n*), **falsehood** (*n*).

undecidable (*adj*) a statement (↑) which cannot be shown to be either true or false, better not used for a statement which has not been shown to be true or false, but where this might in future be shown.

Gödel's theorem in any logical (↑) system a formula can be obtained which cannot be proved either true or false.

intuitive (*adj*) not needing logical (↑) proof or definition, or for which this cannot be given. **intuition** (*n*).

intuitionism (*n*) a theory in which mathematics is regarded as depending on intuitive (↑) concepts rather than on logic (↑) or empirical (p. 255) experiments.

self-evident (*adj*) used of a result which is intuitively (↑) clear, or against which any argument (↑) is unlikely or impossible.

symbolic logic logic (↑) using symbols, especially the symbols for connectives (p. 244). Also known as **propositional calculus**, **propositional logic**, **formal logic**.

Boolean algebra an algebra which has the same rules as the algebra of symbolic logic (↑) using the operations (p. 14) 'and' (p. 245) and 'or' (p. 245); and which is also found in the algebra of sets using the operations intersection and union (p. 12); and also in switching algebra (p. 244).

Boolean duality the duality (p. 256) shown in any Boolean algebra (p. 243), e.g. that between union (p. 12) and intersection in the algebra of sets and that between 'and' (↓) and 'or' (↓) in symbolic logic (p. 243).

switching algebra the Boolean algebra (p. 243) of switches (which turn electricity on and off). Switches in series (one after the other) are like 'and' (↓) in symbolic logic (p. 243) and those in parallel (side by side) are like 'or' (↓).

switching algebra

switches in series corresponding to 'and' in logic

switches in parallel corresponding to 'or' in logic

admission table a table of values defining an operation (p. 14) in Boolean algebra.

sentence (*n*) a statement (p. 243) of something which may be true or false (or possibly undecidable, p. 243). **sentential** (*adj*).

assertion (*n*) a sentence (↑) which is believed to be true, or is to be taken as true. *See also* proposition (p. 250). Symbol: $\vdash$ written before the sentence.

connective (*n*) a word or set of words which join sentences (↑), most commonly *and*, *or*, *if* and *if and only if* (pp. 245, 246); also the word *not* which changes a sentence to its negation (↓).

prime sentence a sentence (↑) with no connectives (↑). Also known as **atomic statement**.

composite sentence a sentence (↑) which is not a prime sentence (↑).

truth value if a statement (p. 243) is true it has truth value 1, if it is false it has truth value 0. In some types of logic (p. 243) fractional values are allowed for undecidable (p. 243) statements.

truth table a table giving all possible truth values (↑) of the prime sentences (↑) making up a composite sentence (↑) and the corresponding truth values of the composite sentence.

truth set the set of truth values (↑) for the prime sentences (↑) which give a true composite sentence (↑).

two term logic the usual form of logic (p. 243) which has two truth values (↑) 0 and 1.

three term logic logic (p. 243) with three truth values (↑), usually true, false and one other such as undecided or meaningless.

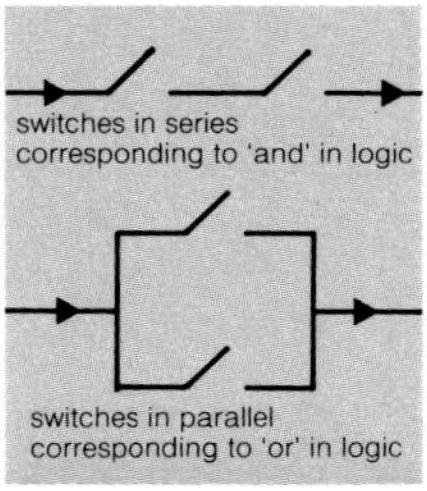

tautology (*n*) a composite sentence (↑) which cannot be false, but which always has truth value (↑) 1, e.g. 'a statement is correct or it is not correct'.

not (*adv*) a connective (↑) which changes a true sentence (↑) into a false one and a false sentence into a true one, e.g. changes 'four is an even number' into 'four is not an even number'. Symbol: ~.

negation
truth table
1 shows the statement is true
0 shows the statement is false

A	$\sim A$
1	0
0	1

negation (*n*) a singulary operation (p. 14) in which a sentence (↑) is changed by adding *not* (↑) to it or removing *not* from it. *The word* negative *is often wrongly used here.* **negate** (*v*). Symbols: $\sim A$, $\rightharpoondown A$, $\tilde{A}$ or A'.

double negation negation (↑) taken twice. In mathematics (unlike sometimes in everyday life) this is always taken to mean the same as if there is no negation. Symbols: $\sim(\sim A)$ etc.

and (*conj*) a connective (↑) which gives a composite sentence (↑) which is true if and only if the two prime sentences (↑) which it joins are both true. Also known as **both** ... **and**. Symbols: $\wedge$, &.

conjunction
truth table

A	B	$A \wedge B$
1	1	1
1	0	0
0	1	0
0	0	0

conjunction (*n*) a binary operation (p. 14) which is the joining of two sentences (↑) A and B by the word *and* (↑) or words of similar meaning. Also known as **logical product**. Symbols: $A \wedge B$, A & B.

or (*conj*) a connective (↑) which gives a composite sentence (↑) which is true if one or the other or both of the two prime sentences (↑) which it joins are true. *Warning*: *or* should never be used in mathematics as it sometimes is in everyday life to mean 'one or the other but not both'. When this meaning is needed the word aut (p. 246) can be used. Also known as **either** ... **or**. *See also* vel (p. 246). Symbol: $\vee$.

disjunction
truth table

A	B	$A \vee B$
1	1	1
1	0	1
0	1	1
0	0	0

disjunction (*n*) a binary operation (p. 14) which is the joining of two sentences (↑) A and B by the word *or* (↑) or words of similar meaning. Also known as **logical sum**, **inclusive disjunction**. *See also* exclusive disjunction (p. 246). Symbol: $A \vee B$.

vel (*conj*) the Latin word for 'or' as defined in mathematics, and sometimes used because *or* in English can have two meanings. *See also* or (p. 245). Symbol: $\vee$.

aut (*conj*) the Latin word for exclusive disjunction (↓) which gives a composite sentence (p. 244) which is true if either, but not both, of its two prime sentences (p. 244) is true. *See also* or (p. 245), vel (↑). Symbol: $\underline{\vee}$.

exclusive disjunction a binary operation (p. 14) which is the joining of two sentences (p. 244) by the word *aut* (↑) or words of similar meaning. Symbol: $A \underline{\vee} B$.

if (*conj*) a connective (p. 244) which gives a composite sentence (p. 244) 'if *A* then *B*' where *A* and *B* are prime sentences (p. 244), which is true if *A* and *B* are both true and also true if *A* is false whether *B* is true or not. Also known as *A* **only if** *B*, *B* **if** *A* and often as *A* **implies** *B*.

conditional (*n*) a binary operation (p. 14) which joins two sentences (p. 244) *A* and *B* by words such as 'if (↑) *A* then *B*'. Also known as **implication**, **material implication**, but these are best not used because they can be mistaken for the relation of implication (↓). Symbol: $A \longrightarrow B$ for 'if *A* then *B*' or $B \longleftarrow A$ for '*B* if *A*'.

imply (*v*) in a conditional (↑) 'if *A* then *B*', *A* is said to imply *B*. (*B* is also said to be implied by *A* or to be inferred, p. 255, from *A*.)

antecedent (*n*) the sentence (p. 244) *A* following *if* (↑) in a conditional (↑) statement (p. 243) 'if *A* then *B*'.

consequent (*n*) the sentence (p. 244) *B* following *then* in a conditional (↑) statement (p. 243) 'if *A* then *B*'.

if and only if a connective (p. 244) which gives a composite sentence (p. 244) 'if and only if *A* then *B*', which is true if *A* and *B* are both true and also if they are both false, but not otherwise. **iff** (*abbr*) is common and often used as a word, *but do not mistake for* if (↑). Symbol: ↔.

biconditional (*n*) a binary operation (p. 14) which joins two sentences (p. 244) *A* and *B* by words such as 'if and only if (↑) *A* then *B*'. Symbol: $A \leftrightarrow B$.

exclusive disjunction (aut)
truth table

A	*B*	*A* aut *B*
1	1	0
1	0	1
0	1	1
0	0	0

conditional
truth table

A	*B*	$A \rightarrow B$
1	1	1
1	0	0
0	1	1
0	0	1

antecedent

if *A*	then	*B*
A	→	*B*
antecedent		consequent
B	if	*A*
B	←	*A*
consequent		antecedent

biconditional
truth table

A	*B*	$A \leftrightarrow B$
1	1	1
1	0	0
0	1	0
0	0	1

nor
truth table

A	B	A nor B
1	1	0
1	0	0
0	1	0
0	0	1

Sheffer stroke function
truth table

A	B	A nand B
1	1	0
1	0	1
0	1	1
0	0	1

nor (*conj*) the connective (p. 244) 'not or' which is written A nor B where A and B are prime sentences (p. 244). Nor gives a composite sentence (p. 244) which is true only when A and B are both false. Symbols: $\downarrow$, $\|$.

nand (*conj*) the connective (p. 244) 'not and' which is written A nand B where A and B are prime sentences (p. 244). Nand gives a composite sentence (p. 244) which is false only when A and B are both true. $A \uparrow A$ is then the same as not A and 'not' (p. 245) can be seen as similar to the other connectives. Symbols: $\uparrow$, $|$.

Sheffer stroke function the function of two prime sentences (p. 244) A and B which is given by $A \uparrow B = C$ (*see* nand, ↑). Also known as **stroke function**.

Sheffer operations the operations (p. 14) which use the connectives 'nor' (↑) and 'nand' (↑).

implication (*n*) a relation (p. 16) between two sentences A and B. If A implies B, then $A \rightarrow B$ is true (*see* conditional, ↑); if A is implied by B, then $A \leftarrow B$ is true. The symbols of implication are now widely used in writing down proofs where each statement (p. 243) follows logically (p. 243) from the one before. **imply** (*v*). Symbols: $\Rightarrow$ for 'implies', $\Leftarrow$ for 'is implied by'.

equivalence (*n*) a relation (p. 16) between two sentences A and B. If A is equivalent to B, then $A \leftrightarrow B$ is true (*see* biconditional, ↑). Equivalence is now used widely in writing down proofs where the argument (p. 243) follows both forwards and backwards. *See also* equivalence relation (p. 18). **equivalent** (*adj*). Symbol: $\Leftrightarrow$.

hence (*adv*) which has the implication (↑) that. Also known as **therefore**, especially in older books. Symbol: $\therefore$, which is now usually replaced by $\Rightarrow$.

because (*adv*) for the reason that. Symbol: $\because$, which is now usually replaced by $\Leftarrow$.

syntactic implication implication (↑) which is based on logic (p. 243) itself and does not depend on the particular meanings of the symbols used.

semantic implication implication (p. 247) which depends on the particular properties of the theory being studied and the meanings given to the symbols being used.

quantifier (*n*) (1) one of the words *all*, *some*, *none* and *not all* or words of similar meaning as used in logic (p. 243); (2) the universal (↓) or particular quantifiers (↓).

universal quantifier the words *for all* or *for every* and words of similar meaning, and also *for none* or *for all not* and words of similar meaning. Symbol: ∀ (meaning 'for all').

exist (*v*) to have being, used of mathematical quantities or concepts which can be found rather than those which are logically (p. 243) impossible.

existential quantifier the words *there is*, *there exists* (↑) or words of similar meaning. Symbol: ∃.

particular quantifier the words *some*, *not all* and words of similar meaning as used in logic (p. 243).

some (*adj*) often used in logic (p. 243) as in 'some men are fathers'. It is often better read as 'at least one'.

universal affirmative a statement (p. 243) of agreement as in 'all *A* are *B*'. *See also* particular affirmative (↓).

particular affirmative a statement (p. 243) of agreement as in 'some *A* are *B*'. *See also* universal affirmative (↑).

universal negative a statement (p. 243) of disagreement as in 'no *A* are *B*'. *See also* particular negative (↓).

particular negative a statement (p. 243) of disagreement as in 'some *A* are not *B*'. *See also* universal negative (↑).

square of opposition a diagram used in logic (p. 243) showing how the quantifiers (↑) are related.

Euler circles a method of showing propositions (p. 250) in logic (p. 243) by means of circles which give the various possibilities. *Do not mistake for* Euler circle, *see* nine-point circle (p. 197).

square of opposition
showing the words used to describe the relations between the statements

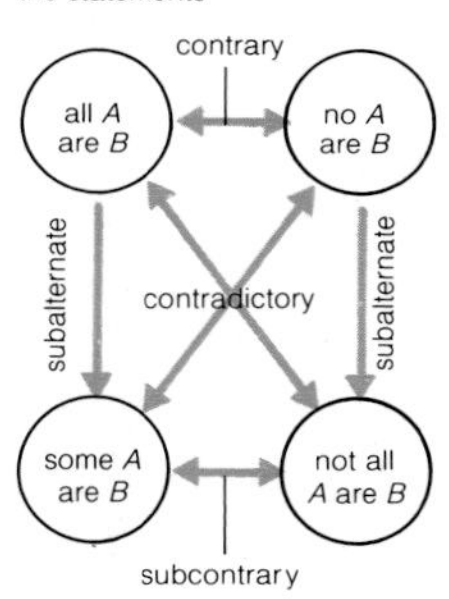

Euler circles

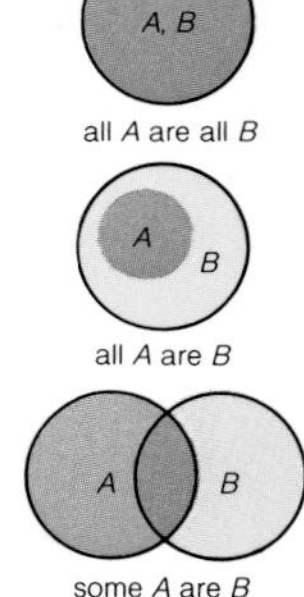

all *A* are all *B*

all *A* are *B*

some *A* are *B*

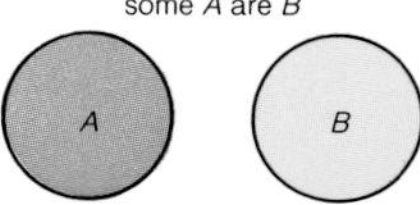

no *A* are *B*

Venn diagrams

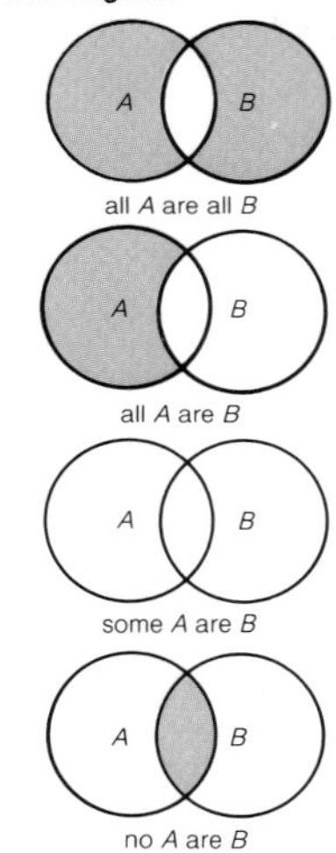

the green areas have no elements

Venn diagram a method obtained from the idea of Euler circles (↑) which includes shading showing the parts of the diagram not concerned in the problem, and which gave rise to the idea of set diagrams (p. 11). Also now widely used with the meaning of set diagram.

necessary condition if a sentence (p. 244) *B* is true only when a sentence *A* is also true, then *A* is a necessary condition for *B*, and we can write $A \Rightarrow B$. *See also* implication (p. 247).

sufficient condition if a sentence (p. 244) *B* is false only when a sentence *A* is also false, then *A* is a sufficient condition for *B*, and we can write $A \Leftarrow B$. *See also* implication (p. 247).

necessary and sufficient condition if sentences (p. 244) *A* and *B* are only true or false together, *A* is said to be a necessary and sufficient condition for *B* and similarly *B* is said to be a necessary and sufficient condition for *A*, and we can write $A \Leftrightarrow B$. *See also* equivalence (p. 247).

condition (*n*) a statement (p. 243) that must be true for a conclusion (p. 251) to be true. Best avoided without *necessary* (↑) and/or *sufficient* (↑).

trivial (*adj*) of little use or importance, though true and needs to be included to complete a statement (p. 243).

contingent (*adj*) used of a conclusion (p. 251) which depends on some necessary condition (↑). *See also* contingency table (p. 128).

denial (*n*) if in any tautology (p. 245) using conjunction (p. 245) and disjunction (p. 245) these operations (p. 14) are interchanged, then the resulting tautology is the denial of the first.

Boolean function an expression in Boolean algebra (p. 243) using the operations (p. 14) 'and' and 'or' and in which the variables have values 0 and 1. *See also* truth values (p. 244).

Boolean polynomial a Boolean function (↓) which is the disjunction (p. 245) of separate conjunctions (p. 245).

disjunctive normal form the canonical (p. 287) form for a Boolean function (↑) in which the terms are separated by disjunctions (p. 245) and each term contains all the variables, e.g. $(A \wedge B) \vee (A \wedge B') \vee (A' \wedge B)$.

converse

statement	A	$\rightarrow$	B
converse	B	$\rightarrow$	A
inverse	$\sim A$	$\rightarrow$	$\sim B$
contrapositive	$\sim B$	$\rightarrow$	$\sim A$

converse (*n*) the converse of the statement (p. 243) '*A* implies (p. 246) *B*' is '*B* implies *A*', e.g. 'all even numbers are divisible by 2' has the converse 'all numbers divisible by 2 are even'.

inverse (*n*) the inverse of the statement (p. 243) '*A* implies (p. 246) *B*' is 'not *A* implies not *B*', e.g. 'all even numbers are divisible by 2' has the inverse 'all uneven numbers are not divisible by 2'. *See also* p. 16.

contrapositive (*n*) the contrapositive of the statement (p. 243) '*A* implies (p. 246) *B*' is 'not *B* implies not *A*', e.g. 'all even numbers are divisible by 2' has the contrapositive 'all numbers not divisible by 2 are uneven'.

proof (*n*) an argument (p. 243) which shows a statement (p. 243) to be true or false; a set of implications (p. 247) showing logically (p. 243) the reasons why a statement is true or false. **prove** (*v*), **provable** (*n*).

proposition (*n*) a statement (p. 243) of a problem which is to be proved. Also known as **assertion**.

enunciation (*n*) a more general statement (p. 243) of the proposition (↑) to be proved at the start of a proof.

predicate (*n*) a proposition (↑) in logic (p. 243) which is formed like a sentence (p. 244) or statement (p. 243) in speech.

propositional function a function in logic (p. 243), the function itself being any possible predicate (↑) and the variable the elements of the set with which the function may be used.

axiom (*n*) a statement (p. 243) which is taken to be true and which is used as one of the starting points of sequences (p. 91) of proofs; a concept which is taken to be true at the start of any branch of mathematics. **axiomatic** (*adj*).

postulate (*n*) an axiom (↑), in geometry also a proposition (↑). **postulate** (*v*).

axiomatics (*n*) the study of parts of mathematics as logical (p. 243) systems which prove the truth or falsity of propositions (↑) obtained from sets of axioms (↑). Also known as **formalism**.

minimum axiom set a set of axioms (↑) which is both a necessary and a sufficient condition (p. 249) for the generation (p. 209) of a mathematical system.

constructive (*adj*) not axiomatic (↑), based on intuitive (p. 243) ideas.

assumption (*n*) a fact which is taken to be true without proof. **assume** (*v*).

hypothesis (*n*) proposition (↑), a statement (p. 243) which may be true or false but is assumed (↑) to be true in an argument (p. 243), a statement which one sets out to prove or disprove. **hypotheses** (*pl*), **hypothesize** (*v*).

primitive proposition an initial assumption (↑) in a mathematical system which cannot or will not be proved. Similar to axiom (↑), but stated at greater length and usually of less importance.

premise (*n*) an assumption (↑) in a logical (p. 243) proof. Also written **premiss**. *See also* syllogism (↓).

conclusion (*n*) the last statement (p. 243) in a proof which repeats the proposition (↑). *See also* syllogism (↓).

syllogism (*n*) a logical (p. 243) argument (p. 243) which has a major (p. 289) premise (↑), a minor (p. 290) premise and a conclusion (↑), e.g. 'all cows have four legs' (major premise), 'this animal is a cow' (minor premise), 'therefore this animal has four legs' (conclusion).

QED *quod erat demonstrandum* 'which was to be shown'. Once placed at the end of a proof in Euclidean geometry (p. 171).

QEF *quod erat faciendum* 'which was to be done'. Once placed at the end of a construction (p. 186) in Euclidean geometry (p. 171).

analytic proposition a proposition (↑) which must, by definition, be true, e.g. 'every even number is divisible by 2'.

synthetic proposition a proposition (p. 250) giving empirical (p. 255) information which cannot be deduced (p. 255) by logical (p. 243) argument (p. 243).

theorem (*n*) a proposition (p. 250) which has been proved and forms an important idea in a branch of mathematics, so that it may be assumed (p. 251) in making further proofs.

strong (*adj*) used of a form of a theorem which is harder to prove but has conditions (p. 249) which do not allow exceptions (p. 288).

weak (*adj*) used of a form of a theorem which is easier to prove but allows exceptions (p. 288) which cannot be decided one way or the other

lemma (*n*) a proposition (p. 250) which is proved separately so that it can then be used in the proof of a theorem.

porism (*n*) a theorem which is obtainable immediately from another theorem; a theorem which shows that where one example of something occurs, then infinitely many examples occur, e.g. Steiner's circles (p. 195). **poristic** (*adj*).

corollary (*n*) a proposition (p. 250) which follows from a theorem which has already been proved but is less important than the theorem, e.g. the exterior angle (p. 177) of a cyclic quadrilateral (p. 193) equals the opposite interior angle (p. 177), which follows from the theorem that the opposite angles are supplementary (p. 175).

rider (*n*) a proposition of less importance than a corollary (↑) which follows from a theorem and is often given as an example or exercise (p. 288) for learners.

extension[2] (*n*) the widening of a theorem or other mathematical concept to make it more general. *See also* pp. 10, 262. **extend** (*v*).

valid (*adj*) correct, used of an argument (p. 243) or proof where only true conclusions (p. 251) are drawn from correct premises (p. 251).

invalid (*adj*) not valid (↑).

fallacy (*n*) a valid (↑) argument (p. 243) from false premises (p. 251) or an incorrect conclusion (p. 251) by false reasoning, usually one where the mistake is hard to find. **fallacious** (*adj*).

theorem
the opposite angles of a cyclic quadrilateral are supplementary
$\hat{B} + \hat{D} = 180°$

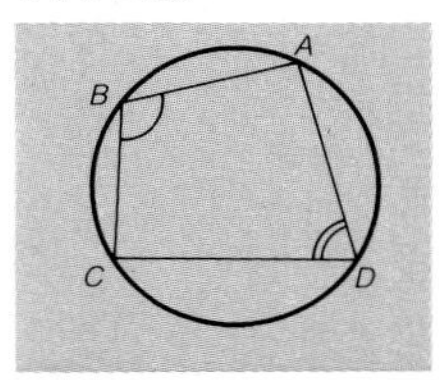

corollary
the exterior angle of a cyclic quadrilateral equals the opposite interior angle
$\hat{B} = A\hat{D}E$

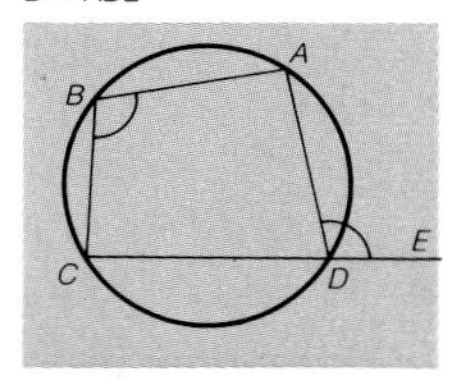

rider
if *ABCD* and *ADEF* are cyclic quadrilaterals and *BAF*, *CDE* are straight lines, then *BC* is parallel to *FE*

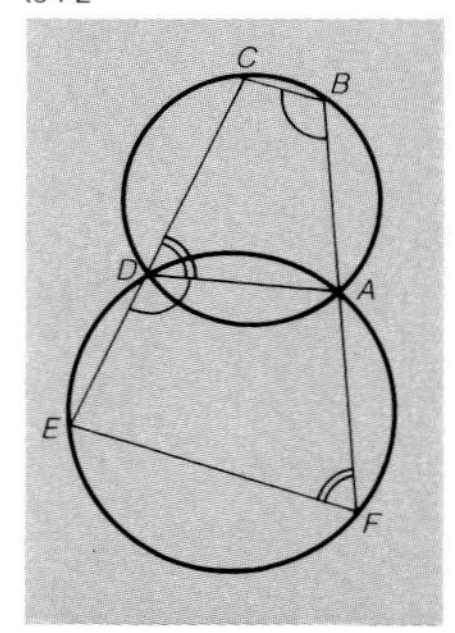

undistributed middle

$A \Rightarrow C$ and $B \Rightarrow C$

falsely used to infer

$A \Rightarrow B$.

A correct argument is

$A \Rightarrow B$ and $B \Rightarrow C$

therefore

$A \Rightarrow C$

undistributed middle a fallacious (↑) argument (p. 243) of the form *A* is *C*, *B* is *C*, therefore *A* is *B*, e.g. 'rectangles are quadrilaterals (p. 177), rhombuses (p. 181) are quadrilaterals, therefore rectangles are rhombuses'.

vicious circle an argument (p. 243) in which the conclusion (p. 251) is assumed (p. 251) in the premise (p. 251) and which is therefore fallacious (↑).

theory (*n*) a set of theorems starting with axioms (p. 250) which covers logically (p. 243) a part of mathematics. **theoretical** (*adj*).

existence theorem a theorem which shows that something exists without finding it.

uniqueness theorem a theorem that shows that something is unique (p. 14), sometimes without finding it or showing how to find it.

consistent (*adj*) used of sets of things, e.g. axioms (p. 250), which can all be true at the same time, and do not lead to false conclusions (p. 251). **consistency** (*n*).

contradictory (*adj*) not consistent. Two statements (p. 243) are contradictory if the truth of one implies (p. 246) the falsity of the other. *See also* square of opposition (p. 248). **contradiction** (*n*), **contradict** (*v*).

contrary (*adj*) used of a statement (p. 243) which is the opposite of a given statement, e.g. 'all squares are rectangles' has the contrary statements 'no squares are rectangles' and 'all squares are not rectangles'. The contrary operation (p. 14) to a given operation is the operation which gives the negation (p. 245) of the given operation. *See also* square of opposition (p. 248).

subcontrary (*n*) *see* square of opposition (p. 248) (diagram).

subalternate (*n*) *see* square of opposition (p. 248) (diagram).

rigorous (*adj*) including everything, making every condition (p. 249) and assumption (p. 251) clear. **rigour/rigor** (*n*).

paradox (*n*) an argument (p. 243) which seems to be consistent (↑) but gives a contradictory (↑) result, e.g. Russell's paradox (p. 254), Zeno's paradox (p. 254). **paradoxical** (*adj*).

antinomy (*n*) two or more statements (p. 243) which disagree although all reached by arguments (p. 243) which seem to be logical (p. 243), e.g. Bertrand's paradox (↓).

Bertrand's paradox an example of antinomy (↑). The probability that a random chord (p. 187) of a circle is longer than the side of the inscribed (p. 198) equilateral (p. 177) triangle is 1/4, 1/3 or 1/2 depending on how the chords are defined.

Bertrand's paradox

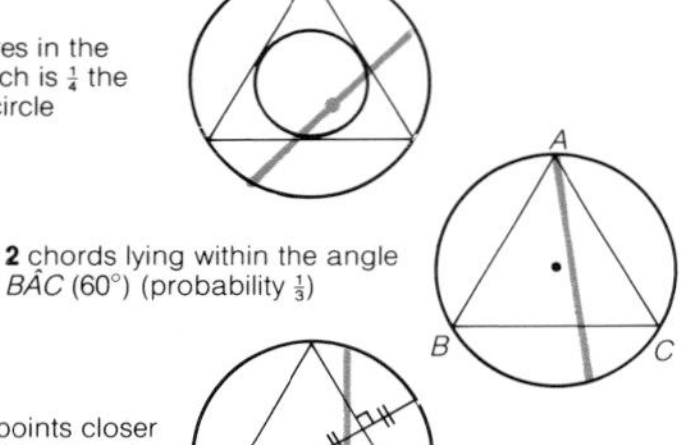

1 chords with centres in the inscribed circle which is $\frac{1}{4}$ the area of the whole circle (probability $\frac{1}{4}$)

2 chords lying within the angle $B\hat{A}C$ (60°) (probability $\frac{1}{3}$)

3 chords with mid-points closer to the centre than the sides of the triangle (probability $\frac{1}{2}$)

Russell's paradox the paradox (p. 253) that if *S* is the set of all sets which do not contain themselves, then *S* must include itself, which is a contradiction (p. 253).

Zeno's paradoxes a series of paradoxes (p. 253) related to concepts in calculus (p. 143), e.g. 'Achilles and the tortoise' showed that a fast runner could not pass a slower one because each time he reached the point where the slower one was, the slower one had moved on further.

ambiguous (*adj*) not clear, able to be understood in two different ways. **ambiguity** (*n*).

conjecture (*n*) a guess, an attempt at a hypothesis (p. 251) which has not been proved or disproved, e.g. Goldbach's conjecture (p. 34).

counterexample a single example which can disprove a conjecture (↑) or hypothesis (p. 251), e.g. the conjecture that all prime numbers (p. 32) are odd is disproved by the counterexample 2. It is important because only one example is needed to do this.

direct proof proof by a sequence (p. 91) of implications (p. 247) or inferences (↓) of the form 'if *A* is true, then *B* is true'.

indirect proof a logical (p. 243) proof not in the form of a direct proof (↑), but using another form of logical argument (p. 243) such as the correct argument 'if *B* is false, then *A* is false, therefore *A* implies (p. 246) *B*'.

reductio ad absurdum a form of indirect proof (↑) which assumes that the proposition (p. 250) is false and shows that this leads to a contradiction (p. 253). Also known as **indirect reduction**, **proof by contradiction**.

inference (*n*) the process of making a logical (p. 243) argument (p. 243) from premises (p. 251) to conclusion (p. 251). *See also* imply (p. 246). **infer** (*v*).

deduction (*n*) the process of making a logical (p. 243) argument (p. 243) from a general case to a particular example, often also used generally for inference (↑). **deduce** (*v*).

entail (*v*) if a statement (p. 243) *B* can be deduced (↑) from a statement *A*, *A* is said to entail *B*. **entailment** (*n*).

reduction (*n*) direct reduction is a method of proof by making changes in an existing argument (p. 243). *Do not mistake for reductio ad absurdum* (↑). *See also* p. 291.

induction (*n*) the process of making an argument (p. 243) from particular examples to a general case, which is not allowed in pure mathematics (p. 8) unless every possibility is considered, but is common in science. Also known as **inductive inference**, **inductive reasoning**. *See also* mathematical induction (p. 256). **inductive** (*adj*). *See also* p. 9.

a priori used of reasoning from cause to effect, and thus of argument (p. 243) by deduction (↑). Also known as **prior** (*adj*).

a posteriori used of reasoning by empirical (↓) methods and argument (p. 243) by induction (↑) or experience. Also known as **posterior** (*adj*).

empirical (*adj*) depending on examples, experience or experiments and not on deduction (↑).

heuristic (*adj*) used of argument (p. 243) by discovery rather than by logic (p. 243). **heuristics** (*n.pl.*).

mathematical induction a logical (p. 243) method of proof which has two parts:

I it is shown that if a proposition (p. 250) is true for some positive integral value of n, say $n = k$, it is also true for another value, usually $n = k + 1$;

II it is shown that the proposition is true for a value of n (usually small) say $n = a$.

The proposition is then true for all values of $n \geqslant a$. *Do not mistake for* induction (p. 255).

proof by exhaustion proof obtained by trying all possible cases.

Hilbert programme 23 problems Hilbert gave in 1900 which mathematicians should try to solve.

dual (*n*) if a theorem remains true when two mathematical concepts in it are changed for each other, then the two concepts and the two theorems are said to be dual, e.g. (1) union (p. 12) and intersection are dual concepts in the algebra of sets, and theorems can be obtained which are duals of each other by interchanging them; (2) some theorems on points and lines, e.g. Pascal's theorem (p. 224) and Brianchon's theorem (p. 224); (3) in three dimensions dual polyhedra (p. 200) are formed by interchanging vertices and faces, e.g. cube and regular octahedron (p. 201), antiprism (p. 203) and trapezohedron (p. 203); (4) dual networks (p. 239) are formed when nodes and regions are interchanged. **duality** (*n*).

dual
two networks: there is a node of one in each region of the other and pairs of intersecting arcs

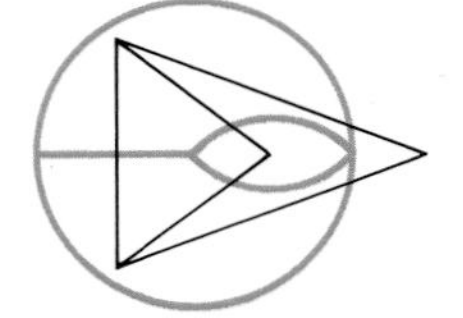

self-dual (*adj*) used of theorems, concepts or configurations which are the same as their duals (↑), e.g. the regular tetrahedron (p. 201).

reciprocal theorems pairs of theorems which are dual (↑).

reciprocal elements the concepts which are interchanged in a pair of dual (↑) theorems.

reciprocal diagrams diagrams in which the lines of one are parallel to those of the other, and regions in one correspond to vertices in the other. This is one sort of duality (↑) and is used in Bow's notation (p. 264).

law (*n*) (1) in pure mathematics (p. 8) a rule which gives a method of logical (p. 243) argument (p. 243); (2) in applied mathematics (p. 8) and science a rule based on empirical (p. 255) experiments. *These two meanings should not be mistaken for each other.*

law of contradiction if A is a sentence (p. 244), A and not A cannot both be true. In symbols $\sim (A \wedge \sim A)$ is a tautology (p. 245).

law of excluded middle if A is a sentence (p. 244), either A or not A is true. In symbols $A \vee \sim A$ is a tautology (p. 245).

law of identity if A is a sentence (p. 244), then A is true implies (p. 246) A is true. In symbols $A \rightarrow A$ is a tautology (p. 245).

de Morgan's laws the laws for sentences A and B:

I not (A and B) = (not A) or (not B);

II not (A or B) = (not A) and (not B),

which are true for any Boolean algebra (p. 243), e.g. in the algebra of sets using union (p. 12) and intersection $(A \cap B)' = A' \cup B'$ and $(A \cup B)' = A' \cap B'$.

absorption laws the laws for sentences A and B:

I A or (A and B) = A;

II A and (A or B) = A,

which are true for any Boolean algebra (p. 243), e.g. (1) in the algebra of sets using union (p. 12) and intersection they are $A \cup (A \cap B) = A$ and $A \cap (A \cup B) = A$, where A and B are any two sets; (2) in logic (p. 243) they are $A \vee (A \wedge B) = A$ and $A \wedge (A \vee B) = A$.

foundations of mathematics the study of logical (p. 243) concepts and the way in which mathematical systems can be built from them.

axiom of choice an axiom (p. 250) in the foundations of mathematics (↑). If S is a set whose elements are other non-empty (p. 11) sets, each pair of which is disjoint (p. 13), then there is at least one set which has exactly one element in common with each of the non-empty sets.

metamathematics (*n*) the study of mathematics from outside, avoiding circular arguments (p. 243).

metatheorem (*n*) a theorem in metamathematics (↑), e.g. Gödel's theorem (p. 243).

kinematics (*n*) the study of the motion of particles or bodies without dealing with the causes of motion.

motion (*n*) movement, the act of moving.

stationary (*adj*) not moving.

particle (*n*) an object which has mass but can be taken to have no size, and thus regarded as a point.

body (*n*) an object which has both mass and size, and which cannot usually be taken as a point.

velocity (*n*) the rate of change of displacement with respect to time. It is a vector quantity and has magnitude (p. 284) and direction. If s is the displacement, t is the time and v is the velocity then $v = \mathrm{d}s/\mathrm{d}t$, or in vector form $\mathbf{v} = \mathrm{d}\mathbf{s}/\mathrm{d}t$. If the velocity is constant, it is simply displacement divided by time. *See also* vector (p. 77).

parallelogram of velocities *see* parallelogram of vectors (p. 81). Similarly **polygon** and **triangle of velocities**.

speed (*n*) velocity (↑) without considering direction. Speed is a scalar (p. 77) quantity.

acceleration (*n*) the rate of change of velocity (↑) with respect to time. It is a vector quantity, and using the symbols given under velocity it is $\mathrm{d}v/\mathrm{d}t$, $\mathrm{d}^2s/\mathrm{d}t^2$ or $v(\mathrm{d}v/\mathrm{d}s)$. In vectors it is $\mathrm{d}\mathbf{v}/\mathrm{d}t$ or $\mathrm{d}^2\mathbf{s}/\mathrm{d}t^2$. **accelerate** (*v*).

retardation (*n*) negative acceleration. Also known as **deceleration**. **retard** (*v*), **decelerate** (*v*).

uniform (*adj*) constant; used especially in motion with speed, velocity (↑) and acceleration (↑), and also with laminas (p. 155) and solids of constant mass per unit area or volume.

equations of uniform motion the five equations which describe the relation between any four of the quantities: displacement, s; initial (p. 289) velocity (↑), u; final velocity, v; uniform (↑) acceleration (↑), a; and time, t, namely:

$$v = u + at,$$
$$s = \tfrac{1}{2}(u + v)t,$$
$$s = ut + \tfrac{1}{2}at^2,$$
$$s = vt - \tfrac{1}{2}at^2,$$
$$v^2 = u^2 + 2as.$$

hodograph
2 is the hodograph of **1**

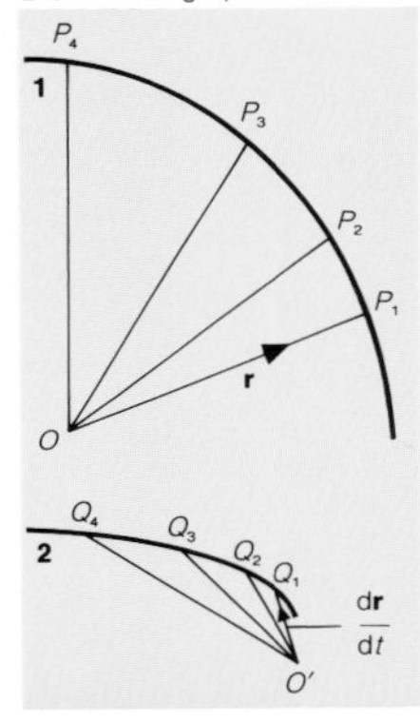

hodograph (*n*) if in a diagram a point *P* has vector **r** at time *t* with origin *O*, and a second diagram is drawn with origin *O*′, then the locus (p. 187) of the point *Q* on the second diagram whose position vector (p. 79) is d**r**/d*t* with respect to *O*′ is the hodograph of *P*.

angular displacement the angle through which a body has turned about a particular axis. The angle is positive if it makes a right-hand screw about the positive direction of the axis, and it is a vector quantity, though it is often treated as a scalar (p. 77) in planar examples.

angular velocity the rate of change of angular displacement. If this is $\boldsymbol{\theta}$ then the angular velocity is the vector $d\boldsymbol{\theta}/dt$.

angular acceleration the rate of change of angular velocity (↑), using the notation given there it is $d^2\boldsymbol{\theta}/dt^2$.

force (*n*) that which can cause a body to accelerate (↑) (or better, to change its momentum, p.268). A body on which no force acts remains stationary (↑) or moves with uniform (↑) velocity (↑). Force is a vector quantity. Also once known as **action**. *See also* vector (p. 77).

parallelogram of forces *see* parallelogram of vectors (p. 81). Similarly **polygon** and **triangle of forces**.

mechanics (*n*) the study of forces and bodies on which they act. The word mechanical is also used to mean related to machines. **mechanical** (*adj*).

statistical mechanics the study of mechanics based on large numbers of small particles and using the theory of probability to predict (p. 290) what will happen.

statics (*n*) the study of forces acting on bodies which are not in motion.

mass (*n*) the quantity of matter in an object; that property of any object which measures its acceleration (↑) for any given force.

gravitational mass the mass as measured by the force acting between two bodies. In a constant gravitational field (p. 267) it is equal to the inertial mass (p. 260).

inertial mass the mass as measured by the inertia (p. 266) of a body.
weight (*n*) that force with which the earth (or another heavenly body) acts on any given body. **weigh** (*v*).
light (*adj*) used of any body or part of a system whose weight need not be considered because it is too small to matter.
heavy (*adj*) used of any body or part of a system whose weight must be taken into account.
reaction (*n*) when two bodies touch or act upon each other, the force of one upon the other. *See also* Newton's laws of motion (p. 266). **react** (*v*).
normal reaction the resolved (p. 80) part of the reaction (↑) perpendicular to the surface between any two bodies.
interaction² (*n*) a force acting between two particles, bodies or systems. *See also* p. 128.
equilibrium (*n*) when all the forces acting on a particle or body cause no change in velocity (p. 258), it is said to be in equilibrium.

reaction
forces between a table and a ball at rest on it: these are equal to the weight of the ball

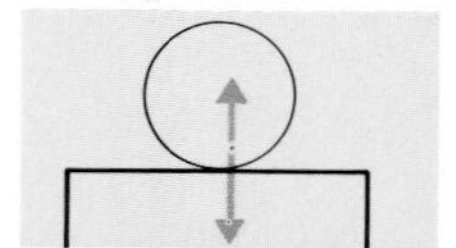

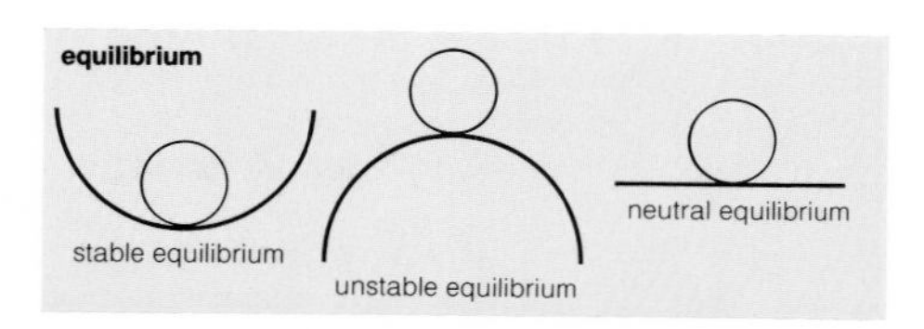

stable equilibrium equilibrium (↑) where after a small displacement a system will return towards its first position.
unstable equilibrium equilibrium (↑) where after a small displacement a system will continue to move away from its first position.
neutral equilibrium equilibrium (↑) where after a small displacement a system remains in equilibrium.
equilibrant (*n*) a force which will, when added to a system, put it into equilibrium (↑). It is equal and opposite to the resultant (p. 80) of the system.
stability (*n*) the state of equilibrium (↑) of a body.

Lami's theorem for three forces in equilibrium
$$\frac{\mathbf{P}_1}{\sin\alpha} = \frac{\mathbf{P}_2}{\sin\beta} = \frac{\mathbf{P}_3}{\sin\gamma}$$

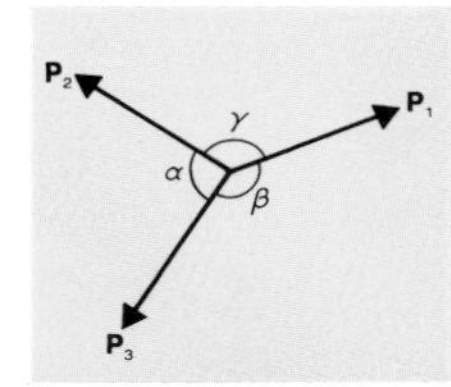

Lami's theorem the sine formula (p. 134) used with a triangle of vectors (pp. 81–2) (usually a triangle of forces). If the three forces are in equilibrium (↑) each is proportional to the sine of the angle between the other two. Also written **Lamy's theorem**.

virtual work the amount of work (p. 269) done when a system is displaced slightly. It is zero when the system is in equilibrium (↑).

rigid body a body whose shape does not change under forces, more exactly a body in which any changes are small enough not to need consideration.

centre of gravity that point of a rigid body (↑) at which the weight may be taken to act.

centre of mass that point of a rigid body (↑) where the mass may be taken to be placed, found by using the formula $\bar{x} = (\Sigma m_i x_i)/(\Sigma x_i)$, where the body is taken to be made of many small masses m_i at distances x_i from a given axis and the centre of mass is at a distance $\bar{x}$ from this axis. By using coordinate axes in turn the position of the centre of mass is found. If the gravitational field (p. 267) is constant, it is the same as the centre of gravity (↑). Also known as **centroid**, **mean centre**. *See also* area under a curve (p. 155), lamina (p. 155), solid of revolution (p. 155), centroid (p. 196).

equivalent forces two sets of forces having the same result when acting on a given rigid body (↑). Also known as **equipollent forces**.

funicular polygon the shape of a light (↑) string whose ends are at two fixed points and to which weights are fixed, or on which forces act at certain points.

funicular polygon for three weights $\mathbf{P}_1$, $\mathbf{P}_2$, $\mathbf{P}_3$

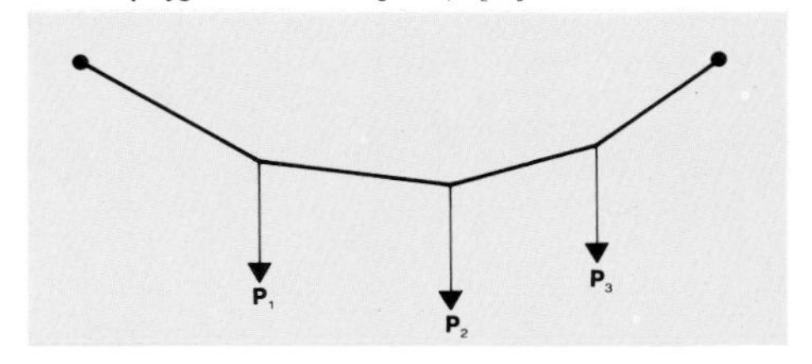

tension (*n*) a force which tries to stretch (↓) a body or make it larger, often used in strings or springs (↓).

spring (*n*) a machine (p. 270) such as a helical (p. 212) wire or a string made of rubber, which when stretched (↓) and let go returns to its original shape.

extension[3] (*n*) the amount by which a spring (↑) is stretched (↓) from its natural length. *See also* pp. 10, 252.

extend (*v*) (1) generally to make longer, used of line segments (p. 167) and curves with the meaning of produce (p. 167) as well as strings and springs (↑); (2) to make more general.

stretch[2] (*v*) to make larger under tension (↑). *See also* p. 237.

compression (*n*) the amount by which a body is made smaller by forces from outside it. **compress** (*v*).

Hooke's law in any spring (↑) the tension (↑) T is proportional to the extension x if this extension is not too great. The formula is $T = \lambda x/l$ where λ is a constant and l is the natural length of the spring.

elastic (*adj*) used of any body which after extension or compression (↑) will return to its first shape.

inelastic (*adj*) not elastic (↑).

elastic limit the stretched (↑) length of a string or spring (↑) beyond which Hooke's law (↑) is no longer true, and it does not return to its first length.

spring rate the constant λ in the definition of Hooke's law (↑).

Young's modulus if the spring rate (↑) λ equals EA where A is the area of the cross-section of the string or other body perpendicular to the tension (↑), then E is a constant for the material called Young's modulus. Also known as **modulus of elasticity**.

stress (*n*) the tension (↑) in a material divided by the area (p. 191) of its cross-section (p. 200) perpendicular to the tension and over which the tension acts, and similarly for a body being compressed (↑). Also known as **tensile stress**.

tension
the forces at the ends of a string or spring: for a 'light' string these are equal

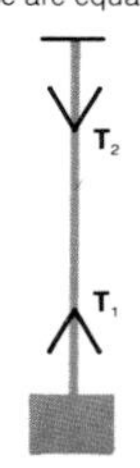

extension
the natural length l of a spring and its extension x under a weight W

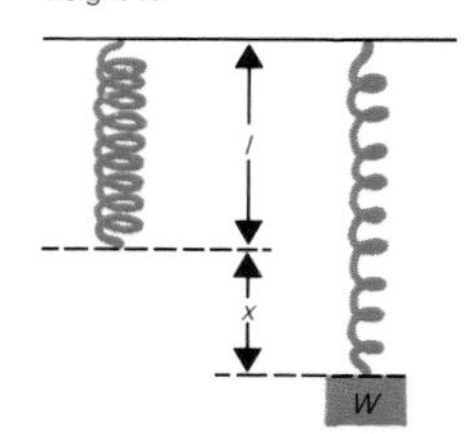

strain (*n*) the extension of a stretched (↑) material, (or compression, ↑, of a compressed material), divided by the unstretched length. Also known as **tensile strain**.

elasticity (*n*) a ratio of stress (↑) divided by strain (↑), e.g. the bulk modulus (↓) and the modulus of rigidity (↓).

bulk modulus the stress (↑) divided by the strain (↑) for an elastic (↑) body which is being compressed (↑).

flexible (*adj*) able to bend without any resistance (p. 265); used mainly of strings.

rod (*n*) a bar, a straight rigid body (p. 261), usually one which can be taken to have no thickness (and sometimes to have no weight).

beam (*n*) a bar, similar to a rod (↑) but often one whose weight must be considered.

thrust
the forces at the ends of a rod under compression

thrust (*n*) the force inside a rod (↑) which acts so as to try to shorten its length. It is the negative of tension (↑).

strut (*n*) a rod (↑) which is in thrust (↑) or compression (↑).

shearing force
between two parts of a beam (shown in different colours): they are equal and opposite

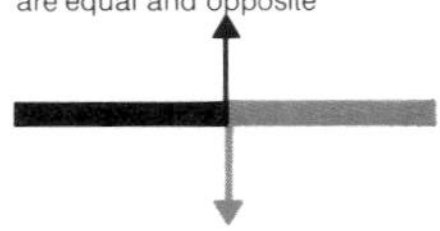

tie (*n*) a rod (↑) or inelastic (↑) string which is in tension (↑).

cantilever (*n*) a rod (↑) or beam (↑) which is not fixed at one end.

shearing force the force inside a beam (↑) or other body at any point which acts between its parts and perpendicular to its length.

bending moment
the couples which are the bending moments for the two parts of a beam (shown in different colours): they are equal and opposite

bending moment the couple (p. 264) inside a beam (↑) (or other body) at any point which acts between its parts.

modulus of rigidity the ratio of the shearing force (↑) per unit area in a solid body divided by the shearing strain (↑) or angle by which the body is stretched (↑) or deformed (p. 242).

hinge (*n*) a joint, usually between two rods (↑) (not often other bodies in mathematics), which can turn with respect to each other. **hinge** (*v*).

freely jointed having hinges (↑) where the friction (p. 265) need not be considered because it is too small.

linkage (*n*), a hinged (↑) system of rods (↑) as in, e.g., a pantograph (p. 186).

trammel (*n*) a machine, usually some form of linkage (p. 263), which causes a point to move along a particular curve.

framework (*n*) a system of rods (p. 263), beams (p. 263), or similar objects joined to make a rigid body (p. 261), as in some bridges.

Bow's notation in a light (p. 260) planar framework (↑) a method of solving the stresses (p. 262) in the rods (p. 263) by giving symbols to the regions between the rods, and using a reciprocal diagram (p. 256) to the diagram of the framework.

method of sections in a light (p. 260) planar framework (↑) a method of solving the stresses (p. 262) in the rods (p. 263) by imagining it to be cut into two parts across not more than three rods.

moment[2] (*n*) the moment of a force is the turning effect of the force about an axis through which the force does not pass and is measured by the product of the force and its perpendicular distance from the axis. If the position vector (p. 79) of a point through which the force passes is **r** and the force vector is **F**, then the moment of the force is the vector quantity **r** × **F**, the direction of this vector being the direction of the moment. *See also* p. 121.

principle of moments if a body is in equilibrium (p. 260), the sum of the moments (↑) of the forces acting on it about any axis is zero. Also known as **law of moments**.

couple (*n*) any system of forces not in equilibrium (p. 260) but which has no resultant (p. 80) force in any direction, especially a pair of equal and opposite forces not in the same straight line (p. 167). Using the symbols under moment (↑) such a system has a total moment of $\Sigma \mathbf{r} \times \mathbf{F}$. If the forces lie in a plane, then the moment of the couple is equal to the sum of the products of the forces with their perpendicular distances from any given axis perpendicular to the plane. For two equal and opposite forces, the magnitude (p. 284) of the couple is the product of one of the forces with the perpendicular distance between them.

framework framework of rods

Bow's notation
the capital letters in **2** correspond to the regions in small letters in **1**. The stress in the rod between *a* and *d* is given by the length *AD* drawn parallel to the corresponding rod, and so on

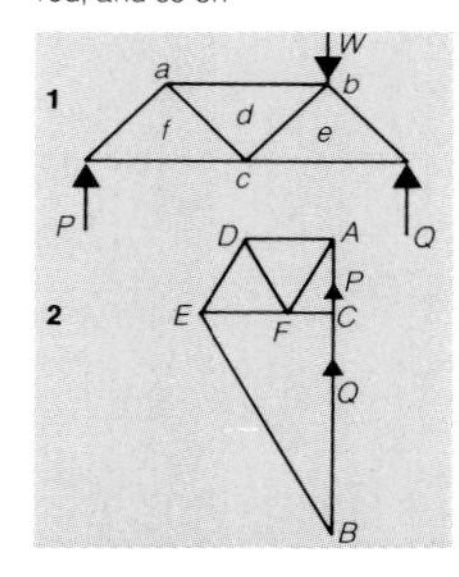

method of sections
if the framework from above is cut as shown, the stresses in the cut rods can be found by resolving and taking moments

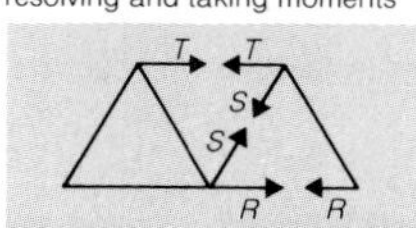

moment
the moment of **F** about *O* is **r** × **F** which is in the direction coming out from the page

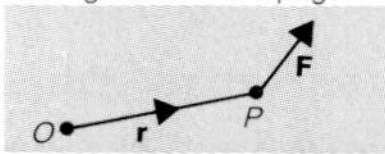

couple
the simplest case: two parallel forces in opposite directions. The direction of this couple is out from the page

torque (*n*) of a couple (↑): its total moment (↑).

wrench (*n*) a system of forces which is equal to a couple (↑), together with a force whose direction is parallel to the axis of the couple.

pitch (*n*) of a wrench (↑): the moment (↑) of its couple (↑) divided by the force along its positive direction.

friction
for a body on a slope

N = normal reaction
F = frictional force
N + F = total reaction; if the body is at rest, this is equal and opposite to the weight
θ = angle of friction

friction (*n*) when two bodies touch, any forces which act against possible sliding or rolling motion between them are said to be caused by friction. **frictional** (*adj*).

static friction friction (↑) between two bodies at rest relative to each other.

limiting friction the maximum value of the static friction (↑) between two bodies.

kinetic friction friction (↑) between two bodies which are sliding relative to each other. It is usually slightly less than the limiting friction (↑), though often taken to be equal to it in exercises (p. 288). Also known as **sliding friction**.

rolling friction friction (↑) between two bodies which roll against each other without slipping rather than sliding.

coefficient of friction the ratio of limiting friction (↑) F to normal reaction (p. 260) R. Symbol: μ, where $F = \mu R$.

angle of friction the angle θ given by $\tan \theta \leqslant \mu$, the coefficient of friction (↑). It is the angle between the normal (p. 188) between the two bodies and the total reaction which is the resultant (p. 80) of the normal reaction (p. 260) and the friction (↑) between them.

resist (*v*) to act against motion as friction (↑) does. **resistance** (*n*).

kinetics (*n*) the study of forces acting on bodies which are in motion.

dynamics (*n*) the study of matter in motion and the forces acting on it.

Newton's laws of motion the three laws on which the study of dynamics (↑) is based:

- I every body remains stationary (p. 258) or in uniform (p. 258) motion in a straight line (p. 167) unless it is acted on by some external (p. 289) force;
- II the rate of change of linear momentum (p. 268) of any body (which is its acceleration, p. 258, if the mass is constant) is proportional to the force acting on it and in the same straight line;
- III if a body *A* makes a force on a body *B*, then the body *B* makes an equal force (the reaction, p. 260) on the body *A* but in the opposite direction. Often stated as 'action and reaction are equal and opposite'.

The second law is often given in simple form as $\mathbf{F} = m\mathbf{a}$ where **F** is the force producing an acceleration **a** in a body of mass *m*.

inertia (*n*) the property of any mass which means that it does not change its motion unless acted on by a force outside it.

gravitation (*n*) a concept which accounts for the motion of bodies by imagining a force between them. **gravitational** (*adj*).

action at a distance used of forces which act without any apparent means, e.g. gravitational (↑) and magnetic (p. 289) forces.

attraction (*n*) action at a distance (↑) towards a given point. **attract** (*v*).

repulsion (*n*) action at a distance (↑) away from a given point. **repulse** (*v*).

gravity (*n*) the gravitational (↑) force on a body due to the earth or some other body. This force is the weight of the body.

Newton's law of universal gravitation the force between two particles of masses m_1 and m_2 at a distance (p. 169) d apart is Gm_1m_2/d^2, where G is a constant, and this force acts on each body towards the other.

gravitational constant the constant *G* in Newton's law of universal gravitation (↑).

inverse square law any law in which one quantity is inversely (p. 250) proportional to the square of the other, e.g. Newton's law of universal gravitation (↑) in which the force is inversely proportional to the square of the distance (p. 169).

gravitational field the space in which a gravitational (↑) force is acting.

centre of attraction the point towards which a gravitational (↑) or other force acts on a given body.

free fall the motion of a body under which the only force is that of gravity (↑).

orbit (*n*) the path of a body moving under a gravitational field (↑). *See also* trajectory (p. 268).

apsis
for an elliptical orbit

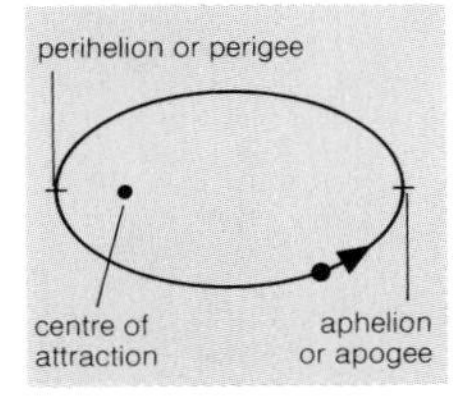

apsis (*n*) one of the points on an orbit (↑) where the body is nearest to or furthest from the centre of attraction (↑). In an elliptical (p. 220) orbit (↑) there are two such points. Also known as **apse**. **apses** (*pl*).

perihelion (*n*) the nearer of the two apses (↑) of an elliptical (p. 220) orbit (↑) to the centre of attraction (↑), especially to the sun (also **perigee**, especially when the earth is the centre of attraction).

aphelion (*n*) the furthest of the two apses (↑) of an elliptical (p. 220) orbit (↑) from the centre of attraction (↑), especially from the sun (also **apogee**, especially when the earth is the centre of attraction).

Kepler's laws laws of motion of a body moving about a centre of attraction (↑) under gravity (↑), e.g. the earth about the sun:

- I the body moves in an elliptical (p. 220) orbit (↑) with the centre of attraction (↑) at a focus (p. 220);
- II the line joining the body to the centre of attraction passes over equal areas (p. 191) in equal times;
- III the square of the time to cover a complete orbit (↑) is proportional to the third power of the semi-axis (p. 215) of the ellipse.

Kepler's laws
second law
if the body takes the same time to move from *A* to *B* as from *C* to *D*, then area *ASB* = area *CSD* where S is the centre of attraction

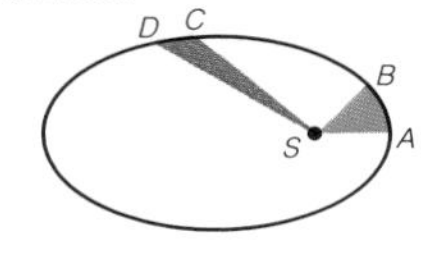

ballistics (*n*) the study of the motion of projectiles (↓).

projectile (*n*) a body moving in a gravitational field (p. 267) under free fall (p. 267), either alone or with some sort of frictional (p. 265) force, e.g. from the air. **project** (*v*). *See also* p. 237.

trajectory (*n*) the path followed by a projectile (↑). Trajectory is similar to orbit (p. 267), but is usually used for paths which are only parts of geometric curves.

range[4] the range of a projectile is the distance (p. 169) between the point from which it is projected (↑) and the point where it lands; usually this is a small enough distance for it to be taken to be a straight line (p. 167), either horizontal or along a slope. *See also* pp. 17, 113, 167.

escape velocity the velocity (p. 258) with which a body must be projected (↑) directly away from another (usually much heavier) body so that gravity (p. 266) will no longer make it return. Used, e.g., of a spaceship leaving the earth.

rocket (*n*) a projectile (↑) which burns material, and is thus changing its mass.

momentum (*n*) the momentum of any body or particle is the product of its mass and its velocity (p. 258). It is a vector quantity. Often known as **linear momentum** to show difference from angular momentum (p. 272).

impulse (*n*) that which causes change in linear momentum (↑), the product of a force and the time for which it acts; used especially for a force like a blow which is large but acts for a very short time. **impulsive** (*adj*).

impact (*n*) the blow when two bodies strike and a large force acts for a short time. Also known as **collision**.

collision (*n*) = impact (↑). **collide** (*v*).

Newton's law of restitution if two bodies A and B collide (↑), and their velocities (p. 258) resolved (p. 80) perpendicular to the plane separating them at the point of impact are u_A and u_B before they collide and v_A and v_B after they collide respectively, all measured in the same direction, then $v_B - v_A = -e(u_B - u_A)$, where e is a constant such that $0 \leqslant e \leqslant 1$.

trajectory

the trajectory of a projectile; when gravity is constant and vertically downwards, this is a parabola

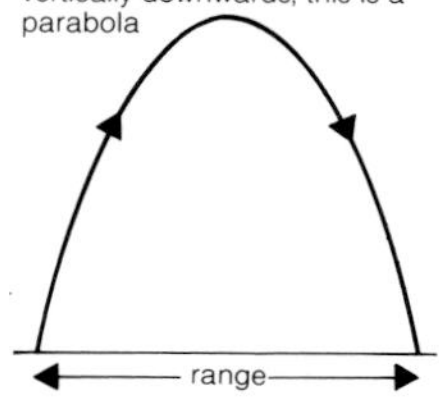

Newton's law of restitution

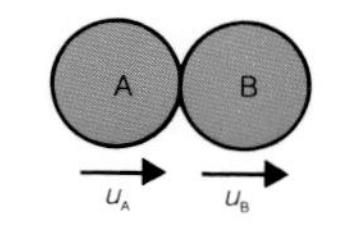

velocities before impact ($u_A > u_B$)

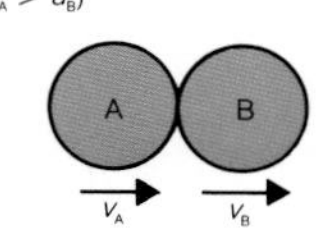

velocities after impact ($v_A < v_B$)

coefficient of restitution the constant e under Newton's law of restitution (↑).

Poisson's hypothesis the hypothesis (p. 251) that if two bodies collide (↑), the ratio of the impulses (↑) on each other before and after the time of maximum compression (p. 262) is as $1 : e$ where e is the coefficient of restitution (↑).

elastic impact impact (↑) in which the coefficient of restitution (↑) e is equal to unity. Similarly **inelastic impact** when $e < 1$.

recoil (*v*) to spring back, as a gun does after firing.

transient (*adj*) acting for a very short time as in an impact (↑). Such a time is sometimes treated as infinitesimal (p. 143).

kineton (*n*) the localized vector (p. 79) which is the product of the mass and the acceleration (p. 258) of a particle at any instant. For a system it is the sum of these vectors. Also known as **linear kineton**, **mass-acceleration product**.

angular kineton if $\Sigma m_i \, d^2\mathbf{r}_i/dt^2$ is the linear kineton (↑) of a system, the vector product (p. 82) $\Sigma \mathbf{r}_i \times (m_i \, d^2\mathbf{r}_i/dt^2)$ is its angular kineton. Also known as **moment of kineton**.

work
the work done by a force is $Fs \cos \alpha$

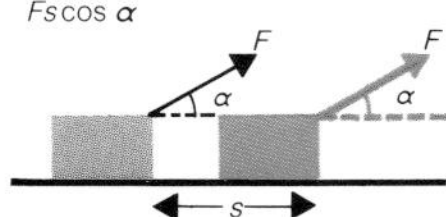

work (*n*) the product of a force and the distance (p. 169) through which it moves in its direction. If the force F moves a body a distance s at an angle α to its direction, then the work done is $Fs \cos \alpha$ or, if F is variable, the integral given by the scalar product (p. 78) of vectors $\int \mathbf{F} \cdot d\mathbf{s}$, where $\mathbf{F}$ is the force vector and $\mathbf{s}$ is the position vector. It is a scalar (p. 77) quantity.

energy (*n*) the ability to produce work (↑), that which a body has as a result of work done on it. It has the same units as work.

kinetic energy that energy (↑) which a body has as a result of its motion. It is $\frac{1}{2}mv^2$ where m is the mass of the body and v is its velocity (p. 258). If the body is rotating it also has kinetic energy $\frac{1}{2}I\omega^2$ where I is its moment of inertia (p. 271) about the axis of rotation (p. 234) through the point for which v is measured and ω is its angular velocity (p. 259) about that axis.

potential energy energy (p. 269) which a body obtains as a result of work (p. 269) done on it against any conservative force (p. 274) such as gravity (p. 266) or the tension (p. 262) in a spring (p. 262). When against gravity it is equal to mgh where m is the mass of the body, g is the acceleration (p. 258) due to gravity and h is the height through which the body has been moved against gravity.

power[2] (*n*) the rate of doing work (p. 269), the rate of change of work with respect to time. Also known as **activity**. *See also* pp. 35, 126.

pressure (*n*) force per unit area (p. 191), measured for a uniform force by dividing the force by the area over which it acts. Sometimes also used unwisely for the total force over the whole of a body.

atmospheric pressure the pressure (↑) at any point on the earth from the weight (p. 260) of the air at that point.

Hamiltonian (*n*) in Newtonian mechanics (p. 282), the sum of the kinetic (p. 269) and potential energies (↑) of a system.

Lagrangian (*n*) in Newtonian mechanics (p. 282), the kinetic energy (p. 269) minus the potential energy (↑) of a system at any given time.

principle of zero activity if a system passes through a position of equilibrium (p. 260), its rate of doing work (p. 269) (or activity, ↑) at that point is zero.

density (*n*) mass per unit volume; when a body has uniform density it is mass divided by volume. *See also* linear (↓) and areal density (↓).

linear density mass per unit length (usually used for uniform and non-uniform rods, p. 263).

areal density mass per unit area (p. 191) (usually used for laminas, p. 155).

homogeneous[2] (*adj*) having constant density (↑). *See also* p. 61.

machine (*n*) an object which changes energy (p. 269), usually kinetic energy (p. 269). Widely used for many simple objects in mechanics using such things as wheels, ropes, levers (↓) and wires.

pulley
a pulley used for lifting a weight

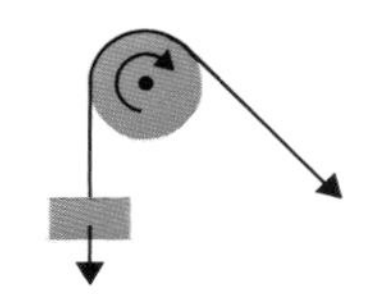

gear
two gears used to alter rate and direction of turning

lever
a rod used as a lever to move a weight

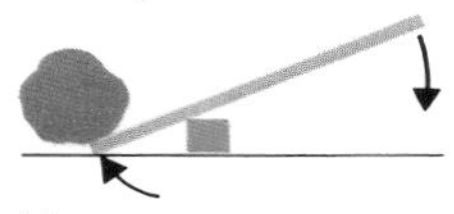

fulcrum
three orders of lever

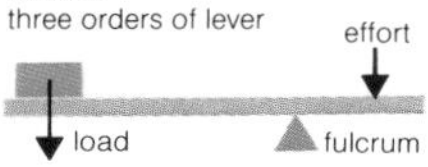

first order: fulcrum between load and effort

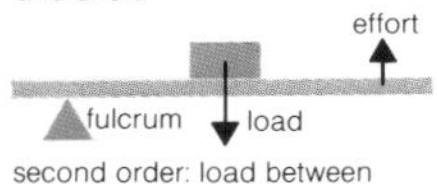

second order: load between fulcrum and effort

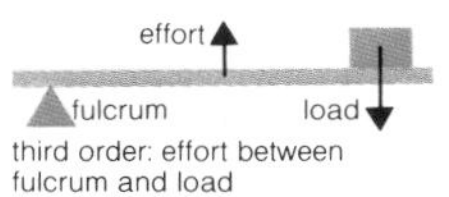

third order: effort between fulcrum and load

perpendicular axis theorem
$I_p + I_q = I_r$ where r is an axis perpendicular to the lamina

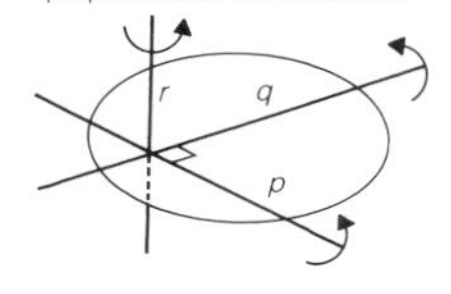

pulley (*n*) a wheel over which a rope or string passes; a machine which changes the direction in which a rope or string pulls.

gear (*n*) a wheel with teeth. A machine with two gears may be used to change speed of turning, and also changes the direction of turning.

lever (*n*) a machine which is a bar with an axis about which it can turn which is used to change forces from one line of action (p. 79) to another.

fulcrum (*n*) the axis about which a lever (↑) turns.

effort (*n*) the force put into a machine.

load (*n*) the force obtained from a machine.

inclined plane a sloping plane which may form part of a machine.

mechanical advantage the ratio of the force put out by a machine to the force put into it, i.e. the load (↑) divided by the effort (↑). Also known as **force ratio**.

velocity ratio in any machine the velocity ratio is the ratio of the distance moved by the effort (↑) in a given time to that moved by the load (↑) in the same time. Also known as **distance ratio**.

efficiency (*n*) the efficiency of a machine or system is the ratio of the energy (p. 269) put out by the machine to that put into it, which is the mechanical advantage (↑) divided by the velocity ratio (↑). It is often given as a percentage.

moment of inertia if a rigid body (p. 261) is made up of point masses m_i, each at a perpendicular distance r_i from a given axis, the quantity $\Sigma m_i r_i^2$ is called its moment of inertia about that axis. For solid bodies the sum becomes an integral of infinitely small particles. In some ways moment of inertia corresponds for angular (p. 173) motion to mass in linear (p. 167) motion. Also known as **rotational inertia**.

perpendicular axis theorem if the moments of inertia (↑) of a lamina (p. 155) about two perpendicular axes through a point O in its plane are I_A and I_B, then the moment of inertia about an axis through O perpendicular to the plane of the lamina is $I_A + I_B$. Also known as **lamina theorem**.

parallel axis theorem if the moment of inertia (p. 271) of a body of mass m about an axis through the centre of mass (p. 261) is I, then the moment of inertia about a parallel axis a perpendicular distance (p. 169) d from the first axis is $I + md^2$.

Routh's rule a rule for remembering the moments of inertia (p. 271) about G of certain uniform solid bodies of mass M with three perpendicular axes of symmetry GX, GY and GZ where G is the centre of gravity (p. 261). If the lengths of the half axes GX and GY are a and b respectively, the moment of inertia about GZ is $M(a^2 + b^2)$ divided by:

- I 3 if the body has edges and corners, e.g. rods (p. 263), rectangular blocks and laminas (p. 155);
- II 4 if the body has edges but no corners, e.g. circular or elliptical (p. 220) discs (p. 192);
- III 5 if the body has no edges or corners, e.g. spheres and ellipsoids (p. 211); for a sphere of radius a about an axis through its centre, it is $M(a^2 + a^2)/5 = \cdot 2Ma^2/5$.

radius of gyration if the moment of inertia (p. 271) of a body of mass m is mk^2 about some axis, then k is its radius of gyration about that axis.

angular momentum of a rigid body (p. 261) about an axis: the moment of inertia (p. 271) about that axis multiplied by its angular velocity (p. 259) about that axis. Also known as **moment of momentum**. *See also* spin (↓).

spin (*n*) rotation; sometimes also used for angular momentum (↑). **spin** (*v*).

impulsive moment of a force, the gain in angular momentum (↑) caused by the force.

flywheel (*n*) a large heavy wheel with high moment of inertia (p. 271) used to make machines rotate more evenly.

gyroscope (*n*) a heavy rotating wheel which turns rapidly and is supported in such a way that it can rotate about three perpendicular axes; often used as an example of a rotating body because the way it behaves is a good example of the theory. It is used to guide ships and as a child's toy.

parallel axis theorem
$I_p = I_a + md^2$ where a is an axis through G, the centre of gravity

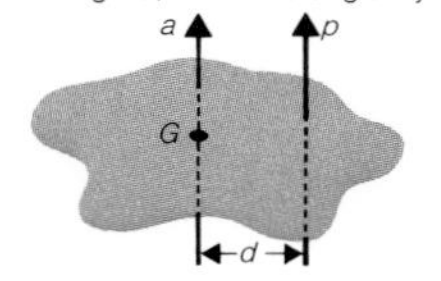

Routh's rule part 1

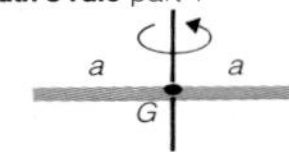

moment of inertia of rod length $2a$ about perpendicular axis through G is $\frac{1}{3}Ma^2$

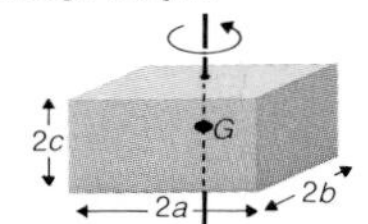

moment of inertia of block about axis shown is $\frac{1}{3}M(a^2 + b^2)$.
Note: the value of c does not matter

Routh's rule part 2

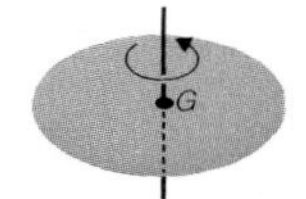

moment of inertia of a disc radius a about perpendicular axis through G is $\frac{1}{4}M(a^2 + a^2) = \frac{1}{2}Ma^2$

gyroscope
the green wheel can rotate about three perpendicular axes, the black ring is fixed

precession (*n*) the property that, if a force is applied to a rotating body, e.g. a gyroscope (↑), at right angles to the axis of rotation (p. 234), then this axis itself moves about a different axis.

space centrode when a lamina (p. 155) moves in its plane, the point in space where the lamina is instantaneously (p. 144) at rest. Interest is often in its locus (p. 187) in space.

body centrode when a lamina (p. 155) moves in its plane, the point in the lamina where it is instantaneously (p. 144) at rest. Interest is often in its locus (p. 187) in the lamina.

centrifugal (*adj*) outwards from a centre along a radius. Used mainly of an imaginary force used at one time to explain motion in a circle which has caused much misunderstanding and is best avoided.

centripetal (*adj*) inwards towards a centre along a radius. Used mainly of forces acting towards a point such as the tension (p. 262) in a string which is swinging or the gravitational (p. 266) force.

banking (*n*) the sloping of roads, etc, round corners so as to provide a centripetal (↑) force which helps vehicles to turn at speed.

banking
a car on a banked road: the reaction has a horizontal component making it easier to go round a bend

central orbit the path of a particle moving freely under a centripetal (↑) force; such a path is a conic (p. 220), e.g. the path of the earth around the sun.

virtual force an imaginary force used to explain acceleration (p. 258) from the point of view of a person who is being accelerated, e.g. centrifugal (↑) force, Coriolis force (↓).

Coriolis force a virtual force (↑) used to explain the force which seems to a person on the earth to act so as to move objects out of their straight paths. This is due in fact to the rotation of the earth which means that paths which appear not to be straight are really so. *See also* Foucault pendulum (p. 275).

d'Alembert's principle a method of changing problems in dynamics (p. 266) into problems in statics (p. 259) by replacing accelerations (p. 258) with virtual forces (↑), and which has often led to misunderstandings.

closed system a system, especially in mechanics, which has no interactions (p. 260) with other objects or systems.

conservation law a law which states that in a closed system (↑) under certain conditions (p. 249) a quantity remains constant. In Newtonian mechanics (p. 282) there are conservation laws of mass, energy (p. 269), linear momentum (p. 268) and angular momentum (p. 272). The last two laws are used mainly with impacts (p. 268). In relativistic mechanics (p. 282) the first two form one law as a result of the mass-energy equation (p. 283) $e = mc^2$.

conservative force a force such that the work (p. 269) done by it in moving a body between two points is the same whatever path is taken.

field² (*n*) a region over which, for every point, a force or other quantity acts in a way which is described by a continuous function. *See also* p. 76.

vector field a field (↑) for which the quantity which is acting is a vector such as force or velocity (p. 258).

scalar field a field (↑) for which the quantity which is acting is a scalar (p. 77) such as temperature (p. 284) or energy (p. 269). Also known as **potential field**.

conservative field a field (↑) defining a conservative force (↑) or another quantity which acts in a similar way. Also known as **lamellar field**.

route independent not depending (p. 102) on the path taken, as in a conservative field (↑).

dissipation (*n*) loss of energy (p. 269) in a system where the conservation law (↑) does not hold good. Such energy is not really lost but appears in different forms, e.g. heat, which are not being considered.

potential (*n*) the potential of a field (↑) at a point P is the quantity $\Sigma(km_i/r_i)$ where m_i are masses in a gravitational field (p. 267) (other quantities in other fields), r_i are their distances from P and k is a constant.

equipotential (*adj*) having the same potential (↑); used usually of lines or surfaces.

pendulum

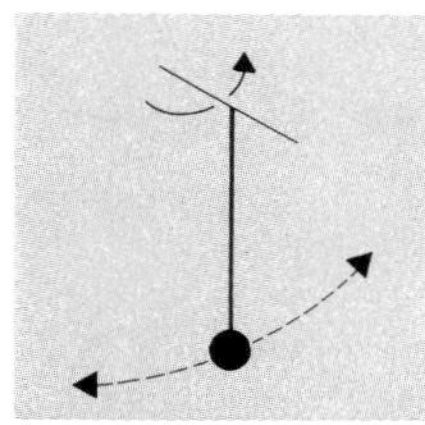

simple pendulum

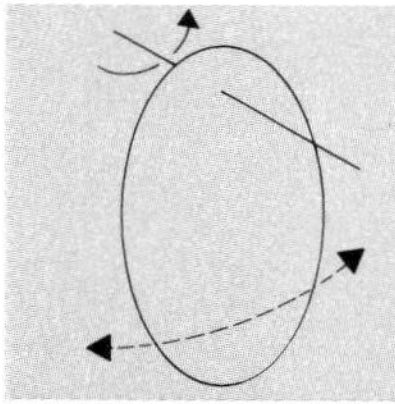

compound pendulum

conical pendulum

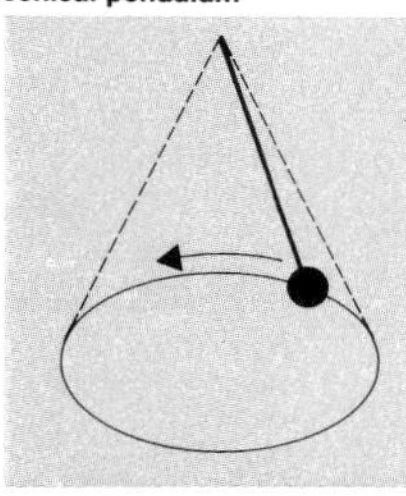

simple harmonic motion
if *P* moves round the circle with constant speed, *Q* moves along the diameter with simple harmonic motion whose amplitude is the radius of the circle

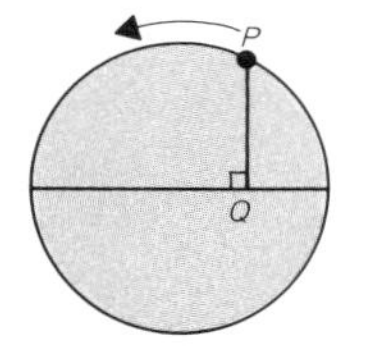

oscillate (*v*) to move backwards and forwards about some central point. *See also* saltus (p. 66). Also known as **vibrate**. **oscillation** (*n*), **oscillatory** (*adj*).

pendulum (*n*) a body which oscillates (↑) about a fixed axis in a gravitational field (p. 267).

simple pendulum an ideal (p. 289) pendulum (p. 275) having a point mass at the end of a light (p. 260), string or rod (p. 263).

compound pendulum a rigid body (p. 261) which oscillates (↑) about a fixed axis in a gravitational field (p. 267).

ballistic pendulum a pendulum (↑) used to measure the velocity (p. 258) of a projectile (p. 268), e.g. a bullet fired from a gun. The projectile sticks to the pendulum whose displacement is measured.

Foucault pendulum a simple pendulum (↑) which swings slowly for a long time. As the earth turns round, the plane in which the pendulum swings stays constant, but to someone on the earth appears to turn, thus giving a reason to imagine the Coriolis force (p. 273).

conical pendulum similar to a simple pendulum (↑) except that the mass, instead of oscillating (↑) in a vertical plane, moves in a horizontal circle so that the string or rod (p. 263) moves in a cone (p. 210).

cycloidal pendulum a simple pendulum (↑) with a flexible (p. 263) string between two arches of a cycloid (p. 225) drawn below its fixed line and which uses the isochrone (p. 226) property of the cycloid.

simple harmonic motion the motion of a point moving in a straight line (p. 167) in which the acceleration (p. 258) of the point is always towards a fixed point on the line and proportional to its distance from it. It is also the component (p. 80) in a fixed direction of the displacement of a point moving with constant speed (p. 258) round a circle. The displacement x is given by $x = a \sin(\omega t + \varepsilon)$, where a, ω and ε are constants and t is the time. An approximate example is the motion of a pendulum (↑) with small oscillations (↑). **S.H.M.** (*abbr*).

amplitude (*n*) the maximum displacement *a* from the fixed point in the equation given under simple harmonic motion (p. 275).

period[3] (*n*) the time of each complete repeat of anything which happens many times, especially used in simple harmonic motion (p. 275) where it is the quantity $2\pi/\omega$ (ω is as given in the equation there). *See also* pp. 39, 133.

periodic (*adj*) used of functions and motions which repeat regularly, e.g. sine function (p. 132).

aperodic (*adj*) not periodic (↑).

periodicity (*n*) the property of having a period (↑) so that $f(x + p) = f(x)$ for all x. The smallest possible value of p is the period.

cycle (*n*) one repetition of a periodic (↑) function or motion, the whole being made up of many repeated cycles.

frequency[2] (*n*) the number of cycles (↑) in any given unit of time, often one second. It is the reciprocal (p. 27) of the period (↑). *See also* p. 106.

angular frequency frequency regarded as the number of rotations in unit time. Also known as **pulsatance**.

phase (*n*) the point in a cycle (↑) at a given time; the quantity $\omega t + \varepsilon$ in the equation of simple harmonic motion (p. 275).

phase constant the value of the phase (↑) at zero time, given by the constant ε in the equation of simple harmonic motion (p. 275). The constant is often taken to be zero. Also known as **phase angle**. *See also* phase angle (↓).

phasor (*n*) a rotating vector of fixed length which can be imagined to generate (p. 209) simple harmonic motion (p. 275).

phase angle the vectorial angle (p. 216) given by a phasor (↑). *See also* phase constant (↑).

lead (*n*) if two periodic (↑) functions have the same period, the difference in the phases (↑) is called the lead or lag. The one which reaches the maximum first is said to lead the other which lags on the first. **lead** (*v*).

lag[2] (*n*) *see* lead (↑) and *also* p. 116. **lag** (*v*).

epoch (*n*) the time of an event (p. 100) measured from some arbitrary (p. 287) origin in time.

amplitude
simple harmonic motion
equation $x = a \sin(\omega t + \varepsilon)$

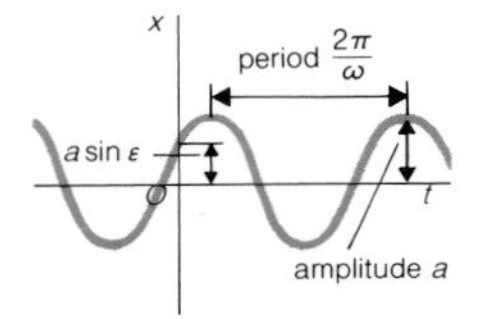

wave (*n*) a periodic (↑) function or motion having a set of cycles (↑) which are usually, but not always, repeats of each other. The commonest example is simple harmonic motion (p. 275).

wavelength (*n*) the length of a complete cycle (↑) of a wave (↑).

wave mechanics the study of the motion of waves (↑), especially that in sub-atomic physics (the physics, p. 290, of very small particles).

wave equation the partial differential equation (p. 159) $\partial^2 y/\partial x^2 = (1/c^2)(\partial^2 y/\partial t^2)$ which describes common types of wave (↑) and whose solution is in the form of a sine wave (↓).

Lissajou's curve

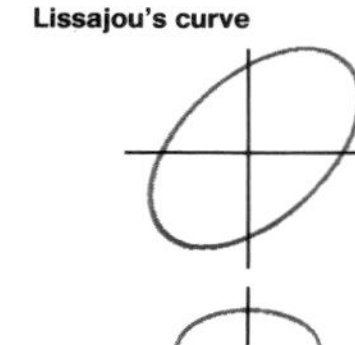

sine wave a wave (↑) whose form is the shape of the graph of the sine function (p. 132).

sinusoidal (*adj*) having the shape of the sine function (p. 132).

Lissajou's curve a locus (p. 187) given by a point having two perpendicular simple harmonic motions (p. 275). Many such curves can be produced by machines.

fundamental (*n*) (1) generally, of any concept or theorem which is the basis (p. 287) of a mathematical theory; (2) the simplest way in which a mechanical system can oscillate (p. 275). **fundamental** (*adj*).

harmonic

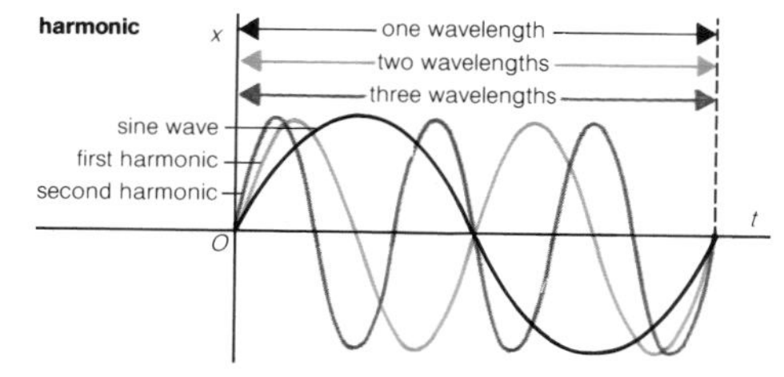

harmonic[2] (*n*) a wave (↑) whose wavelength (↑) is an exact divisor (p. 26) of the wavelength of the fundamental (↑), often used in music. *See also* p. 184. **harmonic** (*adj*).

octave (*n*) in music the harmonic (↑) whose wavelength (↑) is half that of the fundamental (↑); sometimes also used generally for any set of eight elements.

compound harmonic motion motion which consists of two or more simple harmonic motions (p. 275) added together. **harmonic motion** includes both simple and compound types.

damping (*n*) the adding of a resistance (p. 265) to a harmonic motion (↑). **damp** (*v*), **damped** (*adj*).

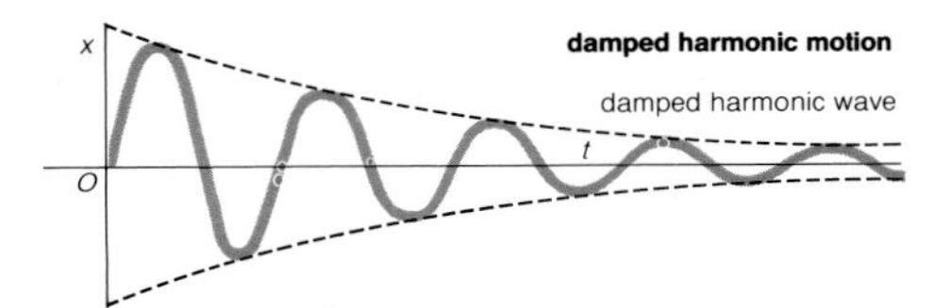

damped harmonic motion a harmonic motion (↑) together with a resistance (p. 265) which causes the amplitude (p. 276) to decrease with time, and usually causes the period (p. 276) to change as well. Sometimes the motion stops oscillating (p. 275) and just dies away.

critical damping damping (↑) which causes an oscillating (p. 275) system to come to rest in the shortest possible time.

overdamping (*n*) damping (↑) in which the resistance (p. 265) is greater than that for critical damping (↑). Similarly **underdamping**.

decrement[2] (*n*) in damped harmonic motion (↑) the ratio of two successive (p. 292) amplitudes (p. 276). *See also* p. 143.

natural frequency the frequency (p. 276) at which a system oscillates (p. 275) when there is no external (p. 289) force which is either large or periodic (p. 276). A natural frequency is usually the fundamental (p. 277) but can be one of the harmonics (p. 277).

free oscillation oscillation (p. 275) of a system at its natural frequency (↑).

forced oscillation oscillation (p. 275) of a system at other than its natural frequency (↑), caused by external (p. 289) forces.

resonance (*n*) the oscillations (p. 275) of large amplitude (p. 276) which are produced when a system is given a periodic (p. 276) force at its natural frequency (↑).

transverse wave a wave (p. 277) in which the motion of any one particle is perpendicular to the direction in which the wave's energy (p. 269) is travelling.

longitudinal wave a wave (p. 277) in which the motion of any one particle is along the direction in which the wave's energy (p. 269) is travelling.

torsional wave a wave (p. 277) in which the motion of any one particle is a rotation about the direction in which the wave's energy (p. 269) is travelling.

travelling wave a wave (p. 277) in which the whole periodic (p. 276) motion travels through the medium (p. 280), e.g. a wave at sea. Also called **progressive wave**.

phase speed the speed (p. 258) at which the phase (p. 276) moves along a travelling wave (↑).

standing wave
the lines show motions of particles in a standing wave

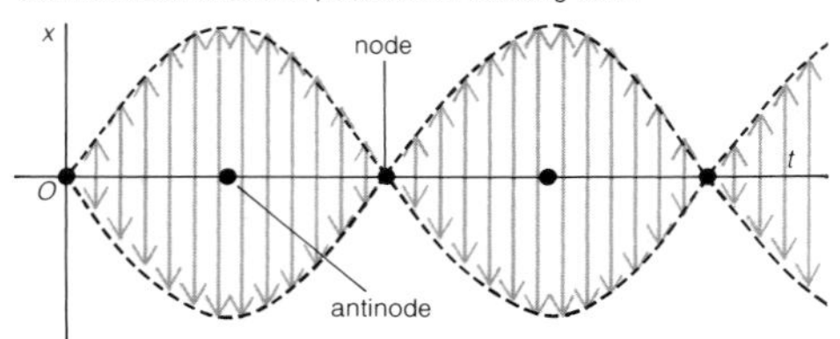

square wave

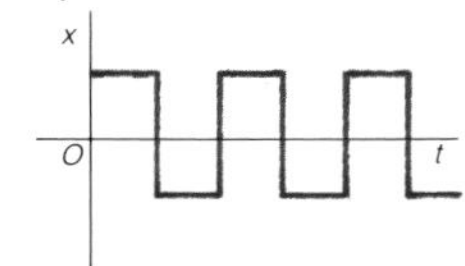

saw-tooth wave

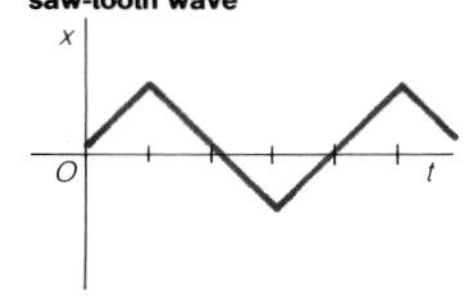

standing wave a wave (p. 277) in which the whole periodic (p. 276) motion does not travel through the medium (p. 280), as in a vibrating (p. 275) string. Also called **stationary wave**.

node[3] (*n*) a point of a standing wave (↑) at which no motion takes place. *See also* pp. 189, 239.

antinode (*n*) a point of a standing wave (↑) where the motion has maximum amplitude (p. 276).

square wave a periodic (p. 276) function having a set of line segments (p. 167) alternately (p. 287) parallel and perpendicular to the axis of the wave (p. 277).

saw tooth wave a periodic (p. 276) function having a set of line segments (p. 167) of alternate (p. 287) gradients (p. 219).

fluid (*n*) that which flows: a liquid or gas.
fluid mechanics the study of forces acting in fluids (↑).
medium (*n*) a fluid (↑) or other substance in which an object is placed, or through which a force or other physical concept acts. Often a resisting (p. 265) medium acting against the motion of bodies through it. The word *medium* is not usually used in mathematics for middle-sized or mean.
hydrostatics (*n*) the study of fluids (↑) at rest in equilibrium (p. 260).
hydrodynamics (*n*) the study of fluids (↑) in motion and the forces acting in them.
aerodynamics (*n*) the study of gases in motion and the forces acting in them.
buoyancy (*n*) the effect of loss of weight which seems to happen to a body placed in a fluid (↑). *See also* upthrust (↓).
upthrust (*n*) the upward force on a body in a fluid (↑) which is caused by the weight of fluid displaced (p. 79) by the body. Also known as **buoyancy**.

upthrust

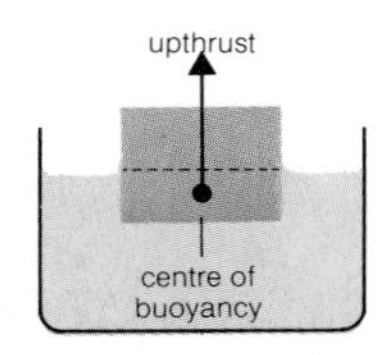

centre of pressure
the centre of pressure of a fluid on a face of a tilted rectangular block placed in it

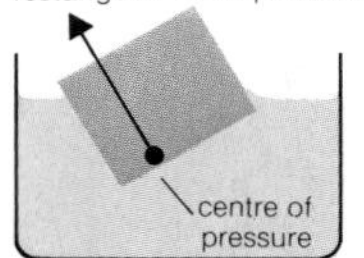

centre of buoyancy the centre of mass (p. 261) of the fluid (↑) displaced (p. 79) by a body placed in it, through which the upthrust (↑) acts.
centre of pressure the point of a plane surface placed in a fluid (↑) through which the resultant (p. 80) of the forces of the fluid acting on the surface passes.
law of flotation a floating body placed in a fluid (↑) displaces (p. 79) its own weight of fluid.
Archimedes' principle if a body is wholly or partly placed in a fluid (↑), the upthrust (↑) equals the weight of fluid which is displaced (p. 79).

metacentre
B_0 is the old centre of buoyancy, B is the new centre of buoyancy and G is the centre of mass

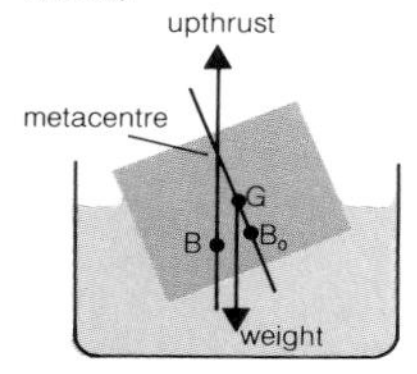

metacentre (*n*) if a floating body is displaced (p. 79) slightly in a plane of symmetry (p. 232) by rotating it so that the mass of liquid displaced is not changed, the point where the vertical through the new centre of buoyancy (↑) meets the line joining the old centre of buoyancy to the centre of mass (p. 261) of the body is the metacentre. If the metacentre is above the centre of mass, the equilibrium (p. 260) is stable (p. 260); if below, it is unstable (p. 260). This is important in building ships.

metacentric height the height of the metacentre (↑) above the centre of mass (p. 261) of the floating body.

terminal velocity the maximum velocity (p. 258) of a body moving under gravity (p. 266) in a fluid (↑).

source
stream lines at a source in a plane

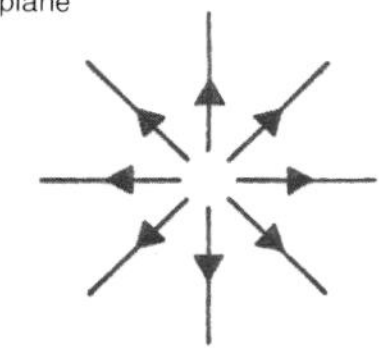

stream line a line in a moving fluid (↑) along whose tangent at any moment in time particles of the fluid on the line are moving.

stagnation point a point in a moving fluid (↑) where the particles have zero velocity (p. 258) and therefore stream lines (↑) may meet.

laminar flow flow which is within a lamina (p. 155) of infinitesimal (p. 143) thickness, and which is therefore smooth flow.

sink
stream lines at a sink in a plane

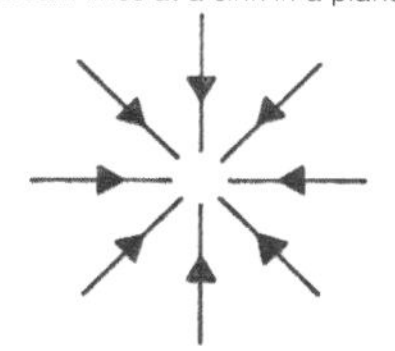

turbulence (*n*) non-laminar flow. Also known as **vorticity**.

source (*n*) a point in a fluid (↑) where fluid is flowing into the system; on a plane surface the stream lines (↑) are radially outwards.

sink (*n*) a point in a fluid (↑) where fluid is flowing out of the system; on a plane surface the stream lines (↑) are radially inwards.

vortex
stream lines at a vortex in a plane

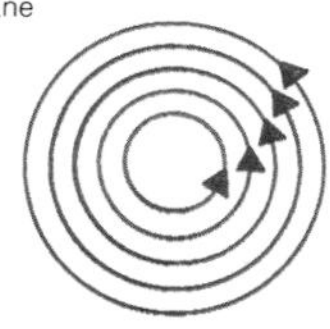

vortex (*n*) a point in a fluid (↑) which is part of a line at any point of which at a given moment in time the particles have only rotation about and movement along the direction of the line. Particles close to a vortex line turn around it. In a plane surface the stream lines (↑) are circles about the point. **vortexes** (*pl*), **vortices** (*pl*).

relativity (*n*) the theory that the laws of physics (p. 290) are not dependent on the particular coordinate system used to describe them. **relativistic** (*adj*).

relativistic mechanics mechanics based on the theory of relativity (p. 281).

Newtonian mechanics mechanics based on Newton's laws of motion (p. 266) and universal law of gravitation (p. 266) rather than on relativity (p. 281). Also known as **classical mechanics**.

general relativity relativity (p. 281) theory which allows coordinate systems to accelerate (p. 258) relative to each other.

special relativity the special case of the theory of relativity (p. 281) which only allows coordinate systems which move with constant velocity (p. 258) relative to each other.

frame of reference a coordinate system which may change with time and which gives the base against which measurements may be taken. Also known as **system of reference**.

inertial frame a coordinate system is an inertial frame with respect to another system if the two systems are at rest or in uniform motion with respect to each other. Also known as **inertial framework**.

principle of equivalence a result of the theory of relativity (p. 281) that gravitational (p. 266) forces are like mechanical forces and gravitation is just a geometrical property of space.

speed of light by the theory of relativity (p. 281) the speed of light is a constant whatever frame of reference (↑) is used. It is approximately 2.998×10^8 metres per second.

rest mass the mass of a body measured by a person at rest relative to it.

relativistic mass the mass of a body moving with velocity (p. 258) v relative to the person measuring it. It is the rest mass (↑) divided by $\sqrt{(1 - v^2/c^2)}$, where c is the speed of light (↑).

proper length the length of an object in the theory of relativity (p. 281) measured in a coordinate system which is at rest relative to the object.

proper time the time between two events (p. 100) in the theory of relativity (p. 281) measured in a coordinate system which moves so that the two events happen at the same point.

Galilean transformation

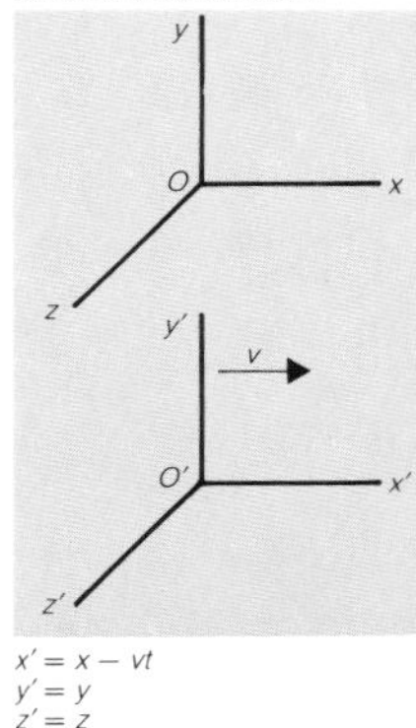

$x' = x - vt$
$y' = y$
$z' = z$
$t' = t$

Lorentz transformation

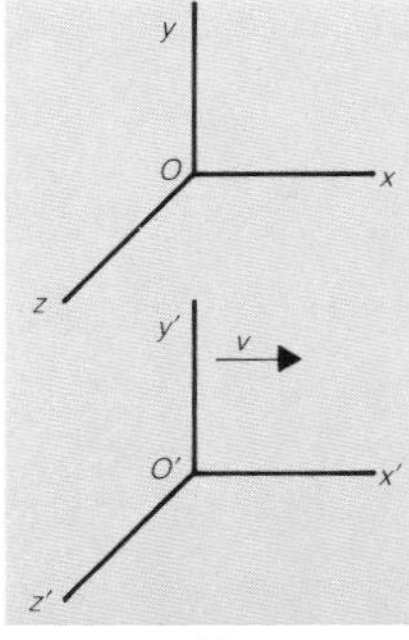

$x' = c(x - vt)/\sqrt{(c^2 - v^2)}$
$y' = y$
$z' = z$
$t' = c(t - vx/c^2)/\sqrt{(c^2 - v^2)}$

mass-energy equation the equation $e = mc^2$ which relates the energy (p. 269) e in a system and the mass m which has been lost to make that energy, and where c is the speed of light (↑). *See also* conservation law (p. 274).

Lorentz-Fitzgerald contraction the length of a body moving with velocity (p. 258) v with respect to a person is made smaller so that it is $\sqrt{(1 - v^2/c^2)}$ times its length at rest relative to that person, where c is the speed of light (↑).

absolute space the single fixed coordinate system for space which is part of the theory of Newtonian mechanics (↑).

Galilean transformation the transformation (p. 54) which allows coordinates to be changed in Newtonian mechanics (↑). If (1) the origins coincide (p. 171) at time $t = 0$; (2) one moves with uniform velocity (p. 258) v relative to the other along the x-axis (p. 215); and (3) the distances are x and x' in the two systems, then $x' = x \pm vt$, the sign depending on the direction of v.

Lorentz transformation the transformation (p. 54) in the theory of relativity (p. 281) corresponding to the Galilean transformation (↑) in Newtonian mechanics (↑). Using the same symbols and axes as for the Galilean transformation, the times t and t' are now different in the two sets of coordinates so that

$$x' = c(x - vt)/\sqrt{(c^2 - v^2)} \text{ and}$$
$$t' = c(t - vx/c^2)/\sqrt{(c^2 - v^2)},$$

where c is the speed of light (↑), and the velocity (p. 258) v remains uniform.

Lorentz covariant used of a law in physics (p. 290) which takes the same form after a Lorentz transformation (↑) has been made.

space-time a four dimensional space, three dimensions being the usual space dimensions and the fourth being time. This is one way of looking at the theory of relativity (p. 281).

world vector a vector with four components (p. 80), used to describe a point in space-time (↑).

world curve a four dimensional curve in space-time (↑).

mensuration (*n*) the study of giving numbers to quantities; the act of measuring.

unit (*n*) a conventional (p. 9) quantity which is used as a basis (p. 287) for mensuration (↑), e.g. metre, hour. *See also* one (p. 23).

measure (*v*) to find the amount of a given unit which describes a quantity. **measurement** (*n*).

magnitude (*n*) size; the number given to any quantity when it is measured in a given unit. *See also* p. 79

direction (*n*) the line towards a given point or along which motion takes place; the distant point towards which a set of parallel lines lead.

conversion (*n*) change between two measurements of the same quantity in different units. **convert** (*v*).

conversion factor the ratio between two measurements of the same quantity in different units, e.g. the conversion factor for changing inches to centimetres is to multiply by 2.54.

calibrate (*v*) to put a scale on something which is then used to measure quantities, e.g. a calibrated straight edge becomes a ruler (p. 185). **calibration** (*n*).

temperature (*n*) that which measures the hotness of bodies.

dimensions of a quantity the units of a quantity given in terms of the fundamental units (↓). In mathematics these are usually mass, length and time, e.g. force is (mass × length) ÷ (time^2).

dimensionless (*adj*) having no units of measurement, being a pure number (used of both quantities and variables).

denominate (*adj*) used of a number having units of measurement, not just a pure number.

dimensional analysis the use of physical dimensions to test the correctness of an equation or to give ideas for a new equation, since the dimensions of the quantities in each term of an equation must correspond.

base unit a unit defined by some physical property and not in terms of other units.

fundamental unit a base unit (↑) in a system of units such as those in the SI system (p. 286).

dimensional analysis

F	$= m$		a
force	= mass	×	acceleration
MLT^{-2}	$= M$	×	LT^{-2}

if dimensional analysis is applied to $F = ma$,

mass has units M

acceleration has units LT^{-2} or (length) ÷ (time)^2

force has units MLT^{-2} or (mass) × (length) ÷ (time)^2

derived unit

	derived unit and symbol	in terms of base units
electrical charge	coulomb (C)	$A\,s^{-1}$
force	newton (N)	$kg\,m\,s^{-2}$
frequency	hertz (Hz)	s^{-1}
power	watt (W)	$kg\,m^2\,s^{-3}$
pressure, stress	pascal (Pa)	$kg\,m^{-1}\,s^{-2}$
work, energy, heat	joule (J)	$kg\,m^2\,s^{-2}$

derived unit a unit defined in terms of some other unit, e.g. units of velocity (p. 258) are defined by units of length divided by units of time.

rationalized unit a unit (especially in electricity and magnetism, p. 289) defined so that irrational numbers (p. 46), usually π (pi, p. 192), do not appear in many common equations. Units of the SI system (p. 286) are rationalized units.

coherent unit a derived unit (↑) obtained from base units (↑) without any numerical multiples being needed.

supplementary unit a dimensionless (↑) unit additional to the usual base units (↑) of a system of units, but which is not a derived unit (↑), e.g. radian (p. 133), steradian (p. 200).

gravitational unit a unit of force, or a unit which is a derived unit (↑) from it, which is defined by means of the earth's gravitational field (p. 267) and based upon weight rather than mass. Becoming disused.

metric system a system of units based upon the metre and the gram, other units being given in powers of ten of these, e.g. the kilogram is 10^3 grams. The SI system (↓) replaces it.

CGS system a system of units based upon the *c*entimetre (10^{-2} metres), the *g*ram and the *s*econd. The SI system (↓) replaces it.

MKS system a system of units based upon the *m*etre, the *k*ilogram and the *s*econd. It became the SI system (↓). Sometimes the ampere was added to make the unit of electric current and give the **MKSA system**.

SI system *Système International d'Unités*. The international system of units based on the seven base units (p. 284) given in Appendix 1 (p. 293) and the two supplementary units (p. 285) of radian (p. 133) and steradian (p. 200). Its units are both coherent (p. 285) and rationalized (p. 285).

SI units units of the SI system (↑).

Imperial units units based on the yard (0.914 4 metres) and the pound (0.453 592 37 kilograms). Imperial units are still used in Britain and elsewhere and are similar to the system still used in the USA. Both are being replaced by the SI system (↑).

kilogram (*n*) the unit of mass in the SI system (↑) equal to a standard (p. 292) mass of metal kept at Sèvres in France. **kg** (*abbr*).

gram (*n*) at first defined as the mass (p. 259) of one cubic (p. 201) centimetre (10^{-6} m^3) of water, now defined as one thousandth part of a kilogram. **g** (*abbr*).

metre (*n*) at first defined as one forty millionth part of the circumference (p. 192) of the earth measured through the poles (p. 205). Now defined from the wavelength (p. 277) of certain radioactive waves (p. 277). **m** (*abbr*).

capacity (*n*) volume, especially used in measuring fluids (p. 280). Not widely used in mathematics.

litre (*n*) a volume which is equal to one thousandth part of a cubic (p. 201) metre and is mostly used to measure capacity (↑) in everyday life. **l** (*abbr*).

accept (*v*) to take to be true, especially of a statistical test. **acceptance** (*n*).

alternate (*adj*) following in turns, correctly used of only two things, e.g. an alternating sequence (p. 92) has terms alternately positive and negative. *See also* alternative (↓).

alternative (*adj*) used of two or more objects, words or elements, any of which may be chosen with the same meaning. Also known as **alternate**, especially in the USA.

analogue (*n*) something which makes an analogy (↓), e.g. an analogue computer (p. 42) is one where some continuously variable quantity, e.g. length, takes the place of the numbers which are fed into it. *Also written* **analog**.

analogy (*n*) once meant ratio, now usually means a similarity in method or in other respects between two problems or examples. **analogous** (*adj*).

apply (*v*) to put to practical use. **applied** (*adj*), **applicable** (*adj*).

approach (*v*) to come close to, tend to, converge (p. 94) to.

arbitrary (*adj*) chosen freely or at random and not by any rule.

assign (*v*) to give, especially give a special value or state (p. 292). **assignment** (*n*).

associate (*v*) to relate or join for a special purpose, e.g. articles in a shop have associated prices.

astronomy (*n*) the study of space, the stars etc.

basis (*n*) that on which something rests, often the starting point for a proof or theory. **basic** (*adj*).

canonical (*adj*) used of expressions or equations in a form which is commonly regarded as usual or standard (p. 292), e.g. the form of equation $y^2 = 4ax$ for the parabola (p. 220).

category (*n*) often used generally for any class (p. 13), set, group (p. 72), etc into which elements may be placed; but now also with particular meaning in algebra (beyond the level of this book). **categorize** (*v*), **categorical** (*adj*).

compare (*v*) to examine the similarities and differences between two or more things, especially the ratio between two quantities or the difference between them. *See also* relative (p. 291).

concept (*n*) an idea, anything thought of in the mind, especially a basic (p. 287) idea in a theory. Also known as **construct.**
consecutive (*adj*) used of a sequence (p. 91) of elements which are next to each other, e.g. 73, 74 and 75 are consecutive integers. *See also* successive (p. 292).
consist (*v*) to be made up of, e.g. the positive integers consist of odd and even numbers.
correct (*v*) to put right, remove mistakes from.
correction (*n*) that which corrects, or which reduces (p. 291) errors (p. 43).
criterion (*n*) a measure or other standard (p. 292) against which things may be tested. **criteria** (*pl*).
demonstrate (*v*) to show clearly, often with the help of examples. **demonstration** (*n*).
develop (*v*) to work out the steps of, as in a proof or theory.
distinguish (*v*) to show the difference between two or more things. **distinguishable** (*adj*).
distribute (*v*) to spread out, often spread out evenly.
double (*v*) to make twice as large, sometimes used for two coincident (p. 171) objects, e.g. a double point. Similarly **triple** (three times), **quadruple** (four times), etc using Latin prefixes (*see* Appendix 2, p. 294), ***n*-tuple** (*n* times).
duplicate (*v*) to double, repeat twice, make a second copy of. Similarly **triplicate** (three), **quadruplicate** (four), etc using Latin prefixes (*see* Appendix 2, p. 294), ***n*-tuplicate** (*n* times).
elementary (*adj*) related to the beginnings of a theory; not usually used in mathematics with the meaning of easy. *See also* element (p. 10).
entity (*n*) a concept, a mathematical idea or quantity which can be given a symbol.
evaluate (*v*) (1) to find the numerical value of; (2) to solve (of an arithmetical problem). **evaluation** (*n*).
example (*n*) a particular case given to show how a general concept may be used.
exception (*n*) that which does not keep to a pattern, is not regular. **exceptional** (*adj*).
exercise (*n*) a problem or piece of work given to a learner to practise a skill.

express (*v*) to state (p. 292) clearly, especially state in symbols.
exterior (*adj*) lying on the outside, especially on the outside of a closed curve (p. 190) or a solid.
external (*adj*) outside.
formal (*adj*) based on axioms (p. 250) from which results are developed (↑) using formal logic (p. 243). Hence **informal**, not formal.
general (*adj*) not special, relating to all or nearly all the cases being considered.
generalize (*v*) to make more general, to widen the ideas being considered to more cases. **generalization** (*n*).
ideal² (*adj*) used of any imaginary concept which cannot be obtained in real life, but is useful to simplify (p. 292) or complete a theory, e.g. a light (p. 260) string, a point mass, a point at infinity. *See also* p. 75.
in terms of using particular symbols or concepts, e.g. a force may be defined in terms of mass and acceleration (p. 258). *See also* term (p. 59).
inconsistent (*adj*) used of statements (p. 243) which cannot all be true at the same time. **inconsistency** (*n*).
initial (*adj*) first, e.g. the initial term is the first of a series (p. 91).
input (*n*) that which is put in, e.g. data put into a computer (p. 42), or a value of the domain (p. 17) given to a function. *See also* output (p. 290).
interior (*adj*) lying on the inside, especially on the inside of a closed curve (p. 190) or a solid.
internal (*adj*) inside.
irregular (*adj*) not regular. **irregularity** (*n*).
large (*adj*) sometimes used to mean large enough for a particular problem to be solved, without saying how large.
magnet (*n*) a piece of iron or other material which draws towards it other pieces of material. **magnetic** (*adj*), **magnetism** (*n*).
major (*adj*) the larger or more important of two.
manifold (*n*) something made up of several parts, as in a vector space (p. 78) which is made up of vectors with a number of components (p. 80), and is a linear manifold.

method (*n*) a way, rule, algorithm (p. 30), e.g. a particular way of proving a given theorem.
minor[2] (*adj*) the smaller or less important of two.
model (*n*) (1) a copy or example, e.g. a model of a solid configuration; (2) a mathematical concept which helps to explain some physical behaviour, e.g. the sine function (p. 132) is a model of a wave (p. 277). Also known as **mathematical model**.
naive (*adj*) not completely mathematically sound, used in cases such as naive set theory where there are paradoxes (p. 253) which cannot be explained by the simple theory.
neglect (*v*) to take no notice of, usually because the effect is too small, or too difficult, to take account of, e.g. in mechanics, friction (p. 265) may be neglected. Also known as **ignore**.
observation (*n*) a single result obtained from an experiment. **observe** (*v*).
output (*n*) that which comes out, the result of an input (p. 289).
paradigm (*n*) an example, especially one which shows the correct or usual way to do something. **paradigmatic** (*adj*).
pattern (*n*) regularity (p. 291); arrangement in a regular way or to a given rule; anything which has parts related in a way which repeats.
physics (*n*) the study of matter and the forces of the natural world. **physical** (*adj*).
predict (*v*) to forecast, tell what will happen or is expected to happen in the future. **prediction** (*n*).
primitive (*adj*) related to the axioms (p. 250) or basis (p. 287) of a theory. *See also* complete primitive (p. 159). **primitive** (*n*).
problem (*n*) a question for which an answer is to be given; an exercise (p. 288) or example.
procedure (*n*) a way of working, a method for solving a problem.
process (*n*) a method or way of reaching a result.
property (*n*) a special fact which results from a particular mathematical concept, e.g. a property of a square is that its diagonals (p. 177) bisect (p. 171) each other.
quantity (*n*) an amount, number; often but not always numbers with units used in measurement.

realization (*n*) that which makes real or actual and is an example of the use of a piece of mathematics; some actual idea which a mathematical concept models (↑). **realize** (*v*). *See also* p. 48.

reduce (*v*) to make smaller, change to another (usually simpler) form, used of equations and expressions, and in topology (p. 238). **reduction** (*adj*). *See also* p. 255.

redundant (*adj*) unnecessary, giving more facts than are needed, e.g. if a parallelogram (p. 181) is defined as a quadrilateral (p. 177) with pairs of opposite sides equal and parallel, *equal* is redundant, because this can be proved from the definition.

regular (*adj*) (1) according to a rule, not exceptional (p. 288); keeping to a pattern; (2) having symmetries, as in a regular polyhedron (p. 200). **regularity** (*n*).

reject (*v*) to take to be false, used especially of a statistical test. **rejection** (*n*).

relative (*adj*) related to, not absolute (p. 146); often used of a quantity given as a difference from another quantity or as a ratio to another quantity. *See also* compare (p. 287).

relevant (*adj*) concerned with the particular thing or problem being dealt with.

represent (*v*) to act for, act as an example of. **representative** (*n*), **representation** (*n*).

respective (*adj*) related in order, e.g. if *a, b, c* and *d* are the respective sides of a quadrilateral (p. 177), they are in that order around the quadrilateral. **respectively** (*adv*), **respect** (*n*).

restrict (*v*) to place conditions upon, e.g. a function may be restricted to a given domain (p. 17). **restriction** (*n*).

rule (*n*) a method for solving a problem which can be used in many different cases.

score (*n*) (1) an amount, number, especially the result of a game; (2) a set of 20 (rare in mathematics).

selection (*n*) choice; in statistics and probability it is usually random and independent. **select** (*v*).

semi- (*prefix*) placed before words to mean half, e.g. semicircle (p. 192).

separate (*adj*) different; not in the same place; not joined together. **separate** (*v*), **separable** (*adj*).

simplify (*v*) to put in an easier form. **simplification** (*n*).

simulation (*n*) the modelling (p. 290) of one system by another; in mathematics usually used of a mathematical model, often but not always involving ideas of random choice. **simulate** (*v*).

slight (*adj*) very small, infinitesimal (p. 143).

small (*adj*) can mean small enough for a problem to be solved, without saying how small.

standard (*n*) (1) that which is fixed as a unit or basis (p. 287) for measurement, usually by convention (p. 9); (2) the usual or commonest way of stating something; a convention.

state (*n*) the position or arrangement of a system at any given time.

state (*v*) to say clearly in words and/or symbols. *See also* p. 243.

succeed (*v*) to come immediately after.

successive (*adj*) taken in order one after the other. *See also* consecutive (p. 288).

superpose (*v*) to place on top of. **superposition** (*n*).

system (*n*) any set of objects, elements or ideas which makes up a whole, or obeys a particular set of rules. *See also* family (p. 90). **systematic** (*adj*).

test (*n*) a method of finding out whether a statement (p. 243) is true or false. **test** (*v*).

transition (*n*) movement from one state (↑) to another.

type (*n*) a kind, sort, e.g. a rational number (p. 26) is a type of number.

typical (*adj*) being an example of its type with no special or unusual properties other than those of which it is an example.

vanish (*v*) to become equal to zero. *Do not mistake for* tend.

verify (*v*) to make sure by examples, experiments or recalculation that something is correct. In mathematics it does not mean to make completely sure. Also known as **check**. **verification** (*n*).

with respect to with regard to, e.g. derive y with respect to x means find dy/dx rather than dy/dt (with respect to t). **wrt** (*abbr*), **wo** (*abbr*).

International System of Units (SI)

Système International d'Unités

Basic Units

There are seven basic SI units of which three are widely used in mathematics:

metre defined from the wavelength of certain radioactive waves.
kilogram defined from a standard mass of metal kept at Sèvres in France.
second defined from the frequency of certain radioactive waves.

UNIT	SYMBOL	MEASUREMENT
metre	m	length
kilogram	kg	mass
second	s	time
ampere	A	electric current
kelvin	K	temperature
mole	mol	amount of substance
candela	cd	luminous intensity

In addition two supplementary units are also defined which are used in mathematics. These are the radian (symbol r) which is a measure of angle and the steradian (symbol sr), which is a measure of solid angle.

Prefixes for SI Units

MULTIPLE		PREFIX	SYMBOL
10^{12} =	1 000 000 000 000	tera-	T
10^{9} =	1 000 000 000	giga-	G
10^{6} =	1 000 000	mega-	M
10^{3} =	1 000	kilo-	k
10^{2} =	100	hecto-*	h*
10^{1} =	10	deka-*	da*
10^{-1} =	0.1	deci-*	d*
10^{-2} =	0.01	centi-*	c*
10^{-3} =	0.001	milli-	m
10^{-6} =	0.000 001	micro-	μ
10^{-9} =	0.000 000 001	nano-	n
10^{-12} =	0.000 000 000 001	pico-	p
10^{-15} =	0.000 000 000 000 001	femto-	f
10^{-18} =	0.000 000 000 000 000 001	atto-	a

* Except for the centimetre these prefixes are not often now used.

Number prefixes

These number prefixes are widely used. Greek prefixes should be used with words from Greek, Latin prefixes with those from Latin, but misuse of the less common prefixes is often found, e.g. septagon for heptagon and nonagon for enneagon. Other variations also occur such as endecagon for hendecagon. Final vowels are often omitted from the prefixes especially before another vowel. The Latin adjectival forms are less often used, but see, for example, the dictionary definitions containing the words singulary, binary and denary. The French form demi- is also used for half.

NUMBER	GREEK	LATIN	LATIN (*adj*)
$\frac{1}{2}$	hemi-	semi-	
1	mono-	uni-	singula-
$1\frac{1}{2}$		sesqui-	
2	di-	bi-	bina-
3	tri-	tri-, ter-	terna-
4	tetra-	quadri-	quaterna-
5	penta-	quinque-	quina-
6	hexa-	sexi-	sena-
7	hepta-	septi-	septena-
8	octa-	octo-	octona-
9	ennea-	nona-	novena-
10	deca-	deci-	dena-
11	hendeca-	undeci-	undena-
12	dodeca-	duodeci-	duodena-
20	icosa-	vigesi-	
many	poly-	multi-	

Index